Understanding Intermolecular Interactions in the Solid State
Approaches and Techniques

Monographs in Supramolecular Chemistry

Series editors:
Philip Gale, *The University of Sydney, Australia*
Jonathan Steed, *Durham University, UK*

Titles in this series:

1: Cyclophanes
2: Calixarenes
3: Crown Ethers and Cryptands
4: Container Molecules and Their Guests
5: Membranes and Molecular Assemblies: The Synkinetic Approach
6: Calixarenes Revisited
7: Self-assembly in Supramolecular Systems
8: Anion Receptor Chemistry
9: Boronic Acids in Saccharide Recognition
10: Calixarenes: An Introduction, 2nd Edition
11: Polymeric and Self Assembled Hydrogels: From Fundamental Understanding to Applications
12: Molecular Logic-based Computation
13: Supramolecular Systems in Biomedical Fields
14: Synthetic Receptors for Biomolecules: Design Principles and Applications
15: Polyrotaxane and Slide-Ring Materials
16: Boron: Sensing, Synthesis and Supramolecular Self-Assembly
17: Porous Polymers: Design, Synthesis and Applications
18: Pillararenes
19: Supramolecular Chemistry at Surfaces
20: Aromatic Interactions: Frontiers in Knowledge and Application
21: Naphthalenediimide and its Congeners: From Molecules to Materials
22: Functional Supramolecular Materials: From Surfaces to MOFs
23: Supramolecular Amphiphiles
24: Co-crystals: Preparation, Characterization and Applications
25: Molecular Gels: Structure and Dynamics
26: Understanding Intermolecular Interactions in the Solid State: Approaches and Techniques

How to obtain future titles on publication:
A standing order plan is available for this series. A standing order will bring delivery of each new volume immediately on publication.

For further information please contact:
Book Sales Department, Royal Society of Chemistry, Thomas Graham House, Science Park, Milton Road, Cambridge, CB4 0WF, UK
Telephone: +44 (0)1223 420066, Fax: +44 (0)1223 420247
Email: booksales@rsc.org
Visit our website at http://www.rsc.org/Shop/Books/

Understanding Intermolecular Interactions in the Solid State

Approaches and Techniques

Edited by

Deepak Chopra
IISER Bhopal, India
Email: dchopra@iiserb.ac.in

ROYAL SOCIETY OF CHEMISTRY

Monographs in Supramolecular Chemistry No. 26

Print ISBN: 978-1-78801-079-5
PDF ISBN: 978-1-78801-308-6
EPUB ISBN: 978-1-78801-516-5
Print ISSN: 1368-8642
Electronic ISSN: 2041-7144

A catalogue record for this book is available from the British Library

The Royal Society of Chemistry is a charity, registered in England and Wales, Number 207890, and a company incorporated in England by Royal Charter (Registered No. RC000524), registered office: Burlington House, Piccadilly, London W1J 0BA, UK, Telephone: +44 (0) 207 4378 6556.

Visit our website at www.rsc.org/books

Printed in the United Kingdom by CPI Group (UK) Ltd, Croydon, CR0 4YY, UK

Foreword

Intermolecular interactions in solids and in particular crystals have been studied over the last couple of decades with renewed interest, as understanding the bonding characteristics in this arena impacts the properties and paves the way for the design of functional materials. In this context, the inspiration provided by crystal engineering concepts like synthons, landscapes and homo- and hetero-meric interactions as design elements provide the building blocks. However, a deeper insight is required in order to quantify and parameterise each intermolecular contact in terms of electron densities, energies and effects due to non-ambient conditions. I find this book addresses all these factors with chapters written by experts in the area and ably edited by Professor Deepak Chopra.

There are nine chapters covering almost all features of evaluating intermolecular interactions in solids and they provide an in-depth understanding of the techniques and methodologies used. The first chapter, *Integrating Computed Crystal Energy Landscapes in Crystal Form Discovery and Characterisation* by S. Mohamed, Khalifa University, Abu Dhabi, UAE, concerns computational methods in crystal structure prediction, exploring the energy landscape of all possible packing arrangement of molecules based on theoretical evaluation of intermolecular contacts in terms of lattice energies. Chapter 2 by Elena Boldyreva describes in detail various aspects of high pressure crystallography with an emphasis on control, generation and propagation of intermolecular interactions under non-ambient conditions. With her vast experience in this area she has outlined the inputs to derive structure—property correlations. The next chapter provides a detailed description of *in situ* cryo-crystallography, exploring non-covalent interactions among molecules that are in general liquids at room temperature. This chapter by

Monographs in Supramolecular Chemistry No. 26
Understanding Intermolecular Interactions in the Solid State: Approaches and Techniques
Edited by Deepak Chopra

Published by the Royal Society of Chemistry, www.rsc.org

Deepak Chopra's group provides details on solvent interactions in guiding molecular arrangements in intermolecular space and provides insights into possible evaluation of polymorphs.

The next three chapters are devoted to the analysis of charge density distributions in molecular crystals, inorganics, organics and macromolecules, providing a quantification of intermolecular interactions in terms of charge density distributions. These methodologies have been well established in the last few decades and the topological analysis using "atoms in molecules" formulation by Richard Bader allows for a systematic evaluation of intermolecular interactions. Interestingly, three young researchers have contributed these chapters, bringing out the significance of experimental high resolution diffraction data in this context. Chapter 4, *Experimental Electron Studies of Inorganic Solids* by Swastik Mondal, describes the charge density analysis of inorganic solids underlying the technical issues of handling both light and heavy elements. Venkatesha R. Hathwar, in Chapter 5 devoted to *Experimental Charge Density Analysis in Organic Solids* describes the methodology in detail, focussing on the determination of the physical and chemical properties of materials. The next chapter, Chapter 6, *Charge Density Studies and Topological Analysis of Hydrogen Bonds in Proteins* by Parthapratim Munshi's group brings in the flavour of charge density analysis on larger biomolecules like proteins. The database and transferability concepts are discussed in detail and the overall approach is supplemented with examples. Chapter 7 deals with the transferability of functional group polarizabilities, a database of immediate relevance in evaluating the dielectric properties of molecular aggregates. Authored by Piero Macchi and his group this chapter offers an approach to computing the linear optical properties of crystals.

Chapters 8 and 9 examine two cognate methodologies, one using NMR as a probe and the other evaluating the usefulness of the Hirshfeld surface and associated fingerprint plots to quantify weak π–hole and σ–hole interactions in a novel visual manner. The group of F. Rossi, P. Cerreia Vioglio, M. R. Chierotti and R. Gobetto present solid state NMR as a complementary technique to X-ray crystallography; with examples for both hydrogen and halogen bonds the authors bring in the relevance of the technique in the evaluation of weak intermolecular interactions.

The selection of authors has been meticulously done by involving both young and senior contributors with an eye for detail and excellence in the subject.

The book serves as a basic text book for beginners as well as a guide and handbook for experts who wish to get a quantitative insight into the bonding characteristics that operate in intermolecular space.

T. N. Guru Row

Preface

The understanding of intermolecular interactions, which has a characteristic signature in all fields of science, has undergone a major renaissance in the past decade. With major developments in technology and computational resources the subject has evolved tremendously with pioneering contributions from Worldwide researchers working in this area. This has been strongly supported *via* parallel developments in the field of crystallography, which is of a quintessential nature, with particular relevance towards the study of various intermolecular interactions in crystals. Both single and powder X-Ray diffraction have had profound contributions towards the subject. Future studies will be directed towards the quantitative assessment of the nature, energetics and topological features associated with interactions, both intra- and intermolecular, *via* different approaches and techniques. Various solid state phenomena, which involve the understanding of the formation of polymorphs, co-crystals and phase transitions in crystals (organic, inorganic, organometallic and macromolecular origin) is also of extreme importance, both from a fundamental and an applied perspective. Cutting-edge research now involves the determination of the crystal structures of compounds, which are liquids at room temperature, *via* the method of *in situ* cryocrystallization and under the application of pressure on a crystal of a compound of interest. The computational procedures involve evaluation of the lattice energies (including partitioning of the individual interaction energies), predicting crystal structures with emphasis on the mapping of the electron density distribution in crystals.

Keeping in mind the above-mentioned features, the current book comprises nine chapters with contributions from active researchers in the field of crystallography and structural chemistry. Chapter 1 focuses on the importance of crystal structure prediction with special emphasis on the role of weak

Monographs in Supramolecular Chemistry No. 26
Understanding Intermolecular Interactions in the Solid State: Approaches and Techniques
Edited by Deepak Chopra

Published by the Royal Society of Chemistry, www.rsc.org

intermolecular interactions in crystals. Chapters 2 and 3 discuss the application of high pressure and *in situ* cryocrystallization, which has enabled the determination of crystal structures and subsequent investigation of interactions. Chapters 4, 5 and 6 highlight the contributions of both experimental and theoretical charge density analysis in understanding the formation of chemical bonds (both intra- and intermolecular) in small molecules, macromolecules and inorganic systems. Chapter 7 delineates the importance of atomic polarizability between molecules in crystals. Chapter 8 deciphers the importance of solid state NMR as an important technique to unravel weak interactions and Chapter 9 explains the role of solid state DFT calculations in the evaluation of interaction energies along with the application of Hirshfeld surface analysis in crystals.

The significant success in the understanding of interactions in solids, as is clearly evidenced by the contents of this book, is expected to drive curiosity amongst the academicians, undergraduate and postgraduate students, researchers, and industrial chemists who work with numerous solid state properties associated with different compounds having various applications. I sincerely hope that the approaches and techniques discussed in the book will be actively pursued by the scientific community, and this endeavor is expected to provide further quantitative inputs towards an assessment of various intra and intermolecular interactions with an improved understanding in the development of new materials.

Deepak Chopra

Acknowledgements

This exercise has been a major academic challenge and I take this opportunity to thank everyone who has motivated me in my professional endeavors. Firstly, I thank my research supervisor, Professor T. N Guru Row who has been a tremendous motivation, and who introduced me to the field of crystallography and intermolecular interactions, followed by my Post-Doc supervisor Professor A. A. Pinkerton at the University of Toledo, USA with whom I learnt the importance of charge density research. I have had numerous interactions with various researchers, especially those working on intermolecular interactions, and in this regard I would like to thank Professor E. A. Arunan, Professor A. Nangia, Professor G. R. Desiraju, and Professor M. A. Spackman. A special thanks to my colleagues, Dr A. R. Choudhury, Dr Saikat Sen and Dr Sukhendu Mandal with whom I have had very long and extended discussions on the subject. I would also like to thank IISER Bhopal for providing excellent facilities and infrastructure, which has enabled our research group to work at the frontiers of organic solid state chemistry. I also thank all the Faculty Members in the Department of Chemistry, IISER Bhopal. Furthermore, I would like to thank all the members of "*Crystallography and Crystal Chemistry Laboratory*" (both past and present) who have contributed immensely towards the development of the subject. I would like to thank Dr Aparna Ganguly, Rowan Frame, Lindsay Stewart, Dr Hannah Aitchison, Katie Morrey, Dr Robin Driscoll and Connor Sheppard in particular, from the Royal Society of Chemistry, who have helped me in the development of the different stages of the book. I would especially like to thank all the contributors who have contributed towards the different book chapters, including the reviewers who have given their precious time to evaluate these and provide their constructive feedback/suggestions towards the improvement of the same.

Monographs in Supramolecular Chemistry No. 26
Understanding Intermolecular Interactions in the Solid State: Approaches and Techniques
Edited by Deepak Chopra

Published by the Royal Society of Chemistry, www.rsc.org

This section cannot be completed unless I especially thank all my family members, in particular, my parents (Pavan Chopra and Bina Chopra), my in-laws (L. R. Gera and Sushila Gera), for their strong motivation, support, and cooperation during this process. I profusely thank my wife Dr Neeta Gera and my son Agasthya Chopra and my new-born daughter Soumyatara Chopra for their patience and tremendous support during this entire journey.

This book is dedicated to my parents.

Contents

Monographs in Supramolecular Chemistry No. 26
Understanding Intermolecular Interactions in the Solid State: Approaches and Techniques
Edited by Deepak Chopra

Published by the Royal Society of Chemistry, www.rsc.org

CHAPTER 1

Integrating Computed Crystal Energy Landscapes in Crystal Form Discovery and Characterisation

S. MOHAMED

Department of Chemistry, Khalifa University of Science and Technology, PO Box 127788, Abu Dhabi, United Arab Emirates
*E-mail: sharmarke.mohamed@ku.ac.ae

1.1 Introduction

Predicting the three-dimensional crystal structure(s) and properties of organic molecules using computational methods has been a long sought-after goal of materials scientists.[1] Sir John Maddox, who served as the Editor of *Nature*, famously proclaimed[2] as far back as 1988 that 'one of the continuing scandals in the physical sciences is that it remains in general impossible to predict the structure of even the simplest crystalline solids from a knowledge of their chemical composition'. Since that time, theoreticians and experimentalists have taken on the challenge of crystal structure prediction (CSP)[3] using a variety of computational algorithms and, whilst significant progress[4–7] has been made over the past two decades, more work remains to be done. Given that the crystal structure dictates all the solid-state properties of a molecule, a computational algorithm for accurately predicting the

Monographs in Supramolecular Chemistry No. 26
Understanding Intermolecular Interactions in the Solid State: Approaches and Techniques
Edited by Deepak Chopra

Published by the Royal Society of Chemistry, www.rsc.org

structures and properties of all the plausible packing arrangements of a molecule would be a significant step forward in efforts to synthesise solid forms of organic molecules with desirable physicochemical properties.[8]

The development of accurate intermolecular potentials for modelling molecular crystal structures has been an extensively studied[9–12] topic of research and an important application of the theory of intermolecular forces.[13] This is reflected in the variety of force fields[14] that are available for modelling organic crystal structures. Indeed, developments in the field of CSP over the past two decades have been underpinned by advances in computing methods for estimating intramolecular and intermolecular energies. Further progress in the accuracy of CSP methods will be driven by advances in methods for estimating and ranking the relative energies of predicted crystal structures. The key motivation for developing CSP methods is to assess the solid-form diversity and properties of molecules, and to support the discovery and characterisation of all experimentally accessible crystal forms. Within the pharmaceutical industry, the failure to characterise the widest possible range of crystal forms of active pharmaceutical ingredients (APIs) during the pre-formulation stage may be detrimental not just to the bottom line of the developer—in the form of lost intellectual property rights that confer market exclusivity—but also potentially to the safety of the patient. The unexpected transformation of crystal polymorphs or the appearance of new (often more stable) polymorphs of drugs following marketing of the product are all challenges that cannot be anticipated by solid state chemists during the drug development process. Ranitidine hydrochloride (Zantac), paroxetine hydrochloride (Paxil) and ritonavir (Norvir) are all examples of pharmaceutical products that have displayed the phenomenon of disappearing polymorphism[15,16] at different stages of the drug development lifecycle. The phenomenon of disappearing polymorphism refers to the well-documented[15,16] observation that a previously characterised polymorph of a molecule is no longer able to crystallise under conditions identical to those previously reported. In the case of ritonavir, the Norvir capsules that Abbot Laboratories marketed in 1996 were formulated to contain exclusively the form I polymorph of the API. However, a more stable form II polymorph with a lower solubility and lower bioavailability appeared. The form II polymorph had an unfavourable bioavailability profile compared to form I and Abbott Laboratories was forced to withdraw Norvir capsules from the market. The subtle interplay of kinetics and thermodynamics in initiating polymorphic phase transitions is a topic of huge significance to pharmaceutical companies, as the example of ritonavir illustrates. The value of computational methods of CSP as applied to APIs is that they can be used to map the range of plausible polymorphs as well as the relative energies of these polymorphs prior to synthesis and formulation, thereby potentially saving the pharmaceutical industry significant amounts of time and capital. Another key motivation for developing CSP methods is in the domain of chiral resolution *via* crystallisation. Enantiomers of a chiral drug display different biological activities within chiral environments. Thus, the ability to ensure the enantiomeric purity of the API during the pharmaceutical development process is a major issue for the industry. A common strategy used to resolve enantiomers is to prepare diastereomeric salts of the chiral API, which unlike the enantiomers possess different physical properties and hence can be separated. CSP methods have already shown potential[17,18] in aiding efforts at chiral resolution *via* diastereomeric salt formation because, unlike experimental screens that tend to be labour-intensive and expensive in resources, computing methods can be used to screen dozens of chiral resolving agents at the click of a button.

Conducting manual solid-form screens of promising drug candidates as a function of a limited range of experimental variables (*e.g.* solvent, temperature, heating/cooling rate) is emerging as a routine crystal engineering endeavour within industrial and academic research laboratories. However, such solid-form screens cannot be exhaustive enough given the infinite number of experimental variables that could potentially be explored, and whilst automated high-throughput screens and parallel crystallisation platforms with temperature controls have facilitated the industrial solid-form screening process in recent decades, there continues to be an unmet need for an automated, faster and cheaper method for crystal solid form discovery and characterisation. *In silico* solid form screening methods such as CSP have been shown to be a useful complement to experimental crystallisation[19] and characterisation[20] efforts. The availability of cheaper and faster computer processors in research institutions over the past three decades has clearly impacted many areas of chemistry[21] and biology.[22] Moreover, the availability of high-performance computing clusters in most research-intensive universities means that the challenge we face today in predicting the crystal structures and properties of organic molecules is not dictated by the speed or storage capacity of our computers, but by the accuracy of our algorithms in helping us understand which of the hundreds of thousands of predicted crystal structures for a given system may be observed under a given set of experimental conditions.

1.2 Computational Methodology for Predicting Molecular Crystal Structures

1.2.1 Overview

Whilst there are many computational algorithms (*e.g.* GRACE,[23] CrystalPredictor,[24] USPEX,[25] Polymorph Predictor,[26] MOLPAK[27]) for performing CSP, each with its own underlying assumptions, strengths and weaknesses, the central challenge that all computational methods of CSP must solve is how to correctly predict the positions and relative orientations of the molecule(s) in the unit cell in addition to the six crystallographic cell parameters defining the unit cell. This is challenging enough for flexible organic molecules that crystallise in structures with one symmetry independent molecule (Z′ = 1) in the asymmetric unit. For crystal structures derived from multiple chemically unique or symmetry independent molecules in the crystallographic asymmetric unit, the additional degrees of freedom relating the relative orientation of the symmetry independent molecules must also be considered in addition to all the other factors mentioned above. Small variations in one or more torsional degrees of freedom in the parent molecule can significantly alter the lattice energy and hence the relative stability of the predicted crystal structure. Thus, the input molecular geometry and how close this is to the true molecular geometry observed in the experimental crystal structure has a significant effect on the relative stability ranking of the experimentally observed structure. Once the input molecular geometry has been defined, it is necessary to search for putative packing arrangements of the parent molecule(s). At this stage, common sense dictates that since approximately 90% of all organic and organometallic structures reported in the January 2016 release of the Cambridge Structural Database (CSD)[28] crystallise in only 13 space groups, the search space for plausible packing arrangements can be limited by space group symmetry constraints and thus it is not necessary to search for hypothetical crystal structures in all 230 crystallographic space groups. Intrinsic molecular symmetry (*e.g.* mirror planes) require special consideration when defining the input molecular structure. The predicted crystal structures must be ranked in order of stability. This is usually done at the initial structure generation stage using some model for estimating the intramolecular energy penalty for conformational changes and the intermolecular forces derived from estimates of the electrostatic and dispersion–repulsion contributions towards the crystal lattice energy. The final output from a typical CSP study is a set of crystal densities and lattice energies for the predicted crystal structures. This information is typically plotted as a scatterplot of density *vs.* lattice energy and this is referred to as the crystal energy landscape (CEL). Figure 1.1 provides a summary of the sequence of steps needed for a successful CSP study leading to a representative CEL using the molecular structure of carbamazepine as an example.

The practical value of CSP is that the wealth of structural information contained in the CEL can provide insight into a range of solid state chemistry problems typically faced by chemists. Indeed, there are several examples where computed CELs have successfully been used to predict and subsequently characterise previously unknown polymorphs[29] of a molecule or warn against the risks of non-stoichiometric channel hydrate formation in salts[30] with a predisposition to crystallising in open hydrogen bond networks. Carbamazepine (CBZ), an API used in the treatment of epilepsy has been the subject of extensive crystallisation screens,[31–33] but until the work of Arlin *et al.*,[34] a fifth polymorph of this API with a catemer hydrogen bond motif between the carbamazepine molecules was not known. Estimating the relative lattice energy of the crystal that results from substituting the

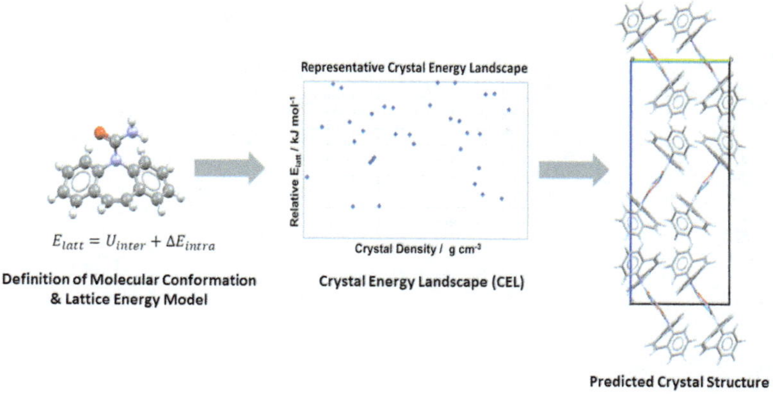

Figure 1.1 Typical steps in a computational CSP study using carbamazepine as a model compound. From left to right, these are: (1) defining the input molecular conformation and appropriate intra- and intermolecular energy model; (2) generating the CEL by computing the range of packing arrangements for the input molecule(s) and optimising the lattice energy as a function of the atomic positions; (3) interpreting the CEL by looking at the properties of the predicted structures and targeting the crystallisation of desired solid forms based on the predictions. The CEL shown is only a representative example and each point on the CEL represents a distinct lattice energy minimum.

molecular structure of CBZ for the molecular structure of 10,11-dihydro-carbamazepine (DHC) in the form II structure of DHC[35] showed that the resulting crystal with a catemer hydrogen bond motif (Figure 1.2) was relatively stable on the CEL. This led to a successful experimental effort to grow single crystals of CBZ form V *via* a templating strategy where crystals of CBZ form V were grown directly onto a single crystal of DHC form II. This work illustrates that *in silico* screens of polymorphism provide a useful complement for targeted polymorph crystallisation and characterisation. The challenge with conventional polymorph screens is that there are many experimental variables to optimise and often a limited timeframe in which to complete the screening and X-ray characterisation work. Thus, while even the most extensive experimental screens of polymorphism conclude at best with the finding of a limited set of novel crystal forms, computed CELs for even the simplest molecules often present the opposite challenge in that they predict too many polymorphs on the static CEL. Nyman and Day[36] recently looked at the static and vibrational energy differences in the crystal structures of 508 polymorphic molecules and concluded that in more than 50% of the cases studied, the lattice energy differences of polymorphs were less than 2 kJ mol^{-1}. This reflects the importance of the intra- and intermolecular energy models used in CSP and the effect this has in the interpretation of the CEL as well as the conclusions drawn regarding the solid-state properties of the molecule(s) in question.

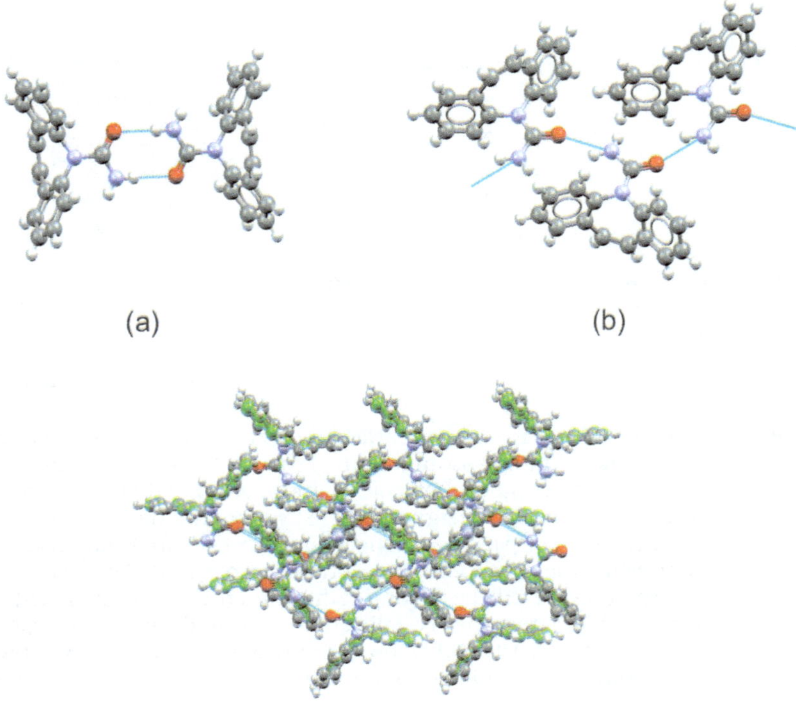

(a) (b)

(c)

Figure 1.2 Illustration of the dimer (a) and catemer (b) hydrogen bond motifs found in the known polymorphs of carbamazepine (CBZ). (c) An overlay of the isostructural dihydrocarbamazepine (DHC) form II (CSD Refcode: VACTAU02) and CBZ form V (CSD Refcode: CBMZPN16) structures showing the catemer hydrogen bond motif common to both structures. DHC form II is coloured by element whilst CBZ form V is coloured green. The root mean-squared deviation for overlaying 15 molecules (ignoring hydrogen atom positions) in the coordination spheres ($RMSD_{15}$) of the VACTAU02 and CBZMPN16 structures was 0.295 Å and the calculated similarity index for the powder X-ray diffraction patterns of the two structures was 0.992.

1.2.2 Searching the Conformational Phase Space and Estimating the Total Crystal Lattice Energy

A successful CSP procedure must be able to estimate and rank the relative stabilities of the predicted crystal structures. The total crystal lattice energy (E_{latt}) can be estimated according to the equation

$$E_{latt} = U_{inter} + \Delta E_{intra}$$

where U_{inter} is the intermolecular lattice energy contribution and ΔE_{intra} represents the energy penalty for conformational changes in the molecule(s) as a function of selected torsional degrees of freedom. The lattice energy is typically reported per molecule so as to allow direct comparison of the relative stabilities of predicted polymorphs with different numbers of molecules in the unit cell (Z). When two or more predicted structures differ in the number of molecules within the asymmetric unit (Z′), assuming the same input molecular conformation is used for all molecules, it is also necessary to scale the calculated lattice energies so as to compare the per-molecule relative energies directly. The challenge with using lattice energies to estimate the relative stabilities of predicted polymorphs is that most models of CSP assume static lattice energies calculated at a nominal temperature of 0 K. Crystal structures, on the other hand, are determined at some finite temperature and whilst there has been an increasing trend towards the characterisation of structures at low temperatures over the past few decades,[37] due to the superior quality structural data obtained at low temperatures as well as the widespread availability of devices for cooling crystals, the computational cost involved in sampling the entropic contributions towards the crystal free energy are often so high so as to make such simulations impractical for most CSP applications. The practical effect of neglecting the entropic contribution is minimized because of the expected cancellations in the entropic contribution as a result of our interest in estimating relative crystal energies rather than absolute crystal energies. For lattice energy estimates obtained from empirically derived force fields, temperature effects would implicitly have been absorbed into the empirical atom–atom intermolecular potentials since the potential parameters were optimized by fitting to experimental lattice parameters and heats of sublimation for a set of known crystal structures. However, the neglect of the entropic and zero-point energy contributions towards the crystal energy can lead to calculated relative energies that are not in accord with experimental stability orders,[38] especially when the errors in the lattice energy model exceed the differences in the relative energies of the predicted structures. DMACRYS[39] can estimate the free energy of the predicted structures assuming rigid body constraints by estimating the phonon frequencies of the lattice energy minima for $\mathbf{k} = 0$. Metadynamics[40,41] represents an alternative approach to sampling the free energy hypersurface, which, unlike the harmonic or quasi-harmonic approximation[42] for estimating phonon frequencies, relies on a time-dependent potential acting on the lattice vectors. Metadynamics has shown potential for producing realistic relative stabilities for the polymorphs of the industrial pigment PR179[43] when the technique is used as the final step for ranking the stabilities of predicted structures.

In defining the lattice energy model, careful consideration must be given to the assumed input molecular conformation for the organic molecule. Thompson and Day surveyed[44] the experimental conformations adopted by flexible single-component organic molecules with no intramolecular hydrogen bonds and compared these experimental conformations with those calculated from global minimisation of the conformational energy. The results of this study indicated that for some molecules the calculated global minimum conformation can be distorted by up to 20 kJ mol^{-1} due to favourable stabilising intermolecular interactions that are achieved in the crystal. Conformers with high energies in the crystal relative to the gas phase global minimum conformation were found to be associated with increases in the relative Connolly[45] surface area ($\Delta A_{\text{Connolly}}$). This was concluded to be important in increasing the potential of the molecule to engage in favourable intermolecular interactions in the crystal. The dimensionality of the conformational search space may be reduced by considering the most likely conformations adopted by similar molecules contained in published crystal structures using the CSD. This can then be used to limit the values that may be adopted by some or all the torsional degrees of freedom in the molecule or to set some torsion angles to specific values and keep them fixed during the conformational search. The accuracy of the final relative crystal lattice energies can be improved by using first principles electronic structure methods such as density functional theory (DFT)[46] to calculate the molecular conformations and, optionally, the final crystal lattice energies. Since re-calculating the molecular conformation at each iterative step during the crystal lattice energy minimisation is computationally expensive, some CSP strategies have adopted a hybrid method[47] where the molecular conformation is calculated *ab initio* using a high quality wavefunction and the intermolecular interactions are calculated using atom–atom potential models. Regardless of which method is used, the choice of input molecular conformation has a significant effect in determining the observed intermolecular interactions as well as the relative energies of the predicted crystal structures.

Alternative methods to the use of empirically derived atom–atom potentials for the estimate of crystal lattice energies do exist. The PIXEL method[48] for example uses the molecular electrostatic potential to derive the four key components for the crystal lattice energy: dispersion, repulsion, coulombic and polarisation terms. The PIXEL method has a much more rigorous theoretical foundation than methods that rely on empirically derived atom–atom potentials, not only because it deconstructs the various contributions towards the lattice energy starting from first principles molecular electrostatic potential calculations, but also because, unlike the atom–atom potential formalism, the PIXEL method does not assume that the atomic nuclear positions are the origin of the intermolecular interactions. Over the past two decades, there has been a significant increase in the number of publications that apply DFT to understand reaction pathways[49–51] and predict molecular properties.[52–54] Within the crystal engineering community, DFT methods have attracted increasing interest for the estimation and ranking of the relative energies of predicted crystal structures.[6,55] It is well known that, unlike empirical force fields, which are inherently limited to estimating the crystal lattice energies of specific types of structures, DFT methods are based on rigorous theoretical foundations and in principle can be applied to any system irrespective of the atom types present or the types of intermolecular/intramolecular interactions seen within the crystal. Apart from the increased computational cost of performing *ab initio* DFT calculations during the structure generation and lattice energy minimisation stages, DFT methods suffer from their inability to account for the long-range dispersion contribution towards the crystal lattice energy. A number of dispersion-corrected density functional theory (DFT-D) approaches[56–59] have been proposed using pairwise dispersion corrections to account for the long-range van der Waals forces. Damping functions are implemented in such DFT-D methods in order to estimate physically realistic values for the dispersion energy at high and low interatomic distances. Since the dispersion contribution towards the crystal lattice energy has a major effect on the long-range order and packing of the molecules in the unit cell, the exact form of the dispersion correction and the damping function used is crucial to the success of DFT-D methods. A number of benchmarking studies have been performed over the years comparing the success of various DFT-D methods in reproducing experimental lattice energies and crystal lattice parameters for a set of well-characterised crystal structures. The X23[60] benchmark set, which builds on the previously reported C21[61] benchmark study, has shown that DFT-D methods are capable of estimating experimental lattice energies with a mean absolute error of 3.92 kJ mol^{-1} assuming the energy model includes the vibrational, exact-exchange and many-body dispersion effects. Subsequent work[62] comparing the performance of a range of DFT-D methods against the results of empirically derived force fields using the X23 benchmark structures has shown that the best DFT-D methods perform better than empirically derived force fields with errors that are on average 2–3 times smaller than those obtained from estimates of crystal lattice energies using anisotropic multipole-based force fields.

1.2.3 Search Methods for Finding Hypothetical Crystal Structures

Most methods of CSP assume that crystallisation is under thermodynamic control and hence proceed to search for minima on the potential energy hypersurface. The key differentiating factor between the various methods is the underlying assumptions behind the search algorithm. Regardless of the method used, once the input molecular conformation and lattice energy model have been defined, the user must specify the number of symmetry-unique (*i.e.* $Z' = 1$ or $Z' > 1$) or chemically distinct molecules (*e.g.* for cocrystals) that are contained within the asymmetric unit. During the search for hypothetical crystal structures, the algorithm is typically sampling the relative orientations and positions of these distinct molecular units whilst keeping the covalent atomic connectivity within each molecular fragment constant throughout the lattice energy minimisation. As such, dynamic effects such as proton disorder are not immediately discernible from rigid body lattice energy minimisations. For systems where there is no barrier for proton transfer, the relative energies of the charged and uncharged states can be estimated by relaxing the atomic positions of the structure as a function of the crystal packing forces using fixed-cell periodic electronic structure calculations.[63]

The CrystalPredictor[24] algorithm is an example of a method that generates plausible packing arrangements of crystal structures by specifying random starting values for the cell parameters within specified bounds subject to space group symmetry constraints. The random starting values are generated using low discrepancy Sobol[64] sequences. Initial packing arrangements with densities that are less than a pre-defined value are rejected, as are packing arrangements with lattice energies higher than a specified cut-off. Structures that satisfy these initial user-defined criteria are subjected to lattice energy minimisation using an atom–atom intermolecular potential model coupled with an atomic point charge model for the electrostatic contributions towards the lattice energy. This is followed by further refinement of the crystal lattice energy using a distributed multipole model for the electrostatic energy derived from fitting to the molecular electrostatic potential. The final step in most CSP studies involves clustering to remove duplicate lattice energy minima.

Results from the most recent blind tests of CSP[4,6,55] suggest that methods based on DFT-D approaches are in general more successful at predicting the target experimental structures than those that rely on empirically derived atom–atom potentials. This does not mean that methods that rely on empirical/semi-empirical force fields will always be inferior to *ab initio* methods, and indeed there have been examples[6,55] from the blind tests where purely *ab initio* based methods for generating and ranking the predicted structures have failed. Moreover, the success of DFT-D methods from the blind tests should be put in proper context since such methods are often computationally very demanding of resources. When discussing the success of various CSP methods, it is important to distinguish between the performance of the search algorithm and the performance of the algorithm used to estimate and rank the crystal lattice energies. These are two different challenges that, when combined, give a clearer picture of the success of the overall CSP approach. The success of the GRACE[23] package, which has been validated in a number of blind tests, shows that an accurate model for estimating the thermodynamic contribution towards the crystal energy can often be enough in finding the experimental structure at or close to the global minimum on the energy hypersurface. The GRACE package relies on using DFT-D energies to generate a tailor-made force field (TMFF) that is unique to the molecule(s) under study. This contrasts with other methods of CSP that assume transferability in the atom–atom potential models used. The TMFF is then used during the structure generation step to sample all the plausible configurations of the molecule(s) within the unit cell. In principle, it is possible to apply DFT-D methods to conduct a final ranking of the stabilities of predicted structures even if the initial structure-generation step employed empirically derived force fields. However, it is important to appreciate that inherent deficiencies in the methodology used to generate the trial structures—such as the neglect of conformational degrees of freedom for flexible molecules—cannot be remedied *via* DFT-D methods even if such methods are used to conduct the final ranking of the crystal lattice energies. Thus, it is important for practitioners of CSP to ensure that the input molecular model used during the structure-generation step is suitable for the system under study. Unlike other methods, the GRACE package simplifies the search for conformational polymorphs because the search algorithm is capable of finding different conformations during the structure generation step without resorting to specifying different input conformations from the outset. The success of DFT-D based CSP methods does come at increased computational cost and this will most likely determine the widespread adoption of such techniques in the future.

1.3 Applications of Computed Crystal Energy Landscapes

1.3.1 Polymorph Screening and Characterisation

One of the original motivations for developing computational methods of CSP was to understand and warn against the risks posed by unexpected polymorphic transformations[65] in organic molecules, particularly with reference to APIs. Although APIs are exposed to extensive manual and/or high-throughput solid form screens as part of the drug development process, there have been cases of disappearing polymorphs[16] even after the characterisation of seemingly stable crystal forms. This suggests that the subject of polymorphism is an ongoing challenge that requires vigilance and a phenomenon that may appear at any time in the lifecycle of a drug regardless of how seemingly exhaustive the polymorph screen may have been. Computational CSP offers the promise of warning against the risks of polymorphic transformations in the parent molecule because the computed CEL contains a wealth of structural information that can be used to map the most significant structural features that distinguish between the most stable lattice energy minima (*i.e.* polymorphs) on the CEL. The likelihood of observing packing polymorphism can be quantified by assessing variations in the intermolecular interactions (*e.g.* hydrogen bonds) between molecules and *via* comparisons of the packing coefficients or crystal densities adopted by the predicted low-energy minima on the CEL. Conformational polymorphs in the parent molecule may be studied by exploring variations in selected torsion angles as a function of the crystal lattice energy and quantifying the relative stability of these conformational polymorphs with respect to the most stable predicted global minimum structure.

Most methods of CSP assume that crystallisation is under thermodynamic control and hence methods for assessing polymorphism also rely on the relative lattice energies of predicted structures. For relatively rigid molecules with no flexible torsional degrees of freedom, the most common method for assessing the risk of polymorphism is to compare the energy of the lattice energy minimum corresponding to the experimental structure with the energy of competitive minima on the CEL. Since rigid molecular structures do not have any torsional degrees of freedom, only the intermolecular energy contribution towards the total crystal energy needs to be optimised. The observation that the experimental structure corresponds to the global minimum in the lattice energy is a good indication that this structure is relatively stable compared to all other possible packings. However, this cannot rule out the finding of more polymorphs in the laboratory since we typically observe many predicted lattice energy minima within a few kJ mol^{-1}. A comprehensive review[66] of the factors behind crystal polymorphism has found that approximately 90% of polymorphs have lattice energies that are within 4 kJ mol^{-1}. For conformational polymorphs, the task of predicting all possible polymorphs is made more challenging by the need to optimise the torsional degrees of freedom during the refinement of the crystal lattice energy—a process that if performed accurately using *ab initio* methods can significantly increase both the accuracy of the final energy rankings and the computational cost of the CSP procedure.

5-methyl-2-[(2-nitrophenyl)amino]-3-thiophenecarbonitrile, also known as 'ROY' for the red, orange and yellow crystals of its polymorphs,[67] has a record seven fully characterised polymorphic structures in the CSD, making it an ideal system for testing and developing computational methods for predicting crystal polymorphs. Recent work[68] applying a multi-stage CSP procedure that considers both the intramolecular and intermolecular contributions towards the crystal lattice energy has successfully predicted all known polymorphs of ROY. However, only five of the seven known polymorphs were found to be in the top 12 lowest energy ranked structures in lattice energy, with the remaining two polymorphs predicted to be at rank 77 and rank 129 on the lattice energy. If the polymorphs predicted to be at rank 77 and 129 were not characterised experimentally, there is a real doubt that any user intervention could have picked them out from the CEL as preferred outcomes of the crystallisation experiment. The CEL of ROY (Figure 1.3) is highly dense with the most stable 150 predicted structures all occurring within 10.7 kJ mol^{-1} of the global minimum. Despite the above challenges, the discovery of most of the known polymorphs of ROY amongst the most stable 12 predicted lattice energy minima on the CEL highlights that a method that considers the energy penalty for varying selected torsional degrees of freedom as part of the lattice energy estimates can lead to a realistic CEL for highly polymorphic systems.

Even after successful characterisation of one or more polymorphs of organic molecules, our reliance on trial-and-error approaches to solid form screens means that it is often not clear if continued crystal form screening for polymorphs is justified. By contrast, successful integration of CSP methods as part of the polymorph screening process has the added value that it can be used to assess the likelihood of finding further polymorphs in screens, often *via* direct assessment of relative energies of predicted and experimental polymorphic structures on the CEL. CSP methods have successfully

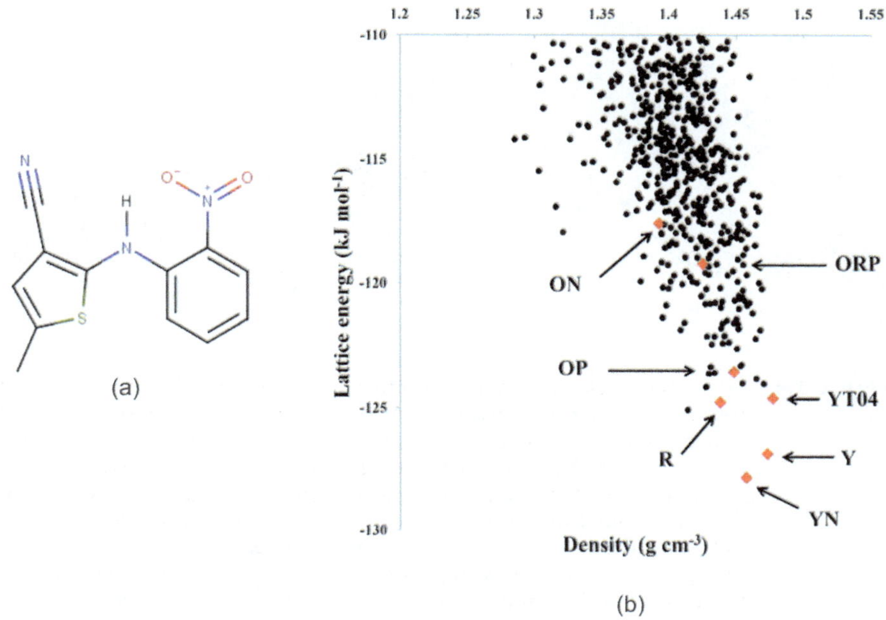

Figure 1.3 (a) Sketch of the molecular structure of ROY and (b) crystal energy landscape for ROY obtained using a comprehensive CSP methodology that considers both the intramolecular and intermolecular contributions to the crystal lattice energy. The red diamonds and labels correspond to predicted structures that match the experimental polymorphs of ROY. The image shown in part (b) is reproduced with permission of the International Union of Crystallography (http://journals.iucr.org/) without any modifications from the source: M. Vasileiadis, A. V. Kazantsev, P. G. Karamertzanis, C. S. Adjiman and C. C. Pantelides, *Acta Crystallogr., Sect. B: Struct. Sci.*, 2012, **68**, 677–685.

been applied to the discovery of a novel high pressure polymorph of dalcetrapib,[69] a cholesterylester transfer protein inhibitor originally developed to raise blood high-density lipoprotein levels. Dalcetrapib has ten torsional degrees of freedom (Figure 1.4) and is thus a very challenging API to model *via* computational methods. Modelling the relative lattice energies of predicted structures on the CEL indicated that the known experimental polymorphs (forms A and B) of dalcetrapib were the most stable at ambient pressures as per lattice energy estimates computed at the DFT level (Figure 1.4) but optimizing the lattice energies of the predicted structures at higher pressures revealed that a previously uncharacterised form C structure was more stable. The results of the CSP procedure were used to successfully target the crystallisation and characterisation of this form C structure in the pressure range 0.02–0.50 GPa. Subsequent experiments revealed that form C is metastable with respect to form A since form C readily converts to form A in solution crystallisation experiments conducted in the pressure range 0.02–0.50 GPa. Crystallisation of form C from the melt at high pressures followed by

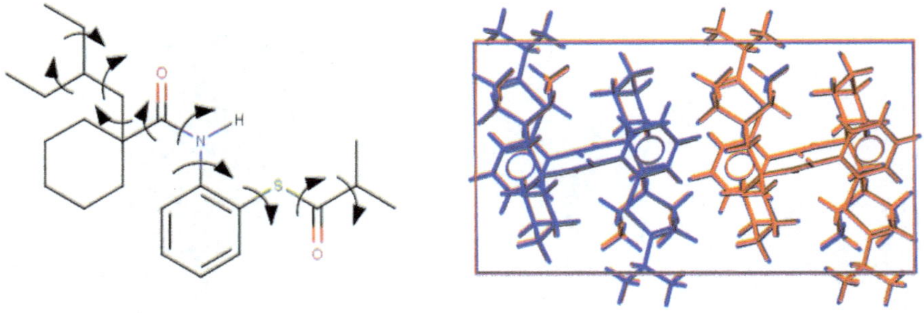

(a) (b)

Figure 1.4 (a) Sketch of the molecular structure of dalcetrapib illustrating all ten torsional degrees of freedom considered during the CSP of this active pharmaceutical ingredient and (b) structural overlays of the experimental crystal structure for form B of dalcetrapib at 100 K (red) with the computer-generated crystal structure 1 (blue). The image shown in part (b) is published under the terms of the CC-BY 4.0 (http://creativecommons.org/licenses/by/4.0/) and has been reproduced without any modifications from the following source: M. A. Neumann, J. van de Streek, F. P. A. Fabbiani, P. Hidber and O. Grassmann, *Nat. Commun.*, 2015, **6**.

subsequent release of the pressure without opening the diamond anvil cell showed that form C could be isolated, but that it converts to form A in a matter of hours. This work shows that computational CSP methods can successfully be integrated with high pressure crystallisation screens to characterise previously unknown polymorphs of highly flexible APIs of the kind typically seen in pharmaceutical drug development pipelines.

1.3.2 Multicomponent Crystal Form Discovery

Multicomponent crystal forms are characterised by two or more chemically distinct molecules or ions in the crystallographic asymmetric unit. Cocrystals, salts, hydrates and ionic cocrystals (ICCs) all fall under the general category of multicomponent crystal forms. The hierarchy[70–72] of hydrogen bond synthons in multicomponent crystal forms suggests that when two molecules form a multicomponent crystal, they adopt predictable hydrogen bonding interactions with each other in the solid state. This complementarity of heterosynthons is generally true for most cocrystals although exceptions are known[73] where cocrystals adopt a mixture of homosynthons and heterosynthons. The carboxylic acid-pyridine heterodimer motif[63] is a typical example of a preferred intermolecular interaction between carboxylic acid and pyridine molecules in cocrystals. Whilst cocrystal heterosynthons have been extensively surveyed using CSD data,[71] and more recently experimental[74,75] and computational[63] approaches, the preferred intermolecular hydrogen bonding interactions in molecular salts, hydrates and ionic cocrystals are yet to be explored in detail. The challenge for CSP methods is that whilst the local structure of the interacting molecules is often predictable to a very high level of accuracy from gas phase *ab initio* electronic structure methods performed with dimers, the long-range order of the molecules inside the crystal quickly becomes very complex to compute accurately as the number of atoms and flexible torsional degrees of freedom in the parent molecules increases.

Cocrystals represent a distinct subset of multicomponent solid forms that are distinguished from salts in that the molecular components are uncharged. Such solid forms have been known for over a century[76] but have witnessed a resurgence of interest in the crystal engineering community over the past decade due to the proven[77–79] utility of cocrystals in modifying the physicochemical properties of the parent molecule. Issa *et al.*[80] tested the hypothesis that cocrystal formation may be predicted on the basis of the lattice energy of the cocrystal being more stable than the sum of the lattice energies of its component molecules (Figure 1.5). This was done by comparing the lattice energies of 12 cocrystals of 4-aminobenzoic acid, eight cocrystals of succinic acid and six cocrystals of caffeine with the sums of the lattice energies of the component molecules. The lattice energies of the cocrystal and component

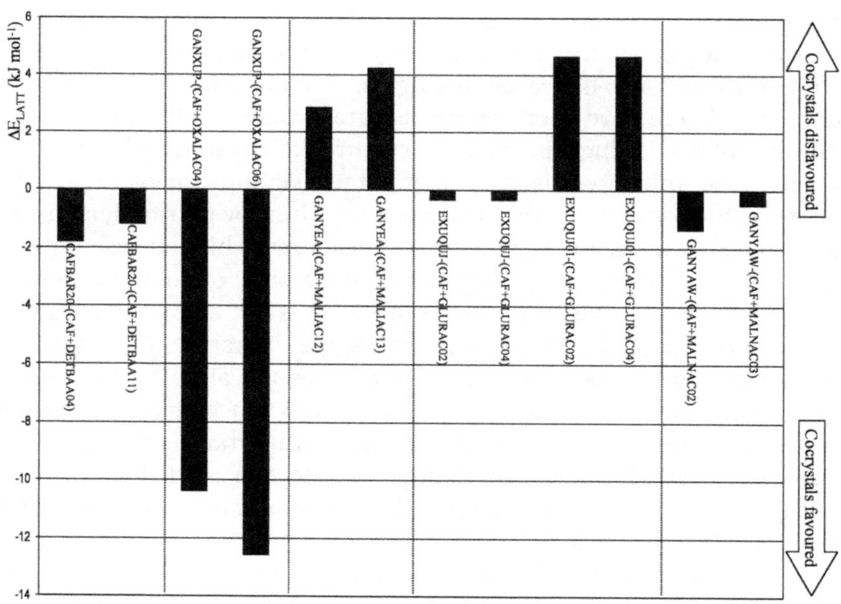

Figure 1.5 Bar chart showing the relative stability of several caffeine cocrystals compared to the sum of the component lattice energies. The cocrystal and the partner crystal structure are denoted by their CSD Refcodes, with each system separated by a vertical line. For ease of comparing the lattice energies, the energies of cocrystals with 2 : 1 stoichiometry are divided by 2 to correspond to one formula unit of caffeine (*i.e.* refer to a 1 : 0.5 cocrystal). Reprinted with permission from N. Issa, P. G. Karamertzanis, G. W. A. Welch and S. L. Price, *Cryst. Growth Des.*, 2009, **9**, 442–453, Copyright 2009 American Chemical Society.

molecules were estimated using an anisotropic atom–atom intermolecular potential derived from the *ab initio* charge distribution of the isolated molecules. The energy penalty for conformational changes in selected flexible torsional degrees of freedom were estimated *ab initio*. Whilst most cocrystal systems were indeed more stable than the sum of their component molecules, the results were not conclusive in suggesting that this method could predict cocrystal formation in all cases. This is because in some instances the relative energy of the cocrystal and its component molecules was found to be too small (*i.e.* <2 kJ mol^{-1}) relative to the inherent errors and assumptions behind the computational model. Of course, not all cocrystals will be thermodynamically preferred outcomes of the crystallisation experiment, so such a method for predicting cocrystal formation based on lattice energy estimates alone will not be valid for all systems.

Hydrate and solvate formation are often serendipitous[81,82] results of the crystallisation process rather than outcomes that are deliberately sought by crystal engineers. Despite this fact, hydrates are an important sub-class of multicomponent crystal forms. Within the pharmaceutical industry, the discovery and characterisation of all solvated forms of drug candidates is an important vehicle for intellectual property creation. Water inclusion into the crystal lattice may create new intermolecular interactions between the API and water molecules, thereby stabilising the hydrated crystal form relative to the anhydrate and hence modifying the solubility of the hydrated drug compound. There have been several reported studies[6,19,83,84] looking at the feasibility of predicting the crystal structures of organic hydrate systems. As is the case for cocrystals, most computational approaches to predicting hydrate formation assume that hydration is driven by thermodynamic considerations. However, attempts to target specific stoichiometries of hydrates as a function of the structural descriptors of the parent and water molecules as well as the properties of the two molecules is not a common feature of most studies reported in the crystal engineering literature. Part of the challenge for crystal engineers is that water is present in almost all organic solvents used in crystallisation experiments and when hydrates are formed, it is very challenging to deconstruct the most important factors that led to hydrate formation. From a computational perspective, modelling the relative intermolecular forces of the organic solvent, water and the parent molecule is a very challenging task.

CSD studies by Aakeröy[85] have shown that salt systems are more prone to crystallising as solvates when compared to cocrystals, reflecting the polarising effect of the charged ions. Computed CELs have shown significant potential in warning against the risks of hydrate formation in pharmaceutically active chloride salts of amantadine and memantine[30] on the basis of the computed solvent accessible volumes in the predicted low-energy structures of the salts (Figure 1.6). The use of calculated solvent accessible volumes in predicted structures alongside traditional lattice energy methods to rank the most likely anhydrate structures that may be susceptible to hydrate formation is an example of the utility of CELs in providing clues as to the likely crystallisation behaviour of candidate molecules. The active ingredients of amantadine

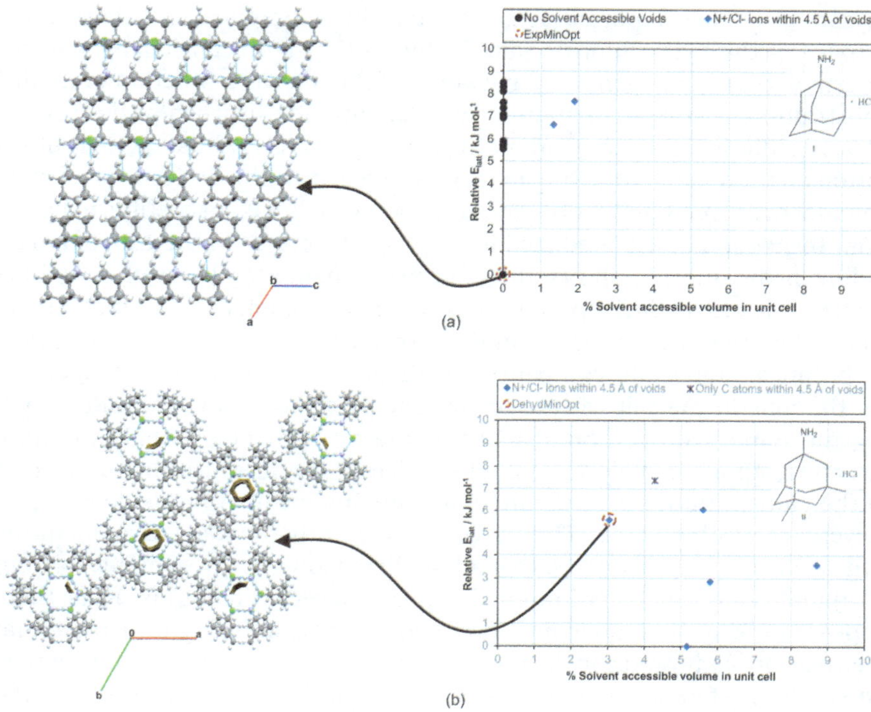

Figure 1.6 Predicted crystal energy landscapes for amantadine hydrochloride (a) and memantine hydrochloride (b) showing that amantadine hydrochloride is less susceptible to hydration when compared to memantine hydrochloride given the lack of many competing low-energy structures with solvent accessible voids within 10 kJ mol^{-1} of the global minimum experimental structure. The lattice energy is plotted relative to the predicted global minimum anhydrate structure for each salt. For the crystal energy landscape (CEL) of amantadine hydrochloride shown in part a, *ExpMinOpt* corresponds to the experimental anhydrate structure of this salt that was calculated using the theoretical model used in the predictions whilst in (b) *DehydMinOpt* corresponds to the calculated lattice energy minimum following dehydration of the channel hydrate structure of memantine hydrochloride. The packing diagrams shown on the left of each CEL corresponds to the anhydrate structure matching the *ExpMinOpt* or *DehydMinOpt*. The CELs shown have been adapted for clarity to show only the predicted structures within 10 kJ mol^{-1} of the predicted global minimum structure for each system. All data correspond to those reported in: S. Mohamed, D. P. Karothu and P. Naumov, *Acta Crystallogr., Sect. B: Struct. Sci., Cryst. Eng. Mater.*, 2016, **72**, 551–561. Reproduced with permission of the International Union of Crystallography (http://journals.iucr.org/).

and memantine have subtle differences in their molecular structures due to the substitution of two methyl groups in the adamantane ring of memantine that are absent in amantadine. However, the hydration profiles of the hydrochlorides of these APIs are very different with amantadine hydrochloride[86] readily crystallising as an anhydrate whilst memantine hydrochloride[87] has to date only been reported as a hydrate. This difference in the hydration profiles of the hydrochloride salts is due to the significant changes in the intermolecular hydrogen bond networks of the ions as a function of methyl substitution on the adamantane ring. Amantadine hydrochloride forms a pleated ribbon hydrogen bond motif [graph set: $C_2^1(4)$] whilst methyl substitution on the adamantane ring in memantine hydrochloride leads to an open channel hydrogen bond motif [graph set: $R_6^3(12)$] that is susceptible to hydration. Inspection of the predicted low-energy structures in the CEL of amantadine hydrochloride shows that the experimentally observed anhydrate structure is energetically preferred as it is approximately 5.6 kJ mol^{-1} more stable than the nearest second ranked structure, and there does not appear to be an energetically competitive hydrogen bond packing arrangement for the ions in the salt with most low-energy structures in the CEL adopting the same hydrogen bond motif as the experimental global minimum structure. Approximately 75% of all predicted structures within 10 kJ mol^{-1} of the global minimum structure of amantadine hydrochloride do not have solvent accessible voids. This is contrasted with the CEL of memantine hydrochloride, which showed that all predicted low-energy structures within 10 kJ mol^{-1} of the global minimum have solvent accessible voids. Thus, there appears to be a thermodynamic preference for crystallising structures that are susceptible to water inclusion in memantine hydrochloride but there is no similar preference in amantadine hydrochloride, in agreement with the observation of a hydrated memantine hydrochloride salt structure.

1.3.3 Structure Solution from Powder X-ray Diffraction Data

Single crystal X-ray diffraction is undoubtedly the gold standard analytical technique for obtaining full 3D crystal structures of organic molecules. However, obtaining suitable single crystals that can be successfully analysed using conventional laboratory or even synchrotron sources of radiation is not always practical. Powder X-ray diffraction (PXRD) techniques therefore represent an important alternative method for obtaining structural data. Solving the full 3D crystal structures of organic molecules from PXRD data is inherently difficult because of peak overlap. Computed CELs offer real value in aiding efforts to solve crystal structures from PXRD data. This is because each lattice energy minimum contained in the CEL is potentially a polymorph that can be observed in the laboratory. Since CELs contain many possible crystal structures, each with full 3D structural information, the predicted structures can be used as starting points to match or even refine against the peak positions and intensities contained in the measured PXRD pattern. This is possible because of matching algorithms that compare the simulated PXRD pattern of the predicted crystal structures with the experimental PXRD pattern. Bond and Jones[88] successfully applied a simulated-annealing CSP methodology as an aid for the structure solution of a series of metallo–organic complexes consisting of Zn, Cu and Ni. Whilst CSP methods are generally good at predicting a set of plausible crystal structures with low energy and high density, discerning which of the many predicted low-energy structures will be observed experimentally is a challenging task and scoring functions based on lattice energy alone are not always reliable. Using CSP to complement structure solution from PXRD data means that initial indexing of the experimental PXRD pattern is not necessary since the lattice parameters and 3D atomic positions are discerned by matching algorithms that directly compare the experimental PXRD pattern against a set of simulated PXRD patterns generated from the predicted structures. The challenge, however, with this approach to structure solution using CSP data is that most algorithms for predicting crystal structures do not consider temperature effects, and hence refer to static lattice energies. Thus, errors are introduced when comparing lattice parameters and cell volumes of computed structures with those expected from the measured PXRD pattern.

Given the very small atomic scattering factor for the hydrogen atom, accurate determination of hydrogen atom positions from PXRD data is difficult. This may necessitate the need to place hydrogen atoms in idealised geometries within the crystal structure. However, small changes in the relative positions of hydrogen atoms can have a significant effect on the intermolecular interactions and packing forces within the crystal, thereby affecting the relative stabilities of crystal structures. Wu *et al.*[89] successfully applied CSP methods to resolve the correct hydrogen atom positions in the experimental structure of the diastereomeric salt 1 : 1 L-ephedrine D-tartrate. The experimental structure of the salt—including all hydrogen atom positions—was initially determined using high-resolution synchrotron PXRD data *via* direct space methods using the FOX[90] algorithm. The experimental structure was indexed and refined in the $P2_1$ space group with 2 formula units in the unit cell. Computational CSP of the plausible crystal structures of the diastereomeric salt confined to the experimental $P2_1$ space group, revealed a range of plausible packings for this salt. Careful inspection of the packing in the predicted global minimum structure and comparison of this structure with that determined from the synchrotron PXRD measurements (Figure 1.7) shows that the closest matching predicted structure displays hydroxyl hydrogen atoms that are positioned to allow interlayer hydrogen bonding interactions whilst the experimental structure does not show this feature. Moreover, the fact that the experimental structure undergoes significant structural distortions upon lattice energy minimisation, whilst remaining relatively unstable when compared to the alternative predicted crystal structures in the same space group, suggested that the hydrogen atom positions in the experimental structure were wrongly assigned. The final Rietveld refinement of the crystal structure of the diastereomeric salt using the closest matching predicted structure from the CEL led to almost identical R_{wp} and χ^2 values, suggesting that the computed structure is a good match for the experimental PXRD data. The good match between the final structural model and the experimental data, coupled with the fact that the computed structure displays sensible interlayer hydrogen bonding interactions, provides a strong basis for confidence regarding the final structural model for 1 : 1 L-ephedrine

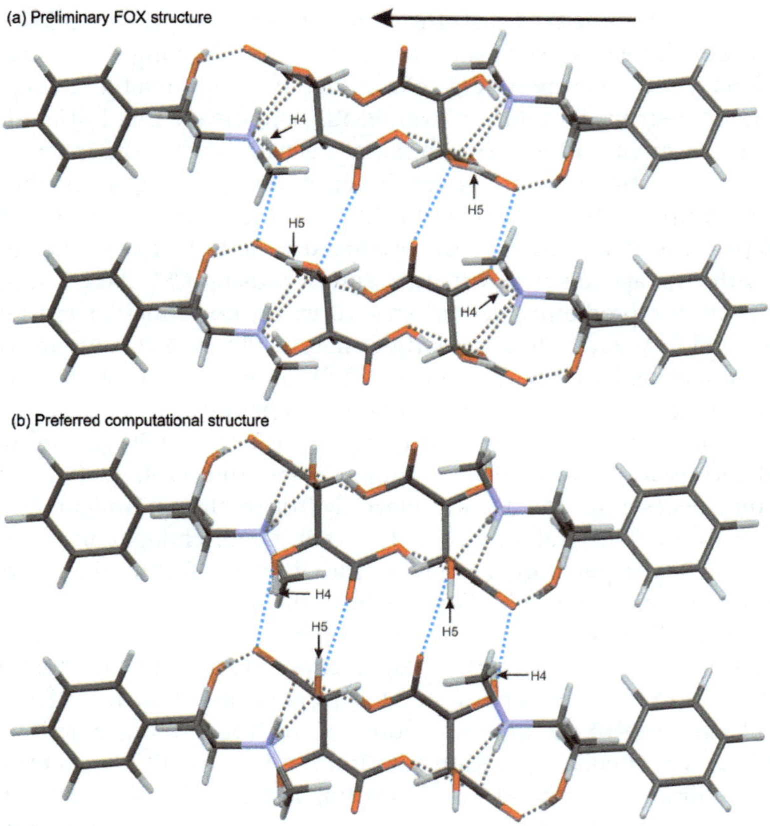

Figure 1.7 Comparison of the crystal packing as viewed down the *b*-axis for (a) the preliminary structure of 1:1 L-ephedrine D-tartrate determined from synchrotron PXRD data with hydrogen atoms placed in idealised geometric positions; (b) the preferred computational structure based on lattice energy calculations that most closely matches the preliminary structure from PXRD data. The hydroxyl hydrogen atoms labelled H4 and H5 have different positions in the two structures. Hydrogen bonds are indicated by dashed lines whilst inter-layer hydrogen bonds are shown in cyan. The large arrow at the top indicates the inter-layer slippage direction following computational relaxation of the crystal lattice energy. Reproduced from ref. 89 with permission from the Royal Society of Chemistry.

D-tartrate. This work highlights the value of integrating CSP methods as part of efforts to determine organic molecular crystal structures from PXRD data.

In recent years, nuclear magnetic resonance (NMR) spectroscopy has also shown potential in aiding the structure solution of powder samples.[91] The basis of the technique is that comparisons of measured ^1H solid-state NMR chemical shifts with those calculated from DFT methods can provide a strong indication of the most likely crystal structure for the powder sample. CSP methods provide a natural means for generating the trial crystal structures needed for the DFT calculations. The correct crystal structure can be identified by comparing the root-mean-squared deviation between the experimental chemical shifts and the chemical shifts of the trial structures generated from CSP methods. The method has already been applied[5,20] to a range of drug molecules and has shown potential in correctly identifying the known crystal structures of cocaine, flutamide and flufenamic acid. Further work by Day and Emsley[92] has shown that the method can also be used to characterise the previously unknown structure of a high molecular weight flexible organic compound starting only with the molecular formula. This provides the foundation for integrating this procedure as part of industrial efforts to characterise the polymorphs of APIs when structure solution *via* single crystal X-ray diffraction methods is not possible.

1.4 CCDC Blind Tests: Assessing Progress in Crystal Structure Prediction Methods (1999–2016)

Since 1999, the Cambridge Crystallographic Data Centre (CCDC) has been organising periodic blind tests[4,6,55,93–95] to assess progress in computational methods for predicting organic crystal structures. This is analogous to the CASP (critical assessment of protein structure prediction)[96] exercises held within the protein structure prediction community. The key motivation behind the CCDC blind tests is to provide a forum for the practitioners of CSP to meet every 3–5 years with the view of objectively assessing the performance of various methods of CSP. Historically, participants were given a list of 3–6 target systems using only the 2D molecular diagram showing the atomic connectivity as an indication of the molecular structure. In recent years, the blind tests have also included the crystallisation conditions needed to successfully target the crystal structure. The participants are usually given 12 months to carry out their computational modelling on some or all the target molecules proposed and are asked to submit the full 3D crystal structure(s) they think would correspond to the experimental structure based on some ranking method (often lattice energy) for deciding amongst the many thousands of predicted crystal structures. In the past, a successful prediction was judged on the basis of the submission of three 'guesses' from the modelling as to what the experimental structure would be, and comparison of these three computed candidate structures with the correct experimental structure held by an independent referee. However, in the most recent sixth blind test of CSP,[55] the organisers decided to permit the submission of a list of 100 predicted structures for each target molecule and optionally a second list of up to 100 structures that may have been generated using a different ranking methodology to the first list. This change in how we assess correct predictions reflects the fact that CELs contain a wealth of information about the crystallisation conditions of the molecule under study and that assigning an arbitrary cut-off of three structures for assessing the accuracy of the CSP prediction neither helps in method development nor in assessing the subtle details about the crystallisation behaviour of the target molecule that can only be discerned by surveying a wider range of structures on the CEL. In recent blind tests, the target molecules chosen have varied from the traditional categories of flexible or polymorphic organic molecules to salts, hydrates, cocrystals and flexible organic molecules that are good analogues of APIs. This gradual evolution in the target molecules proposed in the blind tests reflects the changing interests of crystal engineers and the current focus on the use of intermolecular hydrogen bonding interactions to try and control the crystallisation behaviour of organic molecules in the form of multicomponent crystal forms such as cocrystals.

The target molecules (Table 1.1, labelled **XXII–XXVI**) proposed in the sixth blind test of CSP build on previous blind tests with the notable decision to specify distinct target categories for salt (**XXIV**) and cocrystal (**XXV**) molecular structures in line with growing community interest according to literature reports of experimental[97–100] and computational[30,101–103] studies attempting to understand and control the structures and properties of these multicomponent solid forms. A highly flexible organic molecule (**XXVI**) with several torsional degrees of freedom was also proposed as a target, reflecting the obvious application of CSP methods to molecules of pharmaceutical interest that tend to be flexible. Most participating research groups attempted the simplest category represented by **XXII**, and this structure was successfully generated 12 times. Historically, previous blind tests have shown that whilst the molecular structures of such 'rigid' systems tend to be fairly easy to model using electronic structure methods that can calculate the global minimum conformation with high confidence, the presence of unusual functional groups in the molecular structure (*e.g.* the nitrile and sulphur groups present in **XXII**) can make the CSP of simple rigid molecules very demanding, especially if the atom–atom intermolecular potentials used have not been parameterised on organic molecules with similar functional groups. For **XXIII**, which is a former drug candidate with several torsional degrees of freedom and five known polymorphs consisting of Z′ = 1 (forms A, B and D) and Z′ = 2 (forms C and E) structures, there were some successes by participants in correctly predicting the three polymorphs with Z′ = 1. The experimental structures of all five polymorphs of **XXIII** display an $R_2^2(8)$ dimer hydrogen bond motif between the carboxylic acid groups but there are significant differences in the molecular conformations of form A compared to forms B and D, requiring a robust search algorithm that samples a number of conformations in order to locate and accurately rank the relative energies of these polymorphs. Forms C and E are even more demanding of the search and ranking algorithms as there are added degrees of freedom as a result of these polymorphs comprising Z′ = 2 structures as well as the challenge posed by the two molecules in form E having unique molecular conformations when compared to the conformations adopted by all other polymorphs of **XXIII**. Slurrying experiments confirmed that form A is the most stable polymorph of **XXIII** at 257 K whilst form D is the most stable polymorph at 293 K. The role of temperature in affecting the relative stability of such polymorphs

Table 1.1 Summary of the molecular structures, crystallisation conditions and statistics from the modelling of the target systems studied in the sixth blind test of organic crystal structure prediction methods.[55] The labels (**XXII–XXVI**) for the target molecular structures refer to those used in the blind test. [+] For system **XXII**, the number of CPU hours used in the competition is significantly affected by the resources expended by one group who used approximately 30 million CPU hours on just this system. As a consequence, we show in brackets the total number of CPU hours for **XXII** used by all remaining groups. The information in this table has been obtained from ref. 55. Reproduced with permission of the International Union of Crystallography (http://journals.iucr.org/).

Target system	Crystallisation conditions and remarks	Number of attempted predictions	Number of successful predictions	Total CPU hours used in competition (millions)*
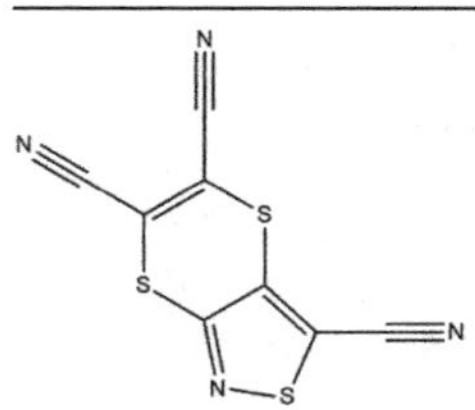 **XXII**	Acetone/water mixture used for crystallisation. Possesses chiral-like character due to potential flexibility of the six-membered ring. However, no chiral precursors were used in the synthesis	21	12	31.72 (1.72)[+]
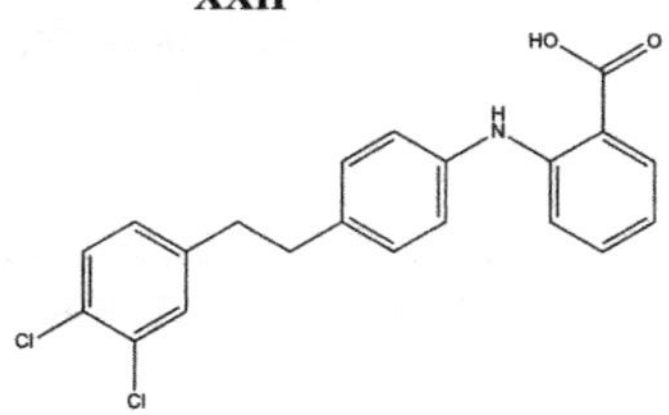 **XXIII**	Five polymorphs (*A–E*) are known. Three (*A*, *B*, *D*) have Z′ = 1, two (*C* and *E*) have Z′ = 2. The most stable polymorphs at 257 and 293 K are both Z′ = 1. Crystallisation conditions include slow evaporation of acetone solution and of ethyl acetate: water mixture	*A*, *B* and *D*: 14; *C* and *E*: 3.	*A*: 4, *B*: 8, *C*: 1, *D*: 3, *E*: 0	1.63
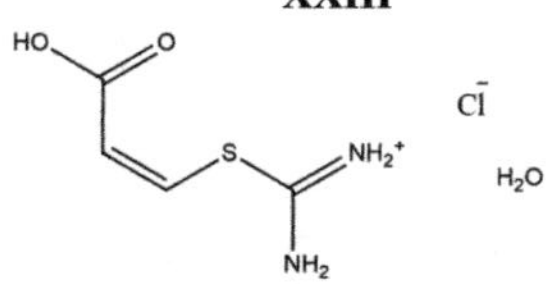 **XXIV**	Crystallised from 1 M HCl solution. The substituents of the C=C double bond are in the *cis*-configuration	8	1	0.71

(*continued*)

Table 1.1 (*continued*)

Target system	Crystallisation conditions and remarks	Number of attempted predictions	Number of successful predictions	Total CPU hours used in competition (millions)*
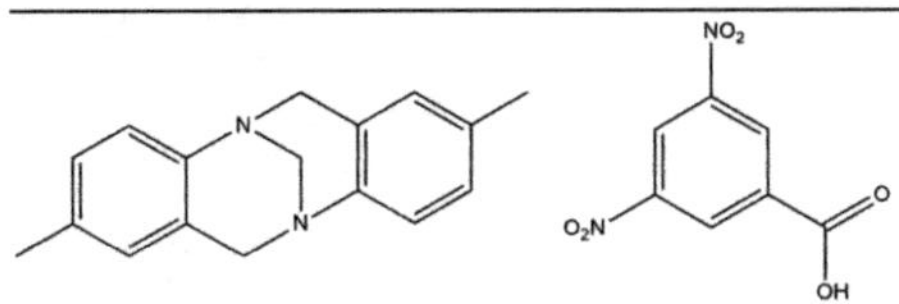 **XXV**	Crystallised from slow evaporation of a methanol solution, which contained a racemic mixture of the enantiomers of Tröger's base	14	5	0.96
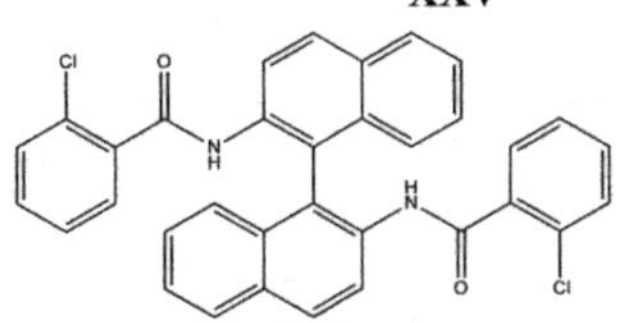 **XXVI**	Crystallised from slow evaporation of a 1:1 mixture of hexane and dichloromethane. No chiral precursors used in the synthesis	12	3	1.52

is a challenge for CSP methods to account for given the static lattice energies employed by most practitioners during the structure generation stage. However, there were some promising signs from this blind test that the use of secondary ranking techniques such as DFT-D methods and estimates of Helmholtz free energies can improve the rankings of predicted structures. In the case of form D, the use of the PBE + MBD functional improved the relative lattice energy ranking of form D to 14th and this was further improved to rank 2 when Helmholtz free energies were estimated at 300 K. The increasing application of DFT methods to rank predicted crystal structures is illustrated by the fact that nearly half of all groups participating in the sixth blind test employed DFT methods, which compares with the fifth blind test where only two groups used DFT energies to rank the predicted structures. In the sixth blind test of CSP, five groups correctly predicted the crystal structure of the cocrystal **XXV** and only three groups correctly predicted the crystal structure of the highly flexible organic molecule **XXVI**. The biggest challenge for most methods of CSP appeared to be for the salt hydrate system **XXIV**. Only eight participants attempted this target system and of those only one group correctly predicted the experimental structure, albeit at rank 2 in energy. This reflects the challenges of modelling a system with three independent species in the asymmetric unit and the difficulty with accurately computing the relative energies of charged systems.

The blind tests have shown that whilst most methods of CSP have varying degrees of success with predicting the crystal structures of different systems, they key differentiating factor between the methods that directly affects their success rate in the blind tests is the structure ranking methodology used and the accuracy of the underlying theoretical models used to rank computed crystal structures. DFT-D methods are increasingly proving[6,104,105] to be better suited in producing CELs where the experimental structure is the most stable structure or amongst the lowest energy ranked structures. The success of the GRACE package,[104,106] which uses a TMFF to carry out the initial structure generation process followed by re-ranking of the low-energy structures produced by the TMFF using a more accurate DFT-D method,[107] illustrates that improved accuracy in structure prediction and ranking may be achieved by moving away from empirical force-field methods for sampling the conformations and lattice energies of organic molecules. The blind tests have also illustrated the importance of sampling the anisotropic environment of molecules during the crystallisation process, as this can have a significant effect on the molecular conformations and packings adopted by molecules. Some methods[7,55] have successfully integrated polarisable continuum models[108] to approximate the polarisation of the molecular charge distribution within the crystal. However, the reliance on lattice energies and the assumption of no entropic contributions towards the crystal growth and stability means that there is still more room for improvement. However, overall, the blind tests demonstrate the value of regular community-wide assessments of the state of the art in a specific area of science. Future progress in CSP methods will be realised by the continued cooperation of experimentalists and theoreticians working at the interface of solid state chemistry.

1.5 Conclusion

Computational methods of CSP are increasingly being applied to a range of solid-state chemistry problems. These include, but are not limited to, those discussed in this chapter, such as assessing the risks posed by crystal polymorphism and quantifying the risks of crystal hydration in molecular organic salts. CSP methods have shown enormous potential in aiding experimental solid-state chemists and crystallographers in their efforts to design, synthesise and characterise the crystal structures of a range of systems such as polymorphs, salts, cocrystals and hydrates. CSP methods are powerful sources of structural information such as the preferred intermolecular hydrogen bonding geometries of cocrystal molecules. DFT-D methods are increasingly proving to be more accurate for ranking the relative energies of predicted crystal structures when compared to the use of empirically derived atom–atom intermolecular potential models. Further progress in the development of accurate theoretical methods for estimating intermolecular energies in molecular crystals at low relative computational cost will be central to the continued advance in the accuracy of CSP methods.

Acknowledgements

Khalifa University of Science and Technology is acknowledged for providing seed funding and access to the high-performance computing cluster that has been instrumental to the research efforts of our group over the past few years.

References

1. G. R. Desiraju, *Nat. Mater.*, 2002, **1**, 77–79.
2. J. Maddox, *Nature*, 1988, **335**, 201.
3. S. M. Woodley and R. Catlow, *Nat. Mater.*, 2008, **7**, 937–946.
4. G. M. Day, T. G. Cooper, A. J. Cruz-Cabeza, K. E. Hejczyk, H. L. Ammon, S. X. M. Boerrigter, J. S. Tan, R. G. Della Valle, E. Venuti, J. Jose, S. R. Gadre, G. R. Desiraju, T. S. Thakur, B. P. van Eijck, J. C. Facelli, V. E. Bazterra, M. B. Ferraro, D. W. M. Hofmann, M. A. Neumann, F. J. J. Leusen, J. Kendrick, S. L. Price, A. J. Misquitta, P. G. Karamertzanis, G. W. A. Welch, H. A. Scheraga, Y. A. Arnautova, M. U. Schmidt, J. van de Streek, A. K. Wolf and B. Schweizer, *Acta Crystallogr., Sect. B: Struct. Sci.*, 2009, **65**, 107–125.
5. E. Salager, G. M. Day, R. S. Stein, C. J. Pickard, B. Elena and L. Emsley, *J. Am. Chem. Soc.*, 2010, **132**, 2564–2566.
6. D. A. Bardwell, C. S. Adjiman, Y. A. Arnautova, E. Bartashevich, S. X. M. Boerrigter, D. E. Braun, A. J. Cruz-Cabeza, G. M. Day, R. G. Della Valle, G.

R. Desiraju, B. P. van Eijck, J. C. Facelli, M. B. Ferraro, D. Grillo, M. Habgood, D. W. M. Hofmann, F. Hofmann, K. V. J. Jose, P. G. Karamertzanis, A. V. Kazantsev, J. Kendrick, L. N. Kuleshova, F. J. J. Leusen, A. V. Maleev, A. J. Misquitta, S. Mohamed, R. J. Needs, M. A. Neumann, D. Nikylov, A. M. Orendt, R. Pal, C. C. Pantelides, C. J. Pickard, L. S. Price, S. L. Price, H. A. Scheraga, J. van de Streek, T. S. Thakur, S. Tiwari, E. Venuti and I. K. Zhitkov, *Acta Crystallogr., Sect. B: Struct. Sci.*, 2011, **67**, 535–551.

7. A. V. Kazantsev, P. G. Karamertzanis, C. S. Adjiman, C. C. Pantelides, S. L. Price, P. T. Galek, G. M. Day and A. J. Cruz-Cabeza, *Int. J. Pharm.*, 2011, **418**, 168–178.
8. M. A. Neumann and M.-A. Perrin, *CrystEngComm*, 2009, **11**, 2475–2479.
9. A. Gavezzotti, *Modell. Simul. Mater. Sci. Eng.*, 2002, **10**, R1.
10. G. Filippini and A. Gavezzotti, *Acta Crystallogr., Sect. B: Struct. Sci.*, 1993, **49**, 868–880.
11. S. L. Price, *CrystEngComm*, 2004, **6**, 344–353.
12. W. T. Mooij, B. P. van Eijck and J. Kroon, *J. Phys. Chem. A*, 1999, **103**, 9883–9890.
13. A. Stone, *The Theory of Intermolecular Forces*, OUP Oxford, 2013.
14. M. Svärd and Å. C. Rasmuson, *Ind. Eng. Chem. Res.*, 2009, **48**, 2899–2912.
15. J. D. Dunitz and J. Bernstein, *Acc. Chem. Res.*, 1995, **28**, 193–200.
16. D.-K. Bučar, R. W. Lancaster and J. Bernstein, *Angew. Chem., Int. Ed.*, 2015, **54**, 6972–6993.
17. P. G. Karamertzanis, P. R. Anandamanoharan, P. Fernandes, P. W. Cains, M. Vickers, D. A. Tocher, A. J. Florence and S. L. Price, *J. Phys. Chem. B*, 2007, **111**, 5326–5336.
18. M. D. Gourlay, J. Kendrick and F. J. J. Leusen, *Cryst. Growth Des.*, 2008, **8**, 2899–2905.
19. D. E. Braun, P. G. Karamertzanis and S. L. Price, *Chem. Commun.*, 2011, **47**, 5443–5445.
20. M. Baias, C. M. Widdifield, J.-N. Dumez, H. P. G. Thompson, T. G. Cooper, E. Salager, S. Bassil, R. S. Stein, A. Lesage, G. M. Day and L. Emsley, *Phys. Chem. Chem. Phys.*, 2013, **15**, 8069–8080.
21. C. J. Manly, S. Louise-May and J. D. Hammer, *Drug Discovery Today*, 2001, **6**, 1101–1110.
22. S. Greengard, *Commun. ACM*, 2014, **57**, 21–23.
23. M. A. Neumann, *J. Phys. Chem. B*, 2008, **112**, 9810–9829.
24. P. G. Karamertzanis and C. C. Pantelides, *J. Comput. Chem.*, 2005, **26**, 304–324.
25. C. W. Glass, A. R. Oganov and N. Hansen, *Comput. Phys. Commun.*, 2006, **175**, 713–720.
26. H. R. Karfunkel and R. J. Gdanitz, *J. Comput. Chem.*, 1992, **13**, 1171–1183.
27. J. R. Holden, Z. Du and H. L. Ammon, *J. Comput. Chem.*, 1993, **14**, 422–437.
28. C. R. Groom, I. J. Bruno, M. P. Lightfoot and S. C. Ward, *Acta Crystallogr., Sect. B: Struct. Sci., Cryst. Eng. Mater.*, 2016, **72**, 171–179.
29. A. T. Hulme, S. L. Price and D. A. Tocher, *J. Am. Chem. Soc.*, 2005, **127**, 1116–1117.

30. S. Mohamed, D. P. Karothu and P. Naumov, *Acta Crystallogr., Sect. B: Struct. Sci., Cryst. Eng. Mater.*, 2016, **72**, 551–561.
31. A. J. Florence, A. Johnston, S. L. Price, H. Nowell, A. R. Kennedy and N. Shankland, *J. Pharm. Sci.*, 2006, **95**, 1918–1930.
32. A. Getsoian, R. M. Lodaya and A. C. Blackburn, *Int. J. Pharm.*, 2008, **348**, 3–9.
33. M. A. O'Mahony, D. M. Croker, A. C. Rasmuson, S. Veesler and B. K. Hodnett, *Org. Process Res. Dev.*, 2013, **17**, 512–518.
34. J.-B. Arlin, L. S. Price, S. L. Price and A. J. Florence, *Chem. Commun.*, 2011, **47**, 7074–7076.
35. W. T. A. Harrison, H. S. Yathirajan and H. G. Anilkumar, *Acta Crystallogr., Sect. C: Cryst. Struct. Commun.*, 2006, **62**, o240–o242.
36. J. Nyman and G. M. Day, *CrystEngComm*, 2015, **17**, 5154–5165.
37. A. E. Goeta and J. A. K. Howard, *Chem. Soc. Rev.*, 2004, **33**, 490–500.
38. S. Mohamed, S. A. Barnett, D. A. Tocher, S. L. Price, K. Shankland and C. K. Leech, *CrystEngComm*, 2008, **10**, 399–404.
39. S. L. Price, M. Leslie, G. W. A. Welch, M. Habgood, L. S. Price, P. G. Karamertzanis and G. M. Day, *Phys. Chem. Chem. Phys.*, 2010, **12**, 8478–8490.
40. A. Laio and M. Parrinello, *Proc. Natl. Acad. Sci. U. S. A.*, 2002, **99**, 12562–12566.
41. R. Martoňák, A. Laio and M. Parrinello, *Phys. Rev. Lett.*, 2003, **90**, 075503.
42. G. M. Day, S. L. Price and M. Leslie, *J. Phys. Chem. B*, 2003, **107**, 10919–10933.
43. T. Zykova-Timan, P. Raiteri and M. Parrinello, *J. Phys. Chem. B*, 2008, **112**, 13231–13237.
44. H. P. G. Thompson and G. M. Day, *Chem. Sci.*, 2014, **5**, 3173–3182.
45. M. Connolly, *J. Appl. Crystallogr.*, 1983, **16**, 548–558.
46. K. Burke, *J. Chem. Phys.*, 2012, **136**, 150901.
47. A. V. Kazantsev, P. G. Karamertzanis, C. S. Adjiman and C. C. Pantelides, *J. Chem. Theory Comput.*, 2011, **7**, 1998–2016.
48. A. Gavezzotti, in *Zeitschrift für Kristallographie - Crystalline Materials*, 2005, vol. 220, p. 499.
49. C. A. Farberow, J. A. Dumesic and M. Mavrikakis, *ACS Catal.*, 2014, **4**, 3307–3319.
50. V. Pallassana, M. Neurock, V. S. Lusvardi, J. J. Lerou, D. D. Kragten and R. A. van Santen, *J. Phys. Chem. B*, 2002, **106**, 1656–1669.
51. L. Yu, X. Pan, X. Cao, P. Hu and X. Bao, *J. Catal.*, 2011, **282**, 183–190.
52. A. Jain, Y. Shin and K. A. Persson, *Nat. Rev. Mater.*, 2016, **1**, 15004.
53. F. Neese, *Coord. Chem. Rev.*, 2009, **253**, 526–563.
54. C.-G. Zhan, J. A. Nichols and D. A. Dixon, *J. Phys. Chem. A*, 2003, **107**, 4184–4195.
55. A. M. Reilly, R. I. Cooper, C. S. Adjiman, S. Bhattacharya, A. D. Boese, J. G. Brandenburg, P. J. Bygrave, R. Bylsma, J. E. Campbell, R. Car, D. H. Case, R. Chadha, J. C. Cole, K. Cosburn, H. M. Cuppen, F. Curtis, G. M. Day, R. A. DiStasio Jr, A. Dzyabchenko, B. P. van Eijck, D. M. Elking, J. A.

van den Ende, J. C. Facelli, M. B. Ferraro, L. Fusti-Molnar, C.-A. Gatsiou, T. S. Gee, R. de Gelder, L. M. Ghiringhelli, H. Goto, S. Grimme, R. Guo, D. W. M. Hofmann, J. Hoja, R. K. Hylton, L. Iuzzolino, W. Jankiewicz, D. T. de Jong, J. Kendrick, N. J. J. de Klerk, H.-Y. Ko, L. N. Kuleshova, X. Li, S. Lohani, F. J. J. Leusen, A. M. Lund, J. Lv, Y. Ma, N. Marom, A. E. Masunov, P. McCabe, D. P. McMahon, H. Meekes, M. P. Metz, A. J. Misquitta, S. Mohamed, B. Monserrat, R. J. Needs, M. A. Neumann, J. Nyman, S. Obata, H. Oberhofer, A. R. Oganov, A. M. Orendt, G. I. Pagola, C. C. Pantelides, C. J. Pickard, R. Podeszwa, L. S. Price, S. L. Price, A. Pulido, M. G. Read, K. Reuter, E. Schneider, C. Schober, G. P. Shields, P. Singh, I. J. Sugden, K. Szalewicz, C. R. Taylor, A. Tkatchenko, M. E. Tuckerman, F. Vacarro, M. Vasileiadis, A. Vazquez-Mayagoitia, L. Vogt, Y. Wang, R. E. Watson, G. A. de Wijs, J. Yang, Q. Zhu and C. R. Groom, *Acta Crystallogr., Sect. B: Struct. Sci., Cryst. Eng. Mater.*, 2016, **72**, 439–459.

56. S. Grimme, *J. Comput. Chem.*, 2006, **27**, 1787–1799.
57. A. Tkatchenko and M. Scheffler, *Phys. Rev. Lett.*, 2009, **102**, 073005.
58. S. Grimme, J. Antony, S. Ehrlich and H. Krieg, *J. Chem. Phys.*, 2010, **132**, 154104.
59. J. Klimeš and A. Michaelides, *J. Chem. Phys.*, 2012, **137**, 120901.
60. A. M. Reilly and A. Tkatchenko, *J. Chem. Phys.*, 2013, **139**, 024705.
61. A. Otero-de-la-Roza and E. R. Johnson, *J. Chem. Phys.*, 2012, **137**, 054103.
62. J. Nyman, O. S. Pundyke and G. M. Day, *Phys. Chem. Chem. Phys.*, 2016, **18**, 15828–15837.
63. S. Mohamed, D. A. Tocher, M. Vickers, P. G. Karamertzanis and S. L. Price, *Cryst. Growth Des.*, 2009, **9**, 2881–2889.
64. I. M. Sobol, *USSR Comput. Math. Math. Phys.*, 1967, **7**, 86–112.
65. J. Bauer, S. Spanton, R. Henry, J. Quick, W. Dziki, W. Porter and J. Morris, *Pharm. Res.*, 2001, **18**, 859–866.
66. A. J. Cruz-Cabeza, S. M. Reutzel-Edens and J. Bernstein, *Chem. Soc. Rev.*, 2015, **44**, 8619–8635.
67. L. Yu, *Acc. Chem. Res.*, 2010, **43**, 1257–1266.
68. M. Vasileiadis, A. V. Kazantsev, P. G. Karamertzanis, C. S. Adjiman and C. C. Pantelides, *Acta Crystallogr., Sect. B: Struct. Sci.*, 2012, **68**, 677–685.
69. M. A. Neumann, J. van de Streek, F. P. A. Fabbiani, P. Hidber and O. Grassmann, *Nat. Commun.*, 2015, **6**, 7793.
70. N. K. Duggirala, G. P. Wood, A. Fischer, Ł. Wojtas, M. L. Perry and M. J. Zaworotko, *Cryst. Growth Des.*, 2015, **15**, 4341–4354.
71. T. R. Shattock, K. K. Arora, P. Vishweshwar and M. J. Zaworotko, *Cryst. Growth Des.*, 2008, **8**, 4533–4545.
72. D. R. Weyna, M. L. Cheney, N. Shan, M. Hanna, L. Wojtas and M. J. Zaworotko, *CrystEngComm*, 2012, **14**, 2377–2380.
73. B. R. Sreekanth, P. Vishweshwar and K. Vyas, *Chem. Commun.*, 2007, 2375–2377.
74. P. Vishweshwar, N. Jagadeesh Babu, A. Nangia, S. A. Mason, H. Puschmann, R. Mondal and J. A. Howard, *J. Phys. Chem. A*, 2004, **108**, 9406–9416.

75. K. Maruyoshi, D. Iuga, O. N. Antzutkin, A. Alhalaweh, S. P. Velaga and S. P. Brown, *Chem. Commun.*, 2012, **48**, 10844–10846.
76. F. Wöhler, *Justus Liebigs Ann. Chem.*, 1844, **51**, 145–163.
77. M. B. Hickey, M. L. Peterson, L. A. Scoppettuolo, S. L. Morrisette, A. Vetter, H. Guzmán, J. F. Remenar, Z. Zhang, M. D. Tawa, S. Haley, M. J. Zaworotko and Ö. Almarsson, *Eur. J. Pharm. Biopharm.*, 2007, **67**, 112–119.
78. A. J. Smith, P. Kavuru, L. Wojtas, M. J. Zaworotko and R. D. Shytle, *Mol. Pharm.*, 2011, **8**, 1867–1876.
79. N. J. Babu and A. Nangia, *Cryst. Growth Des.*, 2011, **11**, 2662–2679.
80. N. Issa, P. G. Karamertzanis, G. W. A. Welch and S. L. Price, *Cryst. Growth Des.*, 2009, **9**, 442–453.
81. K. Shankland, C. K. Leech, S. Mohamed, S. A. Barnett and D. A. Tocher, *Acta Crystallogr., Sect. E: Struct. Rep. Online*, 2007, **63**, o3574.
82. S. Mohamed, S. A. Barnett and D. A. Tocher, *Acta Crystallogr., Sect. E: Struct. Rep. Online*, 2007, **63**, o3575.
83. A. T. Hulme and S. L. Price, *J. Chem. Theory Comput.*, 2007, **3**, 1597–1608.
84. D. E. Braun, R. M. Bhardwaj, J.-B. Arlin, A. J. Florence, V. Kahlenberg, U. J. Griesser, D. A. Tocher and S. L. Price, *Cryst. Growth Des.*, 2013, **13**, 4071–4083.
85. C. B. Aakeröy, M. E. Fasulo and J. Desper, *Mol. Pharm.*, 2007, **4**, 317–322.
86. F. Belanger-Gariepy, F. Brisse, P. D. Harvey, I. S. Butler and D. F. R. Gilson, *Acta Crystallogr., Sect. C: Cryst. Struct. Commun.*, 1987, **43**, 756–759.
87. W.-J. Lou, X.-R. Hu and J.-M. Gu, *Acta Crystallogr., Sect. E: Struct. Rep. Online*, 2009, **65**, o2191.
88. A. D. Bond and W. Jones, *Acta Crystallogr., Sect. B: Struct. Sci.*, 2002, **58**, 233–243.
89. H. Wu, M. Habgood, J. E. Parker, N. Reeves-McLaren, J. K. Cockcroft, M. Vickers, A. R. West and A. G. Jones, *CrystEngComm*, 2013, **15**, 1853–1859.
90. V. Favre-Nicolin and R. Cerny, *J. Appl. Crystallogr.*, 2002, **35**, 734–743.
91. B. Elena and L. Emsley, *J. Am. Chem. Soc.*, 2005, **127**, 9140–9146.
92. M. Baias, J.-N. Dumez, P. H. Svensson, S. Schantz, G. M. Day and L. Emsley, *J. Am. Chem. Soc.*, 2013, **135**, 17501–17507.
93. J. P. M. Lommerse, W. D. S. Motherwell, H. L. Ammon, J. D. Dunitz, A. Gavezzotti, D. W. M. Hofmann, F. J. J. Leusen, W. T. M. Mooij, S. L. Price, B. Schweizer, M. U. Schmidt, B. P. van Eijck, P. Verwer and D. E. Williams, *Acta Crystallogr., Sect. B: Struct. Sci.*, 2000, **56**, 697–714.
94. W. D. S. Motherwell, H. L. Ammon, J. D. Dunitz, A. Dzyabchenko, P. Erk, A. Gavezzotti, D. W. M. Hofmann, F. J. J. Leusen, J. P. M. Lommerse, W. T. M. Mooij, S. L. Price, H. Scheraga, B. Schweizer, M. U. Schmidt, B. P. van Eijck, P. Verwer and D. E. Williams, *Acta Crystallogr., Sect. B: Struct. Sci.*, 2002, **58**, 647–661.
95. G. M. Day, W. D. S. Motherwell, H. L. Ammon, S. X. M. Boerrigter, R. G. Della Valle, E. Venuti, A. Dzyabchenko, J. D. Dunitz, B. Schweizer, B. P. van Eijck, P. Erk, J. C. Facelli, V. E. Bazterra, M. B. Ferraro, D. W. M. Hofmann, F. J. J. Leusen, C. Liang, C. C. Pantelides, P. G. Karamertzanis,

S. L. Price, T. C. Lewis, H. Nowell, A. Torrisi, H. A. Scheraga, Y. A. Arnautova, M. U. Schmidt and P. Verwer, *Acta Crystallogr., Sect. B: Struct. Sci.*, 2005, **61**, 511–527.
96. J. Moult, K. Fidelis, A. Kryshtafovych, T. Schwede and A. Tramontano, *Proteins: Struct., Funct., Bioinf.*, 2016, **84**, 4–14.
97. K. B. Landenberger and A. J. Matzger, *Cryst. Growth Des.*, 2010, **10**, 5341–5347.
98. H. D. Clarke, K. K. Arora, H. Bass, P. Kavuru, T. T. Ong, T. Pujari, L. Wojtas and M. J. Zaworotko, *Cryst. Growth Des.*, 2010, **10**, 2152–2167.
99. S. L. Childs, G. P. Stahly and A. Park, *Mol. Pharm.*, 2007, **4**, 323–338.
100. M. Banik, S. P. Gopi, S. Ganguly and G. R. Desiraju, *Cryst. Growth Des.*, 2016, **16**, 5418–5428.
101. S. Mohamed, *Acta Crystallogr., Sect. E: Crystallogr. Commun.*, 2016, **72**, 1348–1352.
102. S. Mohamed, D. A. Tocher and S. L. Price, *Int. J. Pharm.*, 2011, **418**, 187–198.
103. H. C. S. Chan, J. Kendrick, M. A. Neumann and F. J. J. Leusen, *CrystEngComm*, 2013, **15**, 3799–3807.
104. M. A. Neumann, F. J. Leusen and J. Kendrick, *Angew. Chem., Int. Ed.*, 2008, **47**, 2427–2430.
105. A. Broo and S. O. Nilsson Lill, *Acta Crystallogr., Sect. B: Struct. Sci., Cryst. Eng. Mater.*, 2016, **72**, 460–476.
106. H. C. S. Chan, J. Kendrick and F. J. J. Leusen, *Angew. Chem., Int. Ed.*, 2011, **50**, 2979–2981.
107. M. A. Neumann and M.-A. Perrin, *J. Phys. Chem. B*, 2005, **109**, 15531–15541.
108. J. Tomasi, B. Mennucci and R. Cammi, *Chem. Rev.*, 2005, **105**, 2999–3094.

CHAPTER 2

High Pressure Crystallography: Elucidating the Role of Intermolecular Interactions in Crystals of Organic and Coordination Compounds

E. V. BOLDYREVA

[a]Institute of Solid State Chemistry and Mechanochemistry, Siberian Branch of Russian Academy of Sciences, ul. Kutateladze, 18, Novosibirsk, 630128, Russian Federation; [b]Novosibirsk State University, Department of Solid State Chemistry, ul. Pirogova, 2, 630090, Novosibirsk, Russian Federation
*E-mail: eboldyreva@yahoo.com

2.1 Introduction

The effect of pressure on physical, chemical and biological processes has been known for a long time. Pressure is an important thermodynamic parameter. In nature it can be varied across a larger range than temperature, reaching several thousand atmospheres at the bottom of the oceans and millions of atmospheres in the depths of the Earth and other planets.

Traditionally, high pressure was used mainly by physicists and geologists. The former are interested in the possibility of obtaining a substance in a completely different state, when metals become non-metals and *vice versa*.

Monographs in Supramolecular Chemistry No. 26
Understanding Intermolecular Interactions in the Solid State: Approaches and Techniques
Edited by Deepak Chopra

Published by the Royal Society of Chemistry, www.rsc.org

The latter use high pressure to mimic conditions under which minerals are formed and transformed in nature, and to understand the relation of geochemical reactions to global geological processes.[1] One of the motivations of high-pressure research in the middle of the previous century was the synthesis of artificial diamonds.[2]

The first studies on the effect of pressure on chemical and biological processes were carried out at the beginning of the previous century, at the same time as the first experiments with minerals. For example, denaturation of chicken white under pressure was documented at the beginning of the 20th century.[3] A really dynamic era of research started in the middle of the 20th century, becoming one of the mainstream directions in chemical research by the turn of the 21st century. High pressure is widely used as a tool to obtain new compounds, as well as to obtain new structures of previously known compounds (polymorphs). However, this is not the only way in which pressure can contribute to chemical research. In the context of this book, it is important to note that intermolecular interactions in condensed phases can be probed by varying pressure.[1] This research direction has been developed in relation to the crystals of organic and coordination compounds in particular.

It is a common practice to derive information on intermolecular interactions in crystals based on the analysis of crystal structures by diffraction techniques. It is often preferential to complement these experiments with vibrational spectroscopy, optical spectroscopy, and measurement of magnetic properties, conductivity, or characterization of other physical properties.

Diffraction techniques give information on interatomic distances and angles, *i.e.* on the geometrical parameters that characterize intermolecular interactions. The general packing of chemical species in a crystal also reflects the presence of selected intermolecular interactions and their relative strengths. Precise diffraction data make it possible to visualize the electron density distribution with sufficient reliability to make unambiguous conclusions on intra- and intermolecular bonds and interactions.

Vibrational spectroscopy gives information on the energetic characteristics of interactions: force constants of bonds, wavenumbers of vibrations, the energies of hydrogen bonds, and data on the dynamics of molecular fragments and molecules, along with the presence of order or disorder in crystals. For example, the formation of hydrogen bonds results in qualitatively and quantitatively different changes in the positions, intensities and half-widths of different vibrational bands. Optical spectroscopy provides information on the energies of electronic transitions, which depend on intra- and intermolecular interactions. Magnetic and other physical properties complete the picture.

Even if a crystal structure is analysed at one selected temperature and pressure, one can learn a lot about the intermolecular interactions. The ability to follow structural changes across multiple temperature and pressure points increases the amount of information enormously.

Measuring the anisotropy of structural strain and the absolute value of compressibility can give valuable information on the absolute and relative strengths of various types of intermolecular interactions. Fine-tuning interactions by continuously changing the intermolecular distances provides a unique possibility to quantitatively study structure–property relationships.

The structures formed under selected temperature and pressure conditions, and the structural changes resulting from temperature and pressure variation, can provide insight into the structure-forming role of various types of hydrogen bonds, stacking interactions, van der Waals interactions, *etc.* One can study the structures formed as a result of crystallization of solids originally dissolved in a liquid, or crystals of compounds that are fluid at ambient pressure and crystallize on compression. One can also compare the structures formed as a result of solid-state polymorphic transformations, or over the course of recrystallization at high pressure.

One can compare the response of different crystalline polymorphs of the same compound, of chiral and racemic counterparts, of single- and multi-component crystals containing the same or related chemical species, or of selected series of compounds. The compression—decompression protocol—in particular the rate of increasing and decreasing pressure—the temperature, the choice of hydrostatic fluid, and the presence of other phases can act as potential seeds and can influence the outcome of the transformation.

Additionally, one can follow the effect of pressure on chemical transformations induced by temperature or light, as well as chemical reactions induced by compression. The role of the crystalline environment in intramolecular chemical reactions can be elucidated in this way, since the "reaction cavity" can be modified in a controlled and continuous manner. Pressure can also induce proton and electron transfer between chemical species, phenomena that are directly related to the characteristics of the intermolecular interactions in the structure.

The range of pressure determines the types of processes that can be expected in a crystal. At pressures reaching Mbars (1 GPa $\approx 10^4$ bar), serious changes in the electronic system occur so that inner-core electrons become involved. For pressures between 100 GPa and 1000 GPa, covalent bonds are affected. Below 10 GPa is the region where van der Waals interactions and hydrogen bonds are modified.[4] It is in this pressure range that many phenomena involving anisotropic compression, phase transitions, crystallization and recrystallization in molecular crystals take place. This is the pressure range of primary interest in the context of this book. The various processes related to rearrangement of intermolecular interactions play an important role in crystal engineering, crystal growth and design, as well as in designing molecular materials, pharmaceuticals and controlling solid-state reactivity and biochemical processes. In some cases a few hundreds of atmospheres are sufficient to cause these transformations. Several reviews have been dedicated to this topic over the last two decades.[1,5–37]

2.2 High-pressure Experiments

In order to get precise and reliable data, one needs well-developed and tested experimental techniques. Much effort has been invested in the hardware and software that is required for high-pressure experiments. This includes large facilities, laboratory instruments, devices to create and estimate pressure, and software to process data and extract as much accurate and reliable information as possible. These achievements have been reviewed in books and papers.[1,3,10,21,36–55] In this Section, I give a very concise and general overview for those who are new to high-pressure techniques.

To increase pressure, one should reduce the volume. Pressure is force divided by surface area, $P = F/S$. To create high pressure, one must therefore apply force to a certain area and, should this area be small enough, the pressure attained can be very high. This is the principle used in high-pressure devices. For example, the highest static pressure that has been achieved up to now in a laboratory device (6 Mbar; 640 GPa) was created at the contact of two balls made of a hard material: sintered nanodiamonds.[56]

Hydraulic presses were the first devices used to create high pressure. For a long time the values of pressure that could be reached using these instruments were limited to 3000 atmospheres (≈0.3 GPa). Engineering developments made by Bridgman enabled considerably higher pressure to be obtained, with the new pressure range reaching 10–40 GPa. Bridgman was awarded the Nobel prize in 1946 for his work in this field. He also studied a great variety of substances, minerals, and inorganic and natural materials under pressure in a systematic way, and summarized his observations in several books and reviews that have not lost their value.[3]

A drawback of the instruments used by Bridgman was that the samples could not be studied at high pressure *in situ*, but could only be studied after decompression. This limited the phenomena that could be studied to irreversible or at least partly quenchable processes. A real milestone in the development of high-pressure devices was the invention of miniature devices: diamond anvil cells (DACs).[1,2,10,21,38–56] The centre of such a device (Figure 2.1) contains a pair of diamonds[†] fixed in the backing plates.

Using either screws or gas-driven membranes one can make the two strictly parallel culets of the diamonds approach each other, pressing on a plate inserted between them; a gasket. The compressed gasket transmits pressure onto a hydrostatic fluid in a hole in the gasket and thus onto a solid sample that resides within this fluid. Since diamonds are transparent to a wide range of radiation, including X-rays, and the DACs are small and light, the latter can be installed on many instruments, like spectrometers, diffractometers, or optical microscopes (Figure 2.1b). The structures, spectra and physical properties of samples can be studied at high pressures *in situ*, and the processes can be monitored visually in real time. Membrane DACs make

[†]In some experiments, diamonds can be substituted for sapphires, fianites and some other materials.

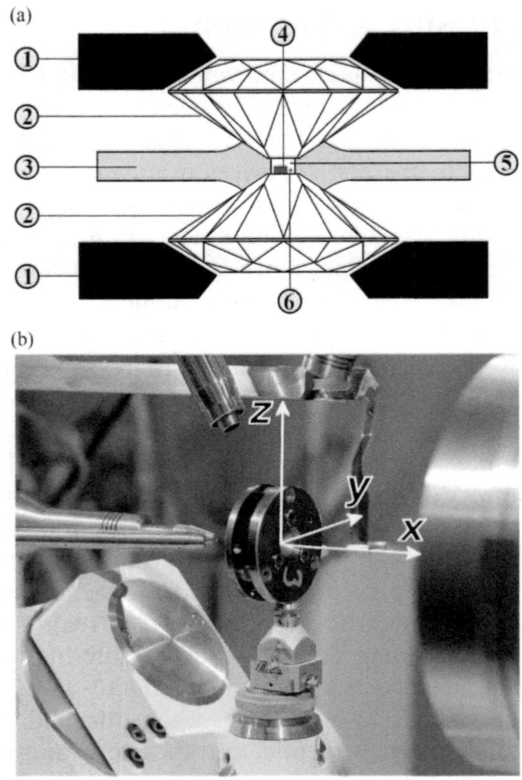

Figure 2.1 The "core" of a diamond anvil cell, DAC, (a) and a DAC mounted at a four-circle diffractometer (b). 1: backing plate, 2: diamonds, 3: gasket, 4: sample, 5: fluid, 6: calibrant. The figure is courtesy of Dr B Zakharov; the photo is courtesy of Dr N Tumanov.

it possible to modify and control pressure in very small steps, continuously, reproducibly and reversibly. Screw-driven DACs are more difficult to operate as far as pressure control is concerned, but a skilled experimentalist can still make steps as small as 0.1 GPa. An important advantage of a membrane cell as compared to a screw-driven one is that the rates of compression and decompression can be modified and controlled.[40,55,57]

Hydrostatic compression is uniform through the whole sample only if a fluid is compressed, or if the solid sample is immersed in a hydrostatic fluid. Different fluids have different limits of hydrostaticity, after which they first become highly viscous and then solidify. The choice of hydrostatic fluid is critical in determining the maximum pressure range that can be achieved in the experiment, but may play additional roles in the high-pressure phenomena observed.[47,55,58–64] One must remember that any hydrostatic fluid is a compound that can interact with the sample, dissolve it, modify its surface properties and even react with it chemically. Even if no visible dissolution or interaction can be detected at ambient conditions, they cannot be excluded as remaining insignificant under pressure. These interactions can greatly complicate experiments. In addition to simply studying increasing pressure, varying hydrostatic fluids in a systematic and controlled way can also contribute to the understanding of the intermolecular interactions in the sample, and their role in crystallization, dissolution, and polymorphic and chemical transformations.

Experiments with DACs require one to reliably and precisely estimate the value of pressure in real time and *in situ*. This value cannot be calculated because it remains unknown which part of the applied force is transmitted to the anvils: inner friction, plastic and elastic deformation of the elements of the DAC and the gasket are difficult to control. The value of pressure must therefore be estimated based on experimental data. The values of absolute pressure are only rarely estimated, since this requires a simultaneous measurement of the bulk modulus and the changes of unit cell parameters *vs.* pressure by combining *in situ* X-ray diffraction with the measurements of the velocity of the acoustic waves propagating through the crystal (Brillouin spectroscopy, inelastic neutron scattering, ultrasonic interferometry).[65–67]

More often, pressure is estimated based on various calibrants, for which the effect of pressure on some properties is known. The calibration curves are not very easy to obtain and are re-considered from time to time, as new data become available and more precise experimental techniques are developed. All the calibrants can be classified as being based on fluorescence, diffraction, optical properties, or vibrational properties (Raman or IR spectroscopic). One may also use calibrants for which the points of pressure-induced phase transitions are well characterized. Generally, pressure can be estimated with a precision of ±0.5 GPa. As examples of fluorescence standards one can mention ruby ($Cr^{3+}:Al_2O_3$), $BeAl_2O_4$ or $Sm^{2+}:BaFCl$. Common diffraction-based standards are ionic halides (NaCl, CaF_2), or covalent oxides (quartz, MgO, $AlPO_4$).[21,46,47,55,68–104]

Pressure can be combined with cooling or heating the pressure cell, as well as with irradiation by light and applying a magnetic field. The range of possible experimental conditions is therefore enormous.[1,10,44,46,47,55,105,106]

Processing data obtained at high pressures is not straightforward.[1,10,21,55,107–112] When interpreting the results, one must always remember that a sample is inside a pressure cell. For diffraction experiments, this means that part of the incident and reflected radiation is shadowed by the cell. Special procedures for data collection are required, and the diffraction pattern must be "cleaned" of any reflections from the gasket, diamond and calibrant. One must also eliminate distortion of the intensities and profiles of reflections. Because of the shadowing effect, only part of the reciprocal space is accessible. To get high-quality data, every reflection is valuable. Therefore, experimentalists often need to introduce corrections and apply other special procedures to "save" weak and corrupted reflections. It is advantageous to use radiation with low wavelengths (Ag or Mo Kα), and to use a pressure cell with a high opening angle, to increase the beam intensity (*e.g.* using synchrotron radiation) and exposure time. Data collection from several crystals is often used in order to increase data completeness. The roles of "manual work" and "human error" remain large. Therefore it is especially important that all primary data are stored and accessible on request. When conducting high-pressure spectroscopic experiments, one needs to take into account that diamonds are not transparent across the entire spectral range. Certain sample vibrations cannot be observed. In such cases diamonds can sometimes be substituted for other materials that are transparent in the radiation range, where diamonds absorb.

The present state of the art in high-pressure research makes it possible, when studying a crystal in a DAC, to solve most of the problems that can be solved for a "free" crystal (in the order of increasing complexity): refine the cell parameters of a known structure, refine a crystal structure if the starting model is known, follow the changes in interatomic distances with increasing pressure, find the cell parameters of an unknown structure, solve a crystal structure without knowing a model *a priori*, follow the changes in molecular conformations, the changes in the intramolecular distances (bond lengths and angles) and even precisely analyse the electron density.[4,107–112]

2.3 Continuous Anisotropic Compression

The pressure response of unit cell parameters and interatomic distances of a crystalline material can lead to correlations in lattice strain and intermolecular interactions. This correlation was first described for the related variable temperature studies by Fedorov[113] and Ubbelohde.[114–117] They suggested that it might be possible to relate the anisotropy of thermal expansion to the different strength of hydrogen bonds in different crystallographic directions. Ubbelohde even used data on the anisotropy of thermal expansion of hydrogen-bonded crystals to locate hydrogen atoms in the structure. Along the same lines, studies of the anisotropy of structural strain and changes in the geometrical parameters of hydrogen bonds with pressure began for several coordination and organic compounds in the late 1990s.[5,118–123]

Temperature and pressure are scalars. When an isotropic material or a crystal belonging to the cubic crystal system expands/contracts with temperature or pressure variation, structural strain can be characterized by bulk compressibility, $\Delta V/V$, % per unit P, and a linear compressibility, $\Delta l/l$, % per unit P. Here, l and V are related as $V = l^3$. A sphere made of an isotropic material will transform into another sphere with a larger (expansion) or smaller (contraction) radius. The response of any non-cubic crystal structure to isotropic action will be anisotropic, and must be described by a strain tensor.[124] On variation of temperature or pressure, a sphere made of an anisotropic material will be transformed into an ellipsoid, known as the strain ellipsoid. The three axes of this ellipsoid that remain mutually orthogonal during strain are termed the principal axes of the strain ellipsoid. To characterize structural strain, one needs to define the orientation of the three principal axes of the strain ellipsoid with respect to the crystallographic axes, and the linear strain along these axes. Linear strain along any other direction can

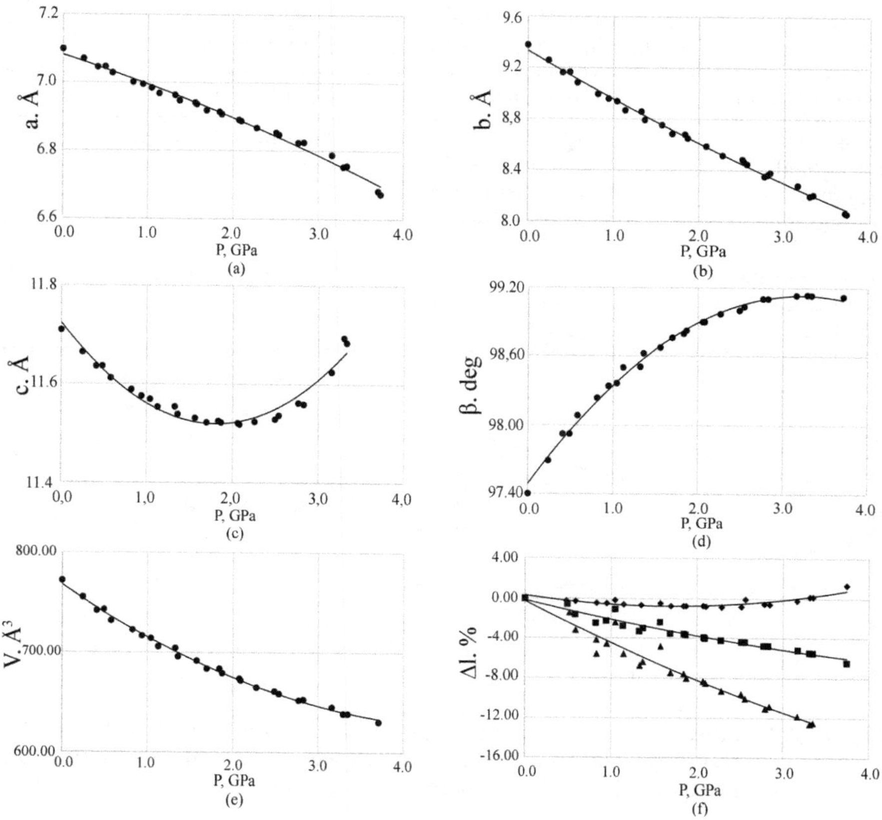

Figure 2.2 (a)–(e) Changes in cell parameters and volume and (f) linear strain in the directions of the principal axes of the strain tensor measured for paracetamol I *versus* hydrostatic pressure. Reproduced with permission of the International Union of Crystallography from E. V. Boldyreva, High-pressure diffraction studies of molecular organic solids. A personal view. *Acta Crystallogr. A*, 2008, **64**(1), 218–231, Copyright 2008 IUCr. http://journals.iucr.org/.

then be calculated by knowing the angles formed by this direction and the principal axes.[124] The strain ellipsoid can be calculated from the experimental pressure-dependent cell parameters[125,126] or directly from changes in the d_{hkl} values.[127] For monoclinic and triclinic systems, the directions of the principal axes do not coincide with the directions of the crystallographic axes, and the curves that describe the changes of cell parameters and linear strain along the principal axes can look very different (Figure 2.2).

According to Neumann's principle,[124] the symmetry group of a physical property must include the symmetry elements of the crystal point group. The symmetry of a physical property must be higher than or equal to the symmetry of the crystal structure. Neumann's principal relates the symmetry of a crystal structure with the possible anisotropy of lattice strain on hydrostatic

Table 2.1 Relations between the orientation of the principal axes of a strain ellipsoid and the symmetry elements in the crystal structures.

Crystal system	Strain ellipsoid	Orientation
Cubic	Sphere	Any one of the axes
Tetragonal	Rotation ellipsoid	∥4
Trigonal	Rotation ellipsoid	∥3
Hexagonal	Rotation ellipsoid	∥6
Orthorhombic		Any one of the axes
Monoclinic	Ellipsoid	An axis 2(⊥m)
Triclinic	Ellipsoid	Any

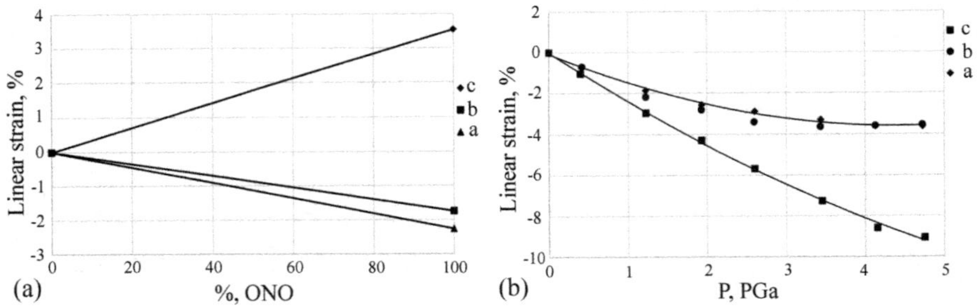

Figure 2.3 Linear strain along the crystallographic axes of $[Co(NH_3)_5NO_2]Cl(NO_3)$ resulting from linkage photoisomerization (a) and hydrostatic compression (b). Based on data from ref. 120 and 128.

compression (Table 2.1). However, the restrictions imposed on the orientation of the principal axes of the strain ellipsoid with respect to symmetry elements of the structure do not say anything about the absolute and relative values of linear strain along these directions: the structure may expand, or be compressed along the same crystallographic direction without violating Neumann's principle. As an example one can compare linear strain in $[Co(NH_3)_5NO_2]Cl(NO_3)$ on hydrostatic compression and as a result of structural compression over the course of the homogeneous linkage photoisomerization. The orientation of the principal axes is the same in both cases (coinciding with the three crystallographic axes), but the relative strain is inverse (Figure 2.3).[118,120,128,129] It is this inversion of relative strain that accounts for the photomechanical effect in crystals of this compound, namely for their bending and jumping.[130]

Moreover, for structures built from more complex units (molecular and ionic-molecular crystals) the linear strain of a structure in a selected direction is not necessarily related to a change in the interatomic distances in a straightforward way. For example, the structure can expand even if the lengths of chemical bonds remain unchanged, or may shorten due to changes in the angles between the bonds, the rotation of molecular fragments, or rotation of entire molecules (Figure 2.4).

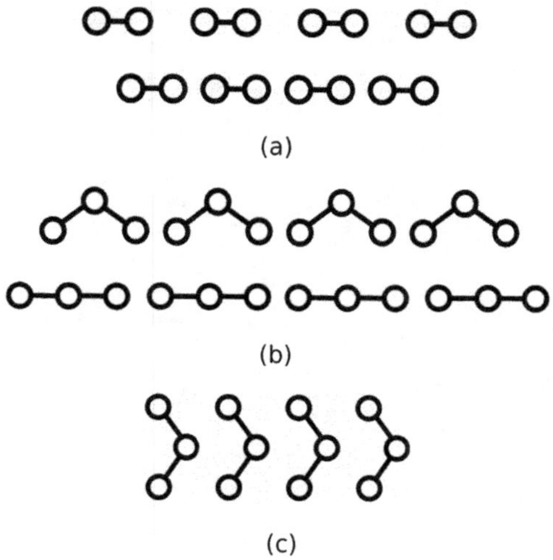

Figure 2.4 Different mechanisms of strain in molecular crystals without changes in the intramolecular bonds: (a) shortening of distances between molecules, (b) changes in valence (or torsion) angles, (c) rotation of molecules. The figure was prepared by Dr B. Zakharov.

Aside from crystal symmetry, the anisotropy of strain on hydrostatic compression is determined by a variety of other factors, including the molecular packing, the nature of hydrogen bonds, as well as the conformational flexibility and shapes of the molecules in the structure. It is informative to compare the pressure-induced strain of crystals across a series of related compounds, of different polymorphs of the same compound, or of the L and DL forms of a compound.[1,5–13,17–19,22–24,27,32] It is of particular interest to compare the anisotropy of strain of the same crystal structure on cooling and on hydrostatic compression corresponding to the same change in the volume. For cubic structures, the coefficients of isothermal compression and of the isobaric thermal expansion are related to each other by a Grüneisen equation.[1,131] In the low-symmetry structures, the strain anisotropy can be radically different if resulting from increasing pressure or from decreasing temperature. The monoclinic structure of $[Co(NH_3)_5NO_2]Cl_2$ (Figure 2.5)[123] is a good example. Differences in the lattice strain anisotropy account for different changes in the interatomic distances corresponding to different types of NH···O and NH···Cl hydrogen bonds in this structure (Figure 2.5b). Similar phenomena—a qualitative difference of the strain anisotropy corresponding to structure compression on cooling and on increasing hydrostatic pressure—have been observed for many other compounds.[1,8–13,17,18,23,35,121–123,132–141]

The bulk compressibilities of polymorphs can be very similar, while the anisotropy of strain can be very different. This reflects the difference in the intermolecular interactions. As an example, one can compare the pressure-induced strain in the polymorphs of paracetamol (Figure 2.6)[132,133,136]: bulk compressibilities are practically the same, whereas the anisotropy of

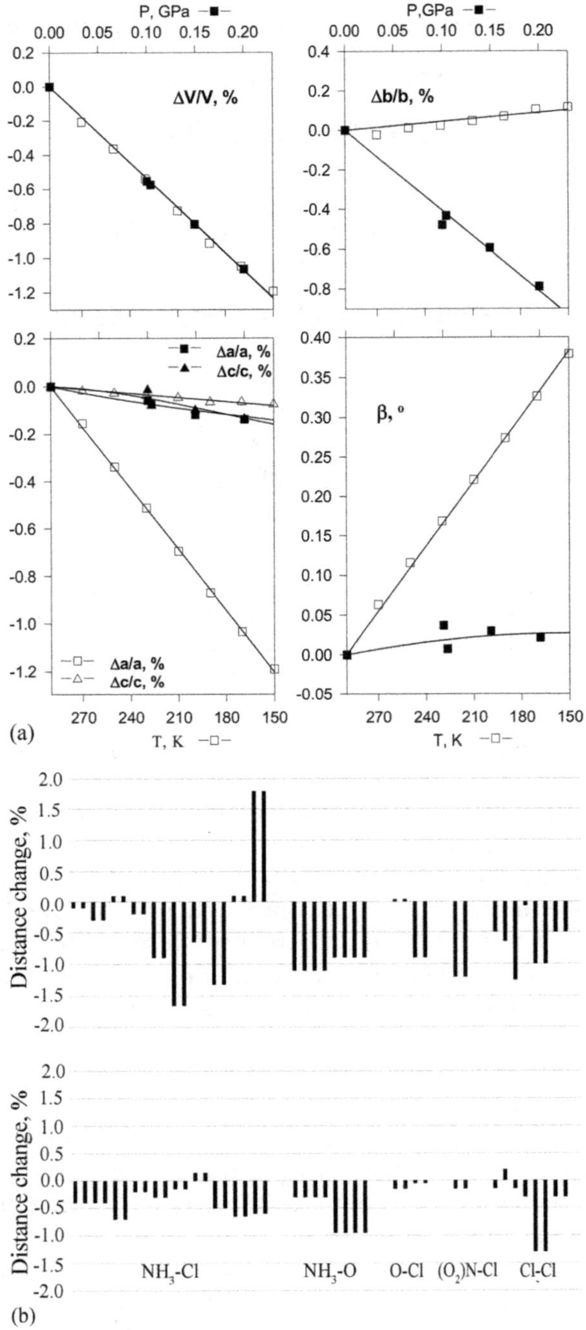

Figure 2.5 Comparison of lattice strain (a) and intermolecular distances (b) corresponding to the same volume change on cooling (empty symbols in (a), upper histogram) and on hydrostatic compression (black symbols in (b), lower histogram). Reproduced with permission of the International Union of Crystallography from E. V. Boldyreva, D. Y. Naumov and H. Ahsbahs, Distortion of crystal structures of some Co^{III} ammine complexes. III. Distortion of crystal structure of $[Co(NH_3)_5NO_2]Cl_2$ at hydrostatic pressures up to 3.5 GPa, *Acta Crystallogr. B*, 1998, **54** (6), 798–808, Copyright 1998 IUCr. http://journals.iucr.org/.

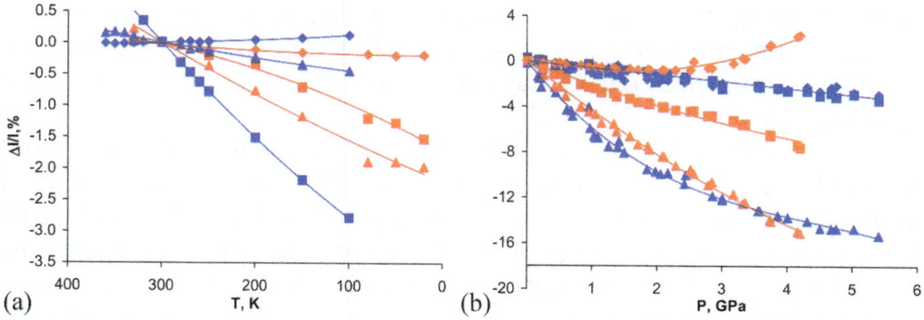

Figure 2.6 The changes in the volume (a) and the linear strain along the three principal axes of strain ellipsoids (b) on hydrostatic compression of the monoclinic (red) and orthorhombic (blue) polymorphs of paracetamol. Reproduced with permission of the International Union of Crystallography from E. V. Boldyreva, High-pressure diffraction studies of molecular organic solids. A personal view. *Acta Crystallogr. A*, 2008, **64**(1), 218–231, Copyright 2008 IUCr. http://journals.iucr.org/.

strain is very different, reflecting the difference in the hydrogen-bonding network in the two structures. The information on the anisotropy of strain on hydrostatic compression can be helpful when optimizing the potentials used to describe the intermolecular interactions in the crystals.[142,143]

In a detailed systematic study, a comparison was done on the strain in each of the four layered polymorphs of the amino acid glycine (α, β, β′, δ forms). This mimics layers in protein structures.[144] Compression along the hydrogen-bonded head-to-tail chains of glycine zwitterions was identical in all structures, whereas strain normal to these chains within a hydrogen-bonded sheet and normal to it depended on the stacking of the layers. This reflected a complex interplay between the compression of different types of hydrogen bonds and an optimizing of van der Waals interactions (minimizing the available free space).[27,135]

Research into the continuous structural strain induced by hydrostatic compression has become very common. As relatively recent examples one can mention the studies of the non-covalent interactions under extreme conditions of high pressure and low temperatures of the isostructural metal–organic networks (4-chloropyridinium)$_2$[CoX_4] (X = Cl, Br),[139] of the giant negative linear compression positively coupled to massive thermal expansion in a metal–organic framework with Ag–Ag interactions,[140] and of the anisotropy of compression for a large class of aromatic diphenyl-1,3,4-oxadiazole molecules.[145] Measurements of the strain anisotropy were used in order to discriminate between attraction and repulsion in Hal–Hal interactions in pressure-frozen CH_2ClI and CH_2I_2.[146] The anisotropy of pressure-induced changes in crystalline structures of organic polymers (polyimides) reflects the structure-forming role of the intermolecular interactions involving aromatic rings.[147]

Precise measurements of linear strain on hydrostatic compression make it possible to follow changes in the relative contributions of different types of hydrogen bonds to structural distortion across different pressure ranges.[148] Interpretation of the results of diffraction studies becomes more reliable if complemented by the data from high-pressure Raman and IR spectroscopy.[1,11,12,20,23–25,27,148,149]

An important application of anisotropic structural strain is to study stacking interactions, or to fine-tune metal–metal interactions in crystals of organic and coordination compounds that have infinite chains of chemical species as structure-forming units or can be considered as coordination polymers.[26,147,150–177] Linear strain in these structures is usually the result of interplay between optimizing intra- and intermolecular interactions. For example, analysis of the high-pressure and low-temperature structures of a series of metal carbonyls shed light on the peculiar behaviour of the metal–metal bond, the most flexible interaction in the molecule. This bond is initially compressed, and subsequently elongated due to increased intracarbonyl repulsion, and eventually compressed again. Theoretical calculations in the solid state allowed the thermodynamic quantities of the transformation to be computed along with prediction of the distortions produced by the external stress. The change in molecular symmetry was shown to be induced by an increase in the high internal energy associated with the staggered carbonyl conformation, which does not allow efficient packing if the crystalline volume is reduced. The eclipsed conformer, though much less stable in isolation, guarantees better entanglement between molecules.[150,151] Another example when a larger distance between metal ions is favourable for denser packing of ligands is provided by $[Gd(PhCOO)_3(DMF)]_n$ compressed to 3.7 GPa and then to 5 GPa.[167]

2.4 Polymorphic Transitions

A crystal structure can adapt to increasing pressure up to a certain limit, after which a structural rearrangement occurs. This can be a truly solid-state transformation without a change in chemical composition, termed a polymorphic transition. This can also be a recrystallization into another polymorph, or a chemical transformation, *e.g.* dehydration of a crystal hydrate, or inclusion of guest molecules from the hydrostatic fluid into the structure to form a host–guest compound or solvate.

Pressure-induced polymorphic transitions in crystals are often isosymmetric, *i.e.* the space symmetry group does not change.[8] This means that it may be impossible to unambiguously distinguish a phase transition from a continuous anisotropic structural distortion without having studied the crystal structure not at two, but at many pressure points. Sometimes the volume change is small, although the linear structural strain in selected crystallographic directions can be significant, compensating each other due to the opposite signs (contraction in one direction and expansion in the other). It is not uncommon that structural strain in some directions is continuous, whereas in the other it is clearly discontinuous. The discontinuity in the volume change *vs.* pressure can be unpronounced, and is only possible to reliably detect by sufficiently

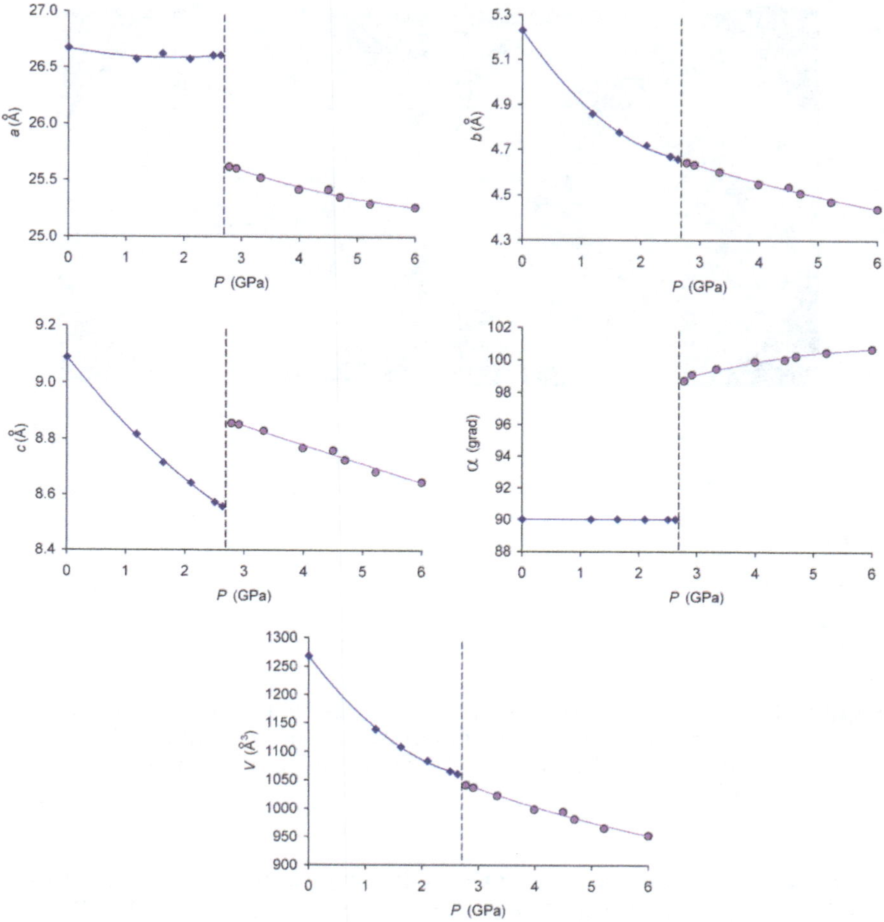

Figure 2.7 The changes in the cell parameters and volume of α-chlorpropamide compressed hydrostatically in its saturated ethanol solution. Reproduced with permission of the International Union of Crystallography from (Y. V. Seryotkin, T. N. Drebushchak and E. V. Boldyreva, Effect of hydrostatic pressure on α-polymorph of chlorpropamide in saturated ethanol solution, *Acta Crystallogr. B*, 2013, **69**, 77–85), Copyright 2013 IUCr. http://journals.iucr.org/.

precise measurements at multiple pressure points. As an example, changes in unit cell parameters and volume during a pressure-induced polymorphic transition in α-chlorpropamide are plotted in Figure 2.7.[178]

Many phase transitions are related to changes in the hydrogen bonding, which are coupled to conformational changes (rotation of molecular fragments), or even to the rotation of the whole molecules. Hydrogen bonds under pressure can shorten, expand, switch over from one group to another, be completely broken or be formed new, get ordered, or become disordered. Different types of hydrogen bonds respond differently to increasing

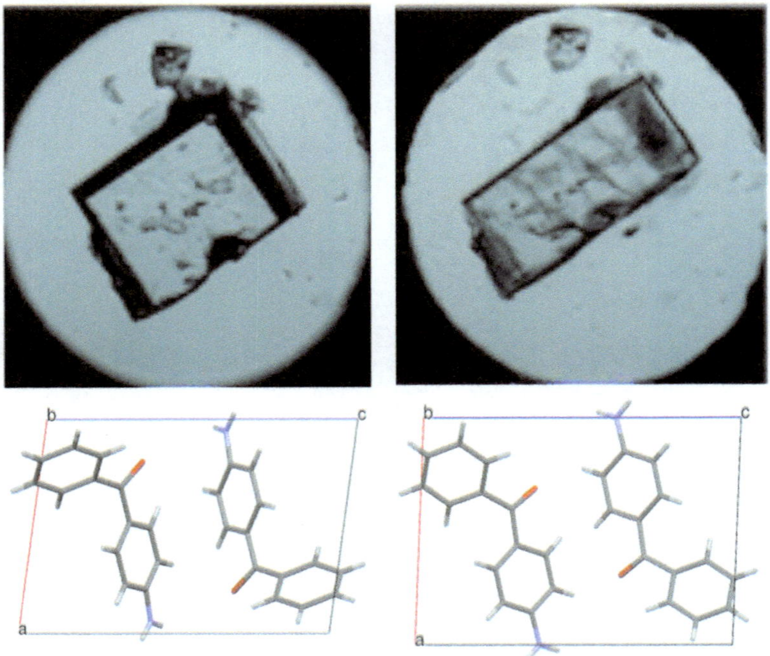

Figure 2.8 The change in the crystal shape corresponding to the conformational re-arrangement of 4-aminobenzophenone on hydrostatic compression. Adapted from ref. 188 with permission from the Royal Society of Chemistry.

pressure. The difference may not only be quantitative (different compression), but also qualitative. For example, selected bonds can be distorted discontinuously, whereas other may be distorted discontinuously. A structural rearrangement can be a result of the cooperative effects between hydrogen bonds of different types, van der Waals interactions and π-stacking interactions.[1,4–13,17–28,32–37,152–214]

In many cases a polymorphic transition is displacive and preserves the main structural framework intact. The hydrogen bonds can act in this case as springs that hold the structural framework together, so that the structure is completely restored on decompression without accumulation of any defects. In some cases even a single crystal is not destroyed in the course of the polymorphic transformation and can go through the phase transition many times without fatigue, even though the crystal shape changes sometimes dramatically (Figure 2.8).[188]

In other cases the transformation is reconstructive and its rate is limited by nucleation of the new phase and its growth. In such cases the transition can be observed at pressures significantly higher than the equilibrium one, or, more often, on decompression from a higher pressure.[1,10,16,36,133,189,193,204] It can be sensitive to the choice of hydrostatic fluid being solvent-assisted and related to a recrystallization, or solid surface–fluid interactions.[178,189–195] Examples when recrystallization facilitates the transition into a more stable form under pressure have been documented for many organic compounds: bromochlorofluoroacetic acid,[196] 3-hydroxy-4,5-dimethyl-1-phenylpyridazin-6-one,[197] imidazole[198] and 4,4′-bipyridine hydrobromide monohydrate.[199] In the latter two cases, high-pressure phases could be obtained only through recrystallization by heating and further slow cooling at definite pressure points. These phases did not form on isothermal compression and transformed spontaneously into other forms on releasing pressure, so they were called hidden polymorphs.[198,199]

The polymorphic transformations can be sensitive not only to the choice of hydrostatic fluid, but also to the choice of the starting polymorph,[178,190,194,195,200–207,215,216] the size of particles, the compression–decompression protocol,[57,133,189,193,204,217] as well as to the presence of seeds of other phases.[194] The ability to vary the compression rate in a membrane-driven pressure cell adds a new perspective to phase diagrams of solids. A particular pressure increase rate may trigger unexpected solid-state transformations, producing otherwise inaccessible phases.[16,57,193,195,217] All of these phenomena are manifestations of what is termed as kinetic control of the polymorphic transformations.[16] At any selected temperature and pressure point a well-defined phase of a compound is thermodynamically stable and should form in accordance with thermodynamics. In reality, however, this is not always the case. A polymorph that is thermodynamically stable at particular conditions cannot be obtained at these conditions unless a special trick is invented. A polymorph that is thermodynamically unstable at particular conditions can be preserved (often for an indefinitely long time) at these conditions. The polymorphs that are formed do not correspond to the absolute minimum of the free energy, but to the minimum barrier separating the new form from the parent one. Based solely on a thermodynamic phase diagram, it is not possible to predict which phase will be formed at a given combination of temperature and pressure. Instead, there exists a complex interplay of nucleation and growth kinetics, alongside thermodynamics, that leads to these unpredictable results.[16] At the microscopic level, the high-pressure phase that is formed (or not formed) in the course of the polymorphic transition is a result of a complex interplay between the response of different types of intermolecular interactions to pressure, the relative role of which changes with pressure. The hydrogen bonds that are normally stronger at ambient conditions (*e.g.* NH···O) can be substituted by the normally weaker CH···O hydrogen bonds at high pressure to obtain more optimum molecular packing.[218] The cooperative effects between hydrogen-bonding and π-stacking interactions are also important.[180]

The role of the starting polymorph in determining which phase will be formed at a selected pressure point can be illustrated by an example of glycine. All polymorphs of glycine have the same structure-forming unit: a head-to-tail chain of zwitterions linked *via* charge-assisted NH···O hydrogen bonds (Figure 2.9). The chains can then form helices (γ-form), or layers stacked in a variety of ways (all other known polymorphs). Depending on the starting polymorph, different phases are formed at the same pressure: α-glycine is

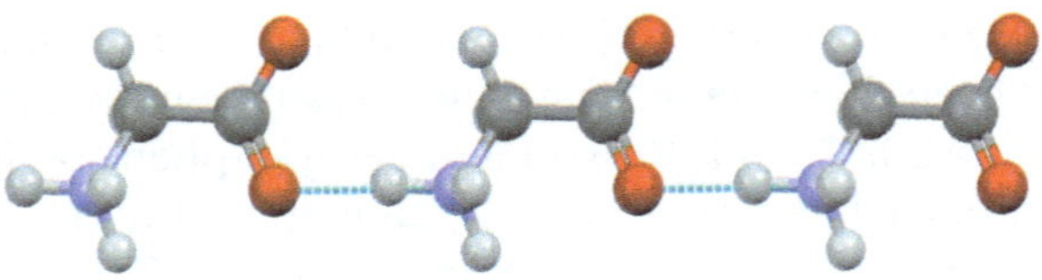

Figure 2.9 The structure-forming unit in glycine polymorphs: a head-to-tail chain of zwitterions linked *via* charge-assisted NH···O hydrogen bonds (shown by dashed lines). Drawn using Mercury software (C. F. Macrae, I. J. Bruno, J. A. Chisholm *et al.*, *J Appl Cryst*, 2008, **41**, 466–70) using (CSD reference code GLYCIN20); all other polymorphs form essentially the same supramolecular synthons.

stable and shows no phase transitions with increasing pressure, β-glycine undergoes a reversible single-crystal to single-crystal transition at 0.76 GPa, and γ-glycine transforms irreversibly from a helical to a layered structure at 3 GPa if pressure is increased rapidly, or at 0.86 GPa if pressure is increased slowly.[200–207,215,216] On decompression, the δ-glycine transforms into another high-pressure form that cannot be obtained on direct compression, namely, the ζ-form.[204] Based on Raman spectroscopy data, the polymorph was supposed to preserve a layered structure.[204] This assumption has been confirmed by diffraction experiments supported by model calculations.[219]

The exact compression–decompression protocol, with particular focus on the rate at which pressure changes occur, can play a decisive role in determining which phases will form at high pressure. In fact, such factors are important to consider whether a phase can be preserved on decompression. Fast compression of the ambient-pressure polymorph I of β-alanine up to 8 GPa allows for the preservation of orthorhombic symmetry in a subtle single-crystal to single-crystal phase transition to form II, even if the sample is kept at this pressure for months. If one leaves the crystal at a lower pressure (6 GPa) for ~15 hours, an incomplete phase transition to monoclinic form V occurs, with a different molecular packing and hydrogen bond network. This incomplete transition conserves a fragment of the initial sample as a single crystal, after which another transition into a finely dispersed form, presumably form IV, is observed. The mechanism of lattice compression for the orthorhombic form at pressures lower than 6 GPa is not equivalent to that at pressures from 6 to 8 GPa. A detailed comparative analysis of deformations occurring during the phase transition from the orthorhombic to monoclinic phase, and during the deformation of the orthorhombic phase from 6 to 8 GPa, showed that further compression of phase II can induce another phase transition, possibly to the same phase V, or another monoclinic form.[193]

The rate at which pressure is increased is very important, though only rarely controlled. L-serine, characterized by a complex high-pressure behaviour with three known polymorphs, can serve as another example. Two parallel paths exist and the composition of the system depends on both the rate with which pressure is increased and the magnitude of pressure increase steps during compression. Phase II can be observed at 5.4 GPa, but only if pressure is increased with rapid, sharp steps such that 5.4 GPa is reached within several hours. This transformation is cooperative in nature ("organic martensite") and preserves a single crystal undamaged. Polymorph II then transforms reversibly into polymorph III at 8 GPa.[57,208–213] A slow and continuous compression, however, favours polymorph IV of L-serine.[57] No direct inter-conversion occurs between polymorphs II and IV. Moreover, phase III originates only from phase II, but never from phase IV (Figure 2.10).[57]

By using state of the art periodic DFT calculations augmented with dispersion corrections and external stress tensor together with gas-phase cluster calculations, the reversible phase transitions of three polymorphs of the crystalline L-serine were thoroughly studied in the pressure range up to 8 GPa. The study has shown that at a macroscopic level the main driving force of the phase transitions is the decrease in the volume of the crystal unit cell, which contributes to the enthalpy difference between the two forms, but not to the difference in their internal crystal energy. At a microscopic level the hydrogen bond overstrain (Figure 2.11) leads to a martensitic-like, cooperative, displacive phase transition into L-serine-II with substantial experimental hysteresis, while no such overstrain was found for the "normal type", atom per atom, reconstructive phase transition into L-serine-III.[213]

Another example of kinetic effects is provided by paracetamol. Single crystals of the monoclinic polymorph I can be compressed to at least 4 GPa without inducing a phase transition.[132] For powder samples of polymorph I, however, a pressure-induced phase transition into polymorph II is observed,[133] which is stable only at elevated pressures.[220,221] Interestingly, this phase transition does not occur immediately on increasing pressure, but only if pressure is first increased to 5 GPa, then subsequently released slowly to lower pressures. Polymorph II is the thermodynamically stable form at high pressure,[220,221] but it cannot be formed if one starts with a single crystal because of kinetic barriers related to the reorientation of every second hydrogen-bonded molecular chain in the structure; nucleation of the new phase is hindered. The pressure protocol itself is of considerable importance, where the powder sample must first be significantly over-pressurized to trigger the transformation, followed by partial decompression to ensure that nuclei of the new phase can actually form and begin growing.[133] This phenomenon can be compared with the overcooling of a melt to give a glass at ambient temperature, which subsequently crystallizes on heating, sometimes requiring temperatures near the melting point. Many polymorphic transitions can be avoided by rapid compression to a pressure much higher than the expected transition point.[57,189,193] One must first increase pressure quickly, keep the sample at high pressure, and then decompress slowly.

An even more complex high-pressure behaviour is shown by an isomer of paracetamol, metacetamol. Like in case of paracetamol, polymorph II of metacetamol[222] is only stable under increased pressure.[223] The I–II–L triple point coordinates for metacetamol TI–II–L = 535 ± 10 K and PI–II–L = 692 ± 70 MPa. However, whereas for paracetamol the enantiotropy under pressure can be foreseen, because the metastable polymorph II is denser, in the case

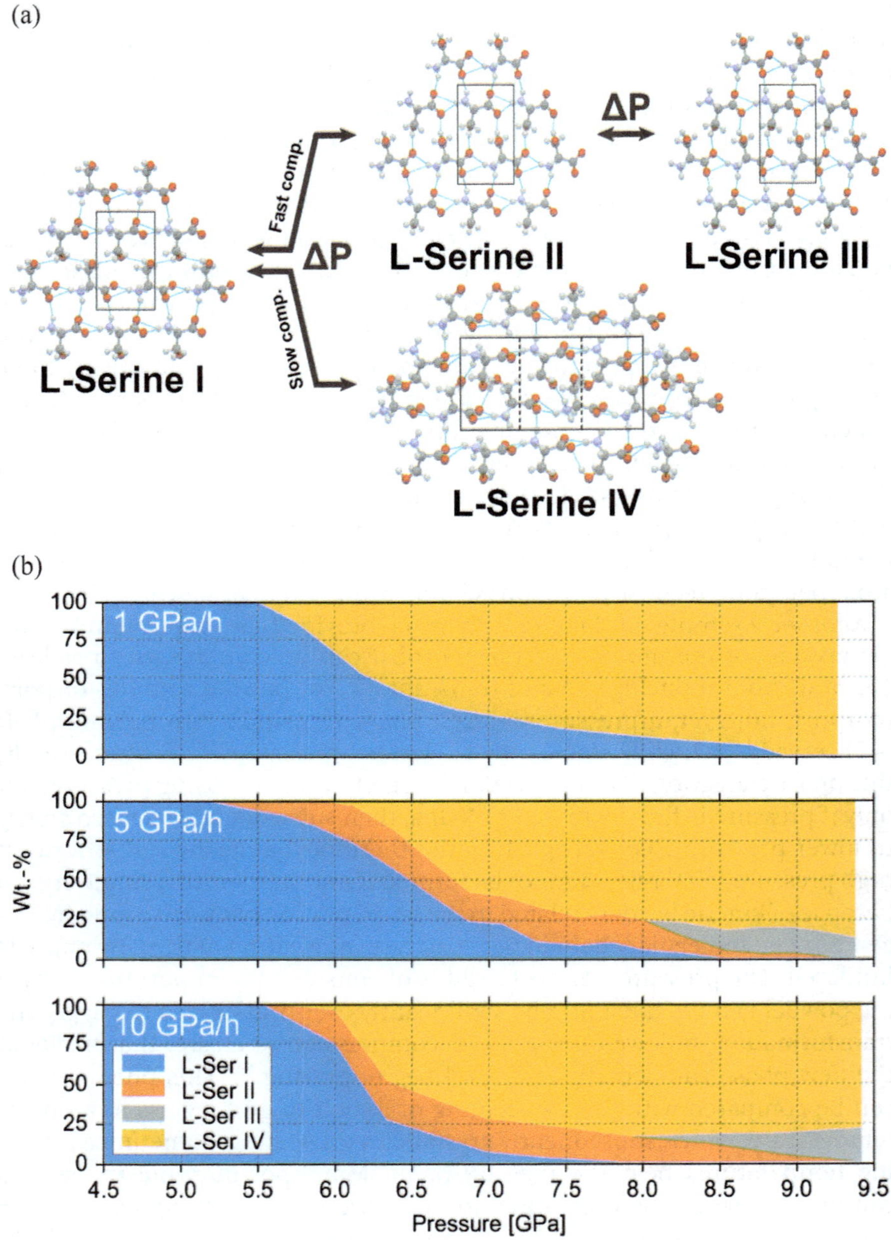

Figure 2.10 (a) Different routes of the polymorphic transformations of L-serine in the crystalline state depending on the pressure increasing rate. (b) Accumulation of different L-serine phases depending on the pressure increasing rate. Reprinted with permission from M. Fisch, A. Lanza, E. V. Boldyreva, P. Macchi and N. Casati, Kinetic control of high-pressure solid-state phase transitions: a case study on L-serine, *J. Phys. Chem. C.*, 2015, **119**(32), 18611–18617, Copyright 2015 American Chemical Society.

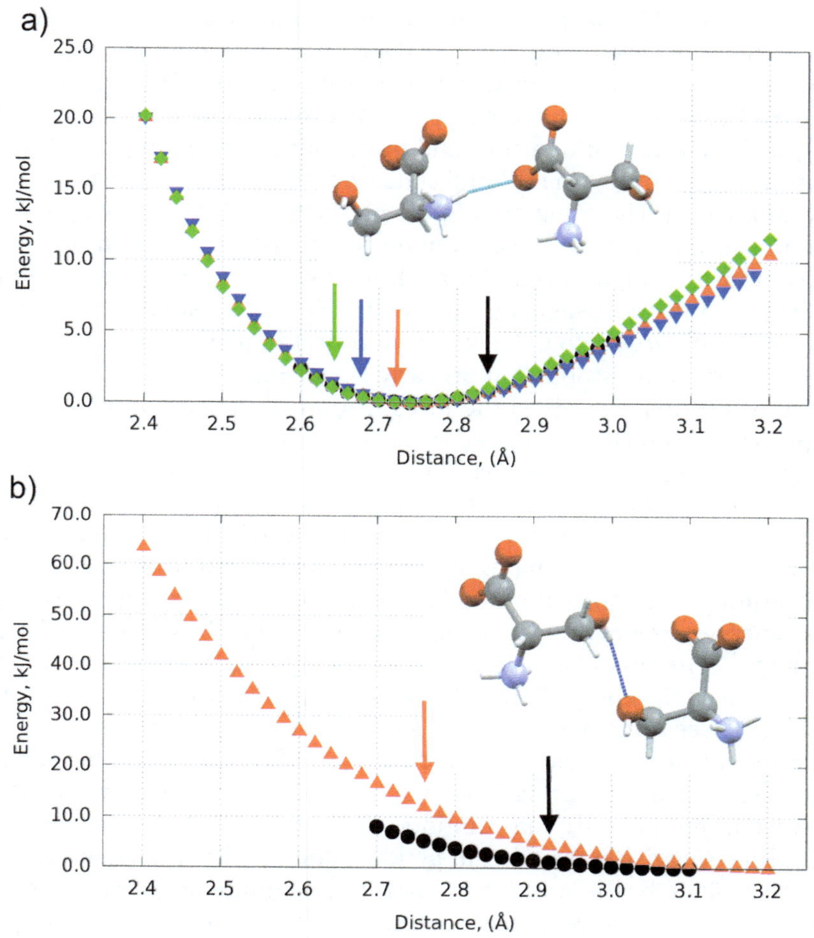

Figure 2.11 Energy wells of pair-wise intermolecular interactions along a NH···OH bond (a) and an OH···OH bond (b) at different pressures. Ambient pressure: black, 4.8 GPa: red, 5.4 GPa: blue, 8.0 GPa: green coloured arrows mark the D···A distance between highlighted dimers at corresponding pressure. An overstrain (red, blue, green arrows (a), red and black arrows (b)) at corresponding pressures is clearly seen. The H bond OH···O disappears after 4.8 GPa. Adapted from ref. 213 with permission from the PCCP Owner Societies.

of metacetamol this is not possible, as the metastable polymorph II is less dense than the stable one.[222] The existence of the stability domain for the less dense polymorph of metacetamol can only be demonstrated by the construction of the topological phase diagram.[223] It is a delicate interplay between the specific volume differences and the enthalpy differences causing the stability domain of the less dense polymorph to be sandwiched between the denser polymorph and the liquid. Metacetamol shares this behaviour with bicalutamide and fluoxetine nitrate.[224,225]

It is also possible that none of the two polymorphs transforms into another on compression, even though one of the two is stable and the other is metastable. An example is provided by polymorphs I and II of tolazamide[226] that were studied using two different fluids as hydrostatic media.[226] The two polymorphs have different densities and molecular packing. Polymorph II converts into polymorph I in the solid state on heating or *via* recrystallization if solvent assisted. The effect of pressure on the two forms and the possibility of a transformation to a denser form on compression have been studied. No phase transitions have been observed in either of the forms in a pentane–isopentane mixture (when no recrystallization is possible). Polymorph II recrystallized partly into a denser polymorph I in methanol at 0.1 GPa, but the transformation stopped at an early stage. Solid-state DFT calculations of the two forms, as well as the investigation of the conformational landscape in the gas phase, were used to rationalize this result. The difference in the intermolecular interactions of tolazamide polymorphs at ambient and higher pressure makes a contribution to the energy cost of II to I transformation. The total barrier that must be overcome to let the polymorphic transition occur includes changing molecular conformation, besides breaking and forming intermolecular bonds. Conformers in tolazamide forms have close energies, the difference being 4.2 kJ mol^{-1}, which is common for polymorphs.[227] One should consider both the energy gap between the conformers, as well as the energy barrier between the two forms, when rationalizing possible phase transitions. The barrier related to molecular motions at high pressure can be expected to be even more difficult to overcome than the energy separating conformations of an individual molecule. Frequently the increase in temperature may diminish this barrier and the transition may occur (like sluggish transitions in pyridine and benzene).[228] However, everything is defined by the strength of intermolecular interactions and their stabilization with pressure. It has been shown for guanidinium nitrate phase III that pressure increases the role of hydrogen bonding, so that it strongly stabilizes the structure.[229] The increase in temperature cannot induce a transformation into a high-pressure phase IV with denser packing. A similar study for tolazamide polymorphs could contribute to understanding the role of intermolecular interactions in stabilizing different crystal structures with increasing pressure.

The role of fluids in pressure-induced polymorphic transitions is another important issue. As mentioned above, hydrostatic conditions cannot be achieved without a fluid phase. The choice of the pressure-transmitting medium is usually rather arbitrary and depends, among other factors, on the technical facilities available at the laboratory. Gas loading is not widely available, and not all groups can work with fluids that are characterized by low boiling points such as pentane-isopentane.[230] A fluid is selected (i) to ensure that hydrostatic conditions are maintained up to the desirable pressure point and (ii) to avoid chemical interaction with the immersed solid, while simultaneously avoiding its dissolution. In regard to the first point, extensive literature data summing up the hydrostaticity limits of various fluids are available.[55,58–63] Recent studies with precise pressure measurements at multiple sites in a pressure cell show that a notable loss of ideal hydrostaticity, and thus the possibility of shear stresses arising in the medium, occur at considerably lower pressures than was previously expected from the broadening of X-ray diffraction peaks or from the measurement of a single pressure point in the cell.[64] In regard to the second point, it is very difficult (if at all possible) to predict *a priori* whether the solubility of an insoluble substance at ambient pressure will increase on compression. It is also difficult to predict surface interactions between a solvent and the crystal material, an interaction that may increase with pressure. Therefore, even if a hydrostatic fluid does not dissolve a solid at ambient conditions, high-pressure dissolution, re-crystallization, and solvate formation remain possible. When compressed in a mixture of ethanol, methanol and water, crystals of L-alanine recrystallize into a mixed solvate with methanol and water. In this solvate, L-alanine zwitterions form a 3D network linked by hydrogen bonds, and solute molecules fill the intermolecular cavities. On decompression, the framework is first preserved, despite desolvation, and a new L-alanine polymorph is formed. However, on complete decompression this new polymorph transforms back to the ambient-pressure polymorph. Solvate/hydrate formation at high pressure has been documented for many pharmaceutical crystals immersed in a fluid, different solvates being formed from different media.[29,231–240] For high-pressure solvates, "over-hydration" seems to be common, when many more solvent molecules can be built into the crystal cell in the high-pressure solvate than is common for low-pressure crystal hydrates. Even if no solvates are formed, the solid–fluid interactions can account for liquid-assisted polymorphic transition or a re-crystallization. This phenomenon is also known at ambient pressure; for example, L-glutamic acid transforms from the stable α-polymorph into a more stable and less soluble β-polymorph.[241–243] No transition has ever been observed on heating a dry crystal of the α-form in dry atmosphere, but the transformation can proceed under humid atmosphere conditions even at ambient temperature, although very slowly (being far from completion even after several months).[241,242] If crystals of the α-polymorph are stored in the saturated solution, they transform rather quickly into the β-polymorph. Interestingly, when crystals of the β-form precipitate, they deposit mainly on the surface of α-crystals rather than crystallizing as single crystals in solution. The two forms fuse into each other to such an extent that their co-existence could be revealed only by powder X-ray diffraction, but not by optical microscopy.[241,242] Without a careful study of the process under different experimental conditions, it is not easy in such case to distinguish between a true solid-state phase transition and a re-crystallization.

A striking example of the effect of the choice of fluid on high-pressure polymorphism is provided by chlorpropamide.[178,191,194,195] This anti-diabetic drug is highly polymorphic. Five of these polymorphs—α, β, γ, δ, and ε—can co-exist at ambient conditions. These forms each have similar hydrogen-bonded chains involving the urea-like core, but differ in the conformation adopted by the phenyl-ring and alkyl tails, and in the relative orientation of the hydrogen-bonded chains. They respond differently to increasing pressure under identical experimental conditions, transforming into different high-pressure phases and at different pressures. The occurrence of phase transitions and the structure of the high-pressure phases depend on the choice of hydrostatic fluid. Dry samples behave completely differently. On tableting, even trace amounts of fluid may be important. In a dry sample, phase transitions in the α-polymorph of chlorpropamide are hindered.[191] In saturated ethanol solution, solvent-assisted polymorphic transition to a new high-pressure phase, α′, takes place.[178] The same starting β-polymorph transforms into different high-pressure forms if immersed in different pressure-transmitting media: He, Ne, paraffin, pentane-isopentane, glycerol-water and ethanol. Moreover, the presence of several crystals of different polymorphs in the pressure cell also leads to non-trivial results. Kinetic barriers to structural rearrangement in the solid-state and of the crystallization at high pressure from solution can be significant and result in formation of metastable materials. Resulting products may simply be those with the lowest kinetic barrier for nucleation. This barrier can obviously be changed in the presence of a seed, but not every polymorph can act as a seed (even if it is denser). Even in the presence of the most stable phase, other crystallization routes to relatively less stable forms remain possible.[194] Thus, in the absence of δ-form seeds, the β polymorph recrystallized into needles of the γ form. Alternatively, in the presence of a δ-form seed, simultaneous slow recrystallization of the β form into both the γ form (new crystals) and the δ form (an increase in the size of seed crystals) was observed. Interestingly, the α polymorph neither acted as a seed in the recrystallization of the β or γ polymorphs, nor did it recrystallize into the denser δ form (Figure 2.12).[194] Based on the relative densities, one might expect three high-pressure recrystallization pathways for β chlorpropamide: β → α, β → γ, or β → δ. However, only the two latter processes were observed. It therefore became important to consider the relative rates with which the required molecular re-organization occurs—sufficient to yield critical nucleation clusters, sufficient only to grow an existing crystal, or so energetically unfavourable that neither nucleation nor growth can occur. One can interpret the observed phenomena having compared molecular conformations and molecular packing in the polymorphs, and assuming that both the conformations of individual molecules and the molecular clusters can be preserved on dissolution. Neither the β → γ nor the β → δ transformations require a fundamental change in the packing of molecular H-bonded chains, whereas the β → α transformation would require that every second chain should be turned "upside down". The β → γ transformation requires some rotation of the phenyl ring, but no major change in the position of the alkyl tail, whereas for both the β → δ and β → α transformations the molecular conformation must change. Thus, one can suggest that the γ form has the lowest nucleation barrier. The newly formed crystals of the γ polymorph were located near the source of chlorpropamide molecules—the β-form crystal. It can be suggested that the molecular conformation and the fragments of the original crystal are preserved in solution for

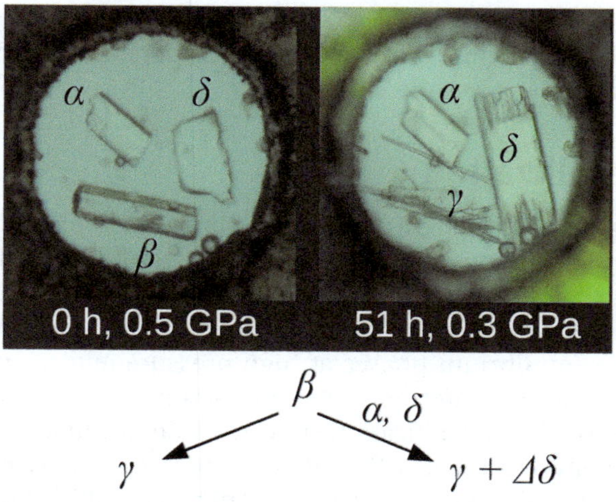

Figure 2.12 The polymorphic transformations of β-chlorpropamide when compressed in a pentane-isopentane mixture in the presence of the crystals of the α and the δ polymorphs: the crystal of the β-form recrystallizes partly into the γ form (new needle-shaped crystals) and the δ form (the present crystal of the δ form grows larger), the crystal of the α form does not change. Reproduced from ref. 194 with permission from the Royal Society of Chemistry.

some time after dissolution, and the nuclei of the γ phase are formed from these fragments. The presence of a seed of the denser δ polymorph, with a different molecular conformation, favours molecular reconfiguration/reorientation, as these molecules can now not form new crystals, but be built into the denser phase where the molecules are already in the required new conformation. The mechanism underpinning the high-pressure polymorphism of this system stems from the fact that the β → δ transformation does not require a change in the relative orientation of the hydrogen-bonded chains with respect to each other, which is in stark contrast to the β → α transition.

The ability of a molecule to preserve its conformation on dissolution from a crystal has been discussed in the literature,[244–246] though it has only been experimentally proven for a few cases to date.[246,247] The "crystalline conformation" can be stabilized both by interactions within the preserved clusters that remain on dissolution and, in some special cases, by interactions with the solvent. The dissolution processes, which do not occur on a "molecule by molecule" basis, but instead as molecular clusters or entire "double layers", have also been discussed, modelled and, in some cases, even observed experimentally.[248] One might expect that the conditions within the confined space of a diamond anvil cell at high pressure would be particularly favourable for preserving molecular clusters on dissolution. This is especially true if the hydrostatic fluid does not readily separate the molecules from each other on dissolution. The transformations in a diamond anvil pressure cell can combine the restrictions of confined crystallization in a small volume with the limitations of convection, the diffusion of molecules through a viscous medium and the difficulties of structural rearrangement of molecules and molecular packing in the solid state.

2.5 Crystallization

One can consider two different cases: crystallization of solids from their solutions, and crystallization of substances that are liquids at ambient pressure but become solid on compression. These experiments are used to obtain new phases, which are metastable at ambient conditions, and quench them, to search for the equilibrium phases at high pressure and to study the structure-forming units and intermolecular interactions in the crystal structures. The outcome of the crystallization process is determined by the interplay of nucleation and crystal growth. Different interactions are responsible for these two processes, and the relative role of different interactions differs with increasing pressure. Therefore, variations of pressure can help to control polymorphism and to obtain metastable forms. Experiments on high-pressure crystallization date back to the early work by Bridgman and are now actively developed by many research groups.

Most often compounds have been crystallized that are liquid under ambient conditions. It is known from early high-pressure experiments that the same liquid can form different polymorphs when cooled or when compressed. Examples include crystallization of benzene,[249–254] ethanol, acetic acid,[255] water,[256,257] acetone,[258] and many other small molecules.[259–263]

Another possibility is to obtain different polymorphs by compressing the same liquid to different pressures. Both of the obtained polymorphs can be thermodynamically stable. However, the situation can be much more complex if kinetic factors come into play such that one or even none of the polymorphs is in fact thermodynamically stable. In addition the sequence of accessing the different pressure points may be important, as well as the compression rate. Different polymorphs are often obtained depending on the "direction of approaching a selected pressure" (*i.e.* on increasing or decreasing pressure).[16,35]

The situation is additionally complicated if the aim is to obtain a high-pressure polymorph as a single crystal. Usually compression of a liquid leads to a polycrystalline sample. In order to obtain a single crystal suitable for X-ray analysis, different strategies can be employed (Figure 2.13).[35] A polycrystalline sample at high pressure can be heated until all crystallites but one is dissolved, after which the heating is stopped and the only remaining crystallite grows. An alternative (yet more experimentally difficult, and hence more rarely used) is to keep temperature constant, but to decrease the pressure until all crystallites but one dissolve. Subsequently, pressure is increased once again to allow the only remaining crystallite to grow. Although the same final (T, P) point is reached in both experiments, their results can be completely different, with different polymorphs formed. Obviously, at least one of them should be metastable. Moreover, the different polymorphs can never

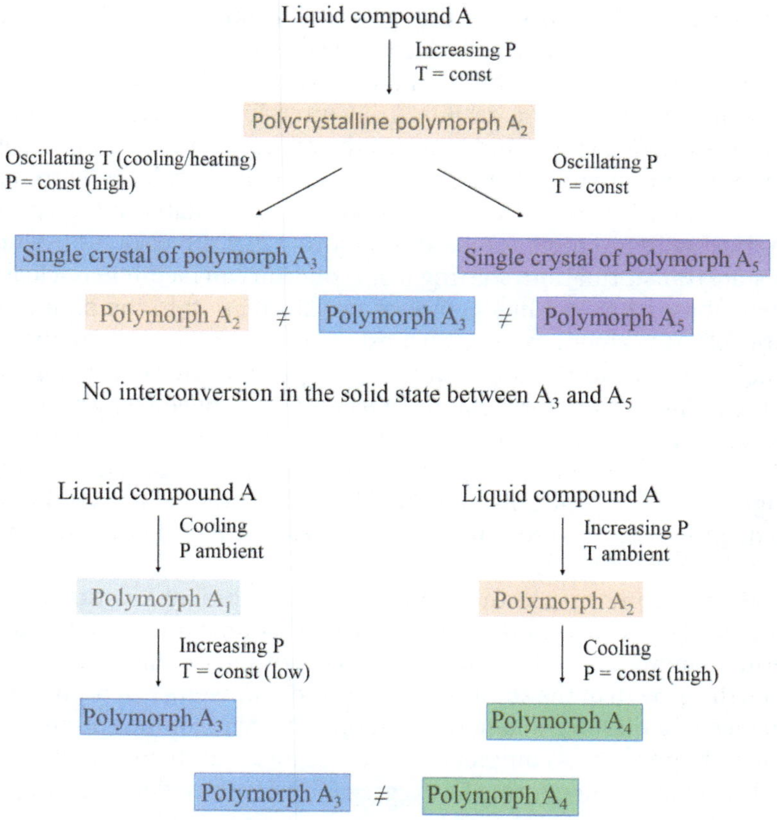

Figure 2.13 Different results of crystallization on variations of temperature and pressure resulting from different experimental protocols. Reproduced with permission of the International Union of Crystallography from G. Resnati, E. Boldyreva, P. Bombicz and M. Kawano, Supramolecular interactions in the solid state, *IUCrJ*, 2015, **2**(6), 675–690, Copyright 2015 IUCr. http://journals.iucr.org/.

interconvert in the solid state. Many examples of this type have been reported in the papers by Katrusiak *et al.*, where such effects were studied systematically.[1,21,23,233–237,260,263–267] A case study is the polymorphism of bromochlorofluoroacetic acid,[196] where pressure affects the balance between secondary intermolecular interactions involving halogen and O atoms. Depending on the protocol with which pressure is increased, either catemers (phase α) or dimers (phase β) were the main structure-forming units in the polymorph formed. Another example is related to comparing molecular aggregation in pyridazine, pyridine and polymorphs of benzene.[228] The interactions governing the molecular arrangement in the series of structures were shown to change gradually from C–H···N to C–H···π hydrogen bonds. High pressures favoured C–H···N interactions over the C–H···π bonds in pyridine, but benzene remained more stable than pyridazine and pyridine.

Another possibility to study the structure-forming role of non-covalent interactions is to consider pressure-induced crystallization of compounds that are soluble in selected fluids under ambient conditions.[14,29,30,268–272] Also in this case crystallization results from interplay between the thermodynamic and kinetic factors, and between the effects of pressure on nucleation and subsequent growth of the crystal nuclei. One possibility is to observe crystallization of the form that is thermodynamically stable at high pressure. In contrast with attempts to obtain this form *via* a direct pressure-induced solid-state transformation, starting from solution can facilitate nucleation of the new phase, thus yielding complete transformation to the high-pressure polymorph. If a kinetic barrier also exists, hindering transformation from the high-pressure to the ambient-pressure polymorph on decompression, then the high-pressure form can be preserved ("quenched") by rapid release of pressure. This "quenched" crystal can then be used to seed mass crystallization at ambient conditions, for example in the pharmaceutical industry. A range of drug polymorphs and their hydrates have been crystallized and subsequently recovered to ambient pressures in this way over the last dec ade.[1,29,30,32,36,231,239,268–272]

Another outcome may be that compression of solutions favours crystallization of form(s) that are metastable at high pressure, but nucleate faster than a thermodynamically stable form. The growth of such a phase can kinetically impede the growth of the stable form. In this circumstance, it is possible that molecules of a dissolved compound adopt different conformations in solution at high pressure as compared with those under ambient conditions. The crystal structure is then formed by aggregation of molecules in the high-pressure conformations, and the lattice energy is optimized for these molecular conformations, not for the ambient-pressure ones.[273] Common methods of crystal structure prediction, which are based on molecular conformations optimized in the gas phase, may not be adequate in such cases. In particular, this approach may become troublesome when crystallization from solution is considered since the molecular conformation in solvent under pressure may be completely different.

An additional factor to consider is that crystal growth at high pressure can be hindered, such that very high values of supersaturation can be achieved. In such cases, crystallization of high-pressure forms is not observed on compression, but instead upon decompression. Rapid increase of pressure to rather high values can help to prevent crystallization even if the sample is kept at this pressure for a very long time.

A special case is the crystallization of high-pressure forms (polymorphs or solvates) on compression of solids immersed in pressure-transmitting fluids. Even if a solid is poorly soluble in the fluid under ambient conditions, its solubility can increase with increasing pressure. In this case, marked recrystallization or a seemingly solid-state (but in fact solvent-assisted) phase transition can take place. The latter can be suspected when the "solid-state" pressure-induced phase transition depends on the choice of pressure-transmitting fluid; either the transition points and/or the structures of the high-pressure form differ. Again, in these situations observation of the high-pressure forms can take place not on increasing pressure, but on decompression, and can be prevented by increasing pressure with sufficient haste. Formation of solvates at high pressure, even for compounds that are reluctant to form solvates at ambient conditions, is another indication that intermolecular interactions in solutions are changed significantly by pressure. This phenomenon seems to be directly related to interactions of proteins with solvents at high pressure.[274]

2.6 Multi-component Crystals

In the last decade ever more research in the field of organic solid-state chemistry is dedicated to crystals that include several chemical species, namely salts, co-crystals, and their solvates. Multi-component crystals are of great interest for the basic understanding of factors that influence the formation of crystal structures from fluid phases under different experimental conditions and their response to variations of temperature and pressure. This is the next level in crystal engineering, since one has a variety of unique fragments that must self-assemble, and one must not only control the formation of a certain polymorph, but also of a multi-component crystal *vs.* several separate mono-component phases. Many of these systems are of practical importance, often in the pharmaceutical industry, as ferroelectrics, non-linear optical materials, or as energetic materials.[1,34,37]

Non-ambient conditions are widely used when studying multi-component crystals in several aspects, for their synthesis, as well as for probing different types of intermolecular interactions by varying temperature and/or pressure. The latter is important in the context of this book.

Multi-component crystals can be studied under variable conditions either in the range of stability of the same phase, or over polymorphic transitions or solvation/desolvation processes.[34–37] In the first case, the main idea is to compare the bulk compressibilities and the anisotropy of strain of a multi-component crystal with those of the phases of individual components. This is especially interesting if some of the structure-forming motifs are the same (or similar) in a salt/co-crystal and the individual component structure;[275–280] one can compare the properties of similar motifs in different crystalline environments.

Another research direction is related to modifying the stability of crystal structures with respect to structural rearrangements. Adding second and third components, one can modify or even radically change the crystal structure and, in this way, also the physical properties of a solid in a controlled way. In particular, a structure can become more robust, or more elastic, more stable with respect to variations in temperature, pressure, and humidity, more soluble, or more compressible. Cases in which the crystal structures of the individual components have the same or similar structural motifs as those of the salts/co-crystals but in the presence of another component have a significant effect on the structural stability are of special interest.

As examples, one can refer to the studies of the effect of pressure on the salts and co-crystals of aminoacids, as compared to the response to the same conditions on crystals of amino acids. Examples when pressure-induced phase transitions are observed for crystals of pure amino acids, but not for their salts, have been provided by cysteinium oxalates. In the crystals of L- and DL-cysteine, the side chains $-CH_2SH$ can form weak hydrogen bonds either with the side chains of the neighbouring molecules (SH···S) or with the terminal carboxy groups (SH···O).[281–284] Crystals of the orthorhombic polymorph of L-cysteine show several phase transitions at pressures between 1 GPa and at 3.9 GPa, and these transitions are related to the rotation of the side chains and the formation of new types of hydrogen bonds.[285,286] Monoclinic polymorph of L-cysteine also undergoes a series of phase transitions with increasing pressure. These transitions are related to a switching over of the hydrogen bonds within the –SH···SH···O– structural motifs.[286] The crystal structure of DL-cysteine is even less stable with respect to increasing pressure and transforms to another polymorph at about 0.1 GPa (an unusually low pressure for a crystalline amino acid).[287] It was then supposed that cysteine-containing crystal structures could be stabilized with respect to phase transitions by measures that reduce the mobility of the side chain in the crystalline environment. This could be achieved by increasing the significance of the side chain –SH group as a participant in intermolecular hydrogen bonds. Through *N*-acetylation one can remove the strong $-NH_3^+$ donor of cysteine, and leave a system without strong charge-assisted interactions. Alternatively, one can co-crystallize cysteine with an acid (oxalic acid) that converts the amino acid to a cation and itself forms a strong anion H-bond acceptor, thus boosting the importance of potential –SH donors. The crystal structures of the three compounds *N*-acetyl-L-cysteine, DL-cysteinium semioxalate, and bis-(DL-cysteinium) oxalate have thus been studied with variation of temperature and pressure. Increasing pressure at least up to 9.5 GPa did not result in any structural phase transitions in *N*-acetyl-L-cysteine and bis-(DL-cysteinium) oxalate. In the case of DL-cysteiniumsemioxalate, increasing pressure caused a phase transition at a much higher pressure (6 GPa), compared to the ranges of pressure-induced phase transitions observed earlier for both monoclinic and orthorhombic L-cysteine (2.5–3.9 GPa and 1.1–2.5 GPa, respectively) or DL-cysteine (0.1–5 GPa). This phase transition had a large hysteresis, and the reverse transformation on decompression was observed at 3.7 GPa. The reverse transformation was accompanied by a change in molecular conformations, as well as by the reorganization of the N–H···O hydrogen bonds in the crystal structure.[288]

The opposite effect is also possible, when a multi-component crystal undergoes a structural rearrangement under milder conditions than the crystals of components do. Examples are provided by salts and co-crystals of glycine. As has been mentioned above, the response of glycine polymorphs to pressure is radically different.[200–207,215,216] It was assumed that the remarkable stability of the α polymorph is provided by the centrosymmetric double layers linked by NH···O hydrogen bonds. As such, if this motif is destroyed by introducing a second component, one can expect that the new crystal structure will be less stable with respect to pressure than α glycine.

In fact, phase transitions have to date been reported for several multi-component crystals containing glycine. Bis-glycinium oxalate undergoes two phase transitions at 1.7 and at 5 GPa, accompanied by a significant rearrangement of hydrogen bonds in the structure and violent crystal fragmentation, as has been evidenced by IR-spectroscopy and X-ray powder diffraction.[289–292] It was supposedly based on the IR-spectroscopy data that the whole hydrogen-bond network is radically re-arranged, so that some NH···O hydrogen bonds are broken, while other, weaker, NH···O hydrogen bonds are formed. This results in the formation of two non-identical glycine molecules per unit cell. At 5 GPa another reorganization of NH···O hydrogen bonds takes place.[289–292] A second salt, 1 : 1 glycinium oxalate, does not undergo structural phase transitions on increasing pressure, although the symmetrization of the OH···O hydrogen bonds occurs at 8 GPa.[292] No phase transitions have been observed in either of the salts on cooling. Co-crystals of glycine with glutaric acid, GGa, undergo a polymorphic single crystal to single crystal polymorphic transition at pressures below 0.1 GPa, a considerably lower transition pressure than has ever been recorded for pure amino acids.[293] The same polymorph is also formed on cooling.[294] The polymorphic transformations on cooling and with increasing pressure are related to the conformational changes in the flexible glutaric acid; every second glutaric acid molecule in the structure changes its shape. The topology of the hydrogen-bonded motifs remains almost the same and hydrogen bonds do not switch to other atoms, although the hydrogen bond lengths do change and some of the bonds become non-equivalent.[293,294] Co-crystals of glycine with other coformers—DL-tartaric and phthalic acids (GTa and GPh, respectively)—behave differently from that with glutaric acid. On cooling, no phase transitions were observed in GTa or GPh, contrasting the situation with GGa. On hydrostatic compression both GTa and GPh underwent reversible phase transformations, accompanied by fracture. In the high-pressure phases the main structural framework is preserved, the number of crystallographically independent molecules in the unit cell increases. In GTa dimers are squeezed together so that some hydrogen bonds get a three-centred character, and the interactions of one of the two glycine molecules changes dramatically.[295]

The high-pressure behaviour was also studied for another multi-component crystal based on glycine, namely, for glycine lithium sulfate.[296] Raman studies and DFT calculations suggested the formation of an intramolecular hydrogen bond at higher pressures. Subsequent to a structural transformation to a new high-pressure phase at 9 GPa, the observed spectral changes in the Raman spectra above 14 GPa indicated possible conformational change of glycine from zwitterionic to a neutral form. At pressures above 18 GPa, the characteristic features in the Raman spectra and the X-ray diffraction patterns suggested transformation to a hydrogen-bond-assisted polymeric phase with intermediate range order.[296]

Another example of making a structure more prone to a high-pressure phase transformation by adding a second co-former, and in this way breaking homomolecular synthons and changing the whole crystal structure radically, is provided by DL-alaninium semi-oxalate monohydrate.[187] DL-alanine itself does not show any phase transitions with increasing pressure.[297] The molecules are linked *via* NH···O hydrogen bonds in a three-dimensional framework, the CH_3- side chains do not form any hydrogen bonds, similarly to the structure of L-alanine, and their rotation does not result in any phase transitions in either L-,[189,298] or in DL- alanine.[297] In contrast, DL-alaninium semi-oxalate monohydrate, in which this homomolecular framework is broken, undergoes a single-crystal to single-crystal phase transition at a pressure between 1.5 and 2.4 GPa.[187] Selected hydrogen bonds involving alaninium cations and crystal water molecules switch over and become bifurcated, whereas the others are compressed continuously. The transition is accompanied by pronounced discontinuities in the cell parameters and volume as a function of pressure, although no radical changes in the molecular packing are induced. Remarkably, the rigid homomolecular chain motifs of alaninum cations and of oxalate anions are preserved with increasing pressure, and no proton transfer is observed.[187]

It is also possible that a structural phase transition in a multi-component crystal (a salt, a hydrate, or a salt hydrate) is observed in the same pressure range as for the crystals of the primary basic component, but the structural mechanism changes. Such an example is provided by a comparison of the effect of pressure on L and DL-serine,[57,148,208–213] L-serine monohydrate,[299] bis-DL-serinium oxalate dihydrate,[300] L-serinium maleate and L-ascorbic acid–L-serine co-crystal.[301] L-serine has a framework crystal structure similar to those of L-alanine, DL-alanine, or the orthorhombic polymorph of L-cysteine. Its side chain, $-CH_2OH$, is involved in the formation of intermolecular OH···O hydrogen bonds that are stronger than the SH···O or SH···S bonds. At ambient pressure, OH···O hydrogen bonds are formed between the $-CH_2OH$ groups of the neighbouring chains of molecules. As pressure increases, a cooperative rotation of the side chains results in a structural phase transition at about 5 GPa, so that new OH···O hydrogen bonds between the side chains and the terminal carboxy groups are formed. As pressure increases further, the side chains rotate at about 8 GPa once again to form a new phase with additional bifurcated hydrogen bonds. The two reversible phase transitions preserve the single crystals intact, and the interface propagates rapidly through the crystal along one of the crystallographic directions. This is possible since the main structure is held by highly elastic NH···O hydrogen bonds between the terminal amino and carboxy groups, which are preserved, while the side chains rotate. The layered structure of DL-serine, in which the $-CH_2OH$ side chains already form OH···O bonds to the carboxy groups at ambient pressure, does not undergo any structural phase transitions with increasing pressure up to at least 8.6 GPa. However, non-monotonic distortion of selected hydrogen bonds does occur in the pressure range below 1–2 GPa, a change in the compression mechanism is seen at 2–3 GPa, and evidence for formation of bifurcated NH···O hydrogen bonds is seen between 3 and 4 GPa.[148]

The crystal structure of L-serine monohydrate (L-serine monohydrate-I) at ambient pressure contains H-bonded layers of zwitterionic serine molecules linked by H bonds to water molecules. The water molecules act as donors to oxygen atoms on the L-serine carboxylate and alcohol groups in separate layers. This phase remains stable from ambient pressure to 4.5 GPa, although notable reduction in the interlayer distance is observed. On increasing the pressure to 5.2 GPa the structure transforms to a high-pressure polymorph, L-serine monohydrate-II. Across the transition the serine interlayer distances reduce further and the water molecules rotate so that both donor interactions are made to the same serine layer. The serine ammonium group adopts an eclipsed conformation, reconfiguring the H bonding within the serine layers. The disruption of H bonding as water is pushed into the serine layers suggests that a similar process may occur as a first step in the pressure-denaturation of proteins. Though the water molecules become coordinatively saturated with respect to H bonding, and interlayer serine–serine Coulombic interactions are strengthened, model calculations suggested that overall the intermolecular interactions are weaker in phase II than phase I. According to the authors of the work, the lattice enthalpy becomes more negative through the transition as a result of the smaller PV term applying to the more efficiently packed phase-II structure.[299]

Another multi-component crystal containing serine, namely bis-DL-serinium oxalate dihydrate, with a three-dimensional network of hydrogen bonds linking oxalate anions, serinium cations and water molecules, shows a reversible phase transition also at about 4–5 GPa, resulting in the formation of multiple disordered domains because of the reorientation of molecules and molecular fragments, which no longer hold together as robustly as homomolecular H bonds hold molecules in L- or DL-serine.[300]

Pressure is well known to influence the proton transfer between the molecules in mono-component molecular crystals. Both the symmetrization and, oppositely, an increase of ionicity of the hydrogen bonds (OH···O or NH···N) have been reported in the literature.[1,4–13,19–27,32–37,189,302] An increase in the ionicity of the intermolecular NH···O bonds between the terminal amino and carboxy groups with increasing pressure has been reported for L-alanine.[189] A bright example of a complete intermolecular hydrogen transfer has been reported for oxalic acid dihydrate. This results in the formation of a salt from the original molecular hydrate.[291,303,304] Interestingly, this transfer takes place only if the α polymorph is compressed, and was not observed for the β polymorph.[304] As other examples of the systems for which the interplay between the structural transformations and solid-state acid–base (proton transfer) reactions at high pressure has been studied in relation to the paraelectric and ferroelectric properties, one can mention the co-crystals of phenazine with chloranilic, bromanilic, and fluoranilic acids,[305] or of co-crystals of anilic acids and 2,3-di(2-pyridinyl)pyrazine,[306] where hydrogen bonds are polarized and neutral molecules are converted into ionized species on cooling or on increasing pressure.

2.7 Pressure-induced Reactions and Effect of Pressure on Photo- and Thermo-chemical Transformations

The main effects of pressure on (molecular) crystals are summarized in Figure 2.14. Many of these phenomena are not only related to the anisotropic compression and polymorphic transformations, but also to chemical reactions. When pressure is applied, the molecules inside a crystal undergo significant changes in their stereoelectronic properties, often related to charge transfer, valence isomerization, and formation of new chemical bonds.[4,23,26,32,118–123,128,129,149–177,306–310] The change in colour is the most obvious visually observed manifestation of such effects termed piezochromism. The most interesting are the changes that enhance the reactivity of systems that would otherwise be rather inert at ambient conditions. The proton migration mentioned in the previous section is also a chemical reaction giving a salt from neutral components. Charge disproportionation, in general, is favoured by increasing pressure, enabling better packing.[311] Another manifestation of this phenomenon is de-aromatization of aromatic compounds under pressure.[112,312] In aromatic compounds, molecular stability originates from the resonance between two electronic configurations. The resonance energy can be decreased in molecular crystals on application of pressure. For example, syn-1,6:8,13-biscarbonyl[14] annulene, an aromatic compound at ambient conditions, gradually localizes one of the resonant configurations on compression.[312] This phenomenon is evident from the molecular geometries measured at several pressures and from the experimentally determined electron density distribution at 7.7 GPa; the experimental observations have been validated by periodic DFT calculations.

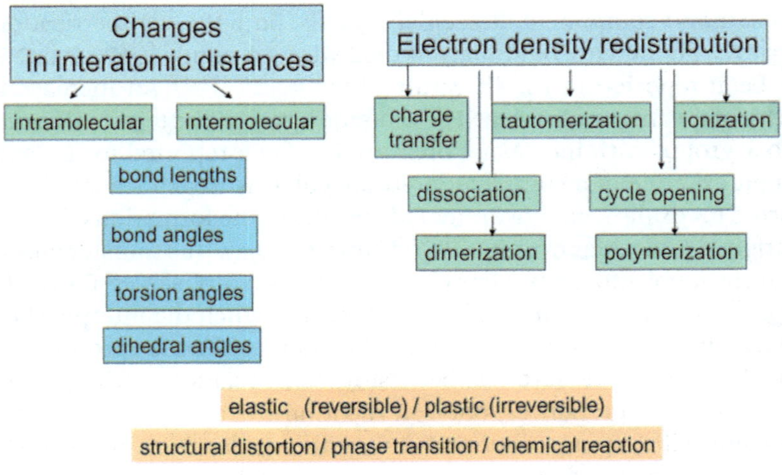

Figure 2.14 The main effects induced in solids by high pressure.

Other chemical transformations related to bond rearrangement include isomerization, dimerization, polymerization, charge-transfer processes,[313–333] dehydration of crystal hydrates,[334] desolvation and reverse solvation of metal–organic frameworks (MOFs), or coordination polymers, often related to structural rearrangement of the host MOF.[176,192,334–351] Quite often a bond rearrangement transformation is cooperative, similar to what is observed during polymorphic transitions. In fact, there is no sharp border line between a solid-state structural phase transition and a solid-state reaction. The ability to induce all of these transformations by increasing pressure, rather than by heating or irradiation, is a bright illustration that pressure can have a very significant effect both on intra- and intermolecular interactions in solids.[352–355] This is the field that has attracted much attention in recent years and is developing very quickly.

Fluid–solid interactions under high pressure are of special interest. Hydrostatic conditions cannot be achieved without a hydrostatic medium—a fluid phase, *i.e.* a gas or a liquid. The same fluid can not only enable the hydrostaticity of compression, but also participate in a pressure-induced process involving the solid phase, which can be a phase transition, a recrystallization (as has been described in the previous Sections 2.4 and 2.5), or a chemical reaction giving a "host–guest" compound, when the molecules of the pressure-transmitting fluid penetrate into the solid.

The importance of selecting a pressure-transmitting fluid is well understood in high-pressure mineralogy, where the hydrostatic media are considered as being either penetrating, or non-penetrating.[356] This approach is especially common in relation to microporous materials, zeolites, which are characterized by an "open" system of interconnected channels and cavities, or channelled minerals.[357] For example, when present in the hydrostatic medium (even in trace quantities), molecules of H_2O can enter zeolite channels upon compression; this process is reversible.[356–358] The so-called "pressure-induced over-hydration"[356–358] can cause either volume increase (abrupt as in[359,360] or gradual as in[361]), or simply a decrease in compressibility as compared with the non-hydrated species, due to the site occupancy increase of the already existing water sites.[362,363] It is important to note, however, that this pressure-induced hydration effect is not always observed.[364,365] Gases used as pressure-media can also penetrate into the solid on compression. This can result in a decreased compressibility of the material,[366–369] or also in the structural transformations, as for cristobalite,[370] or vitreous silica[371] compressed in helium.

For organic, metal–organic and coordination compounds the studies of the role of fluids in pressure-induced solid-state transformations have started relatively recently.[190–195,236–240,271,272,338–350,372,373] In addition to the effects of fluids on phase transitions and recrystallization (Sections 2.4 and 2.5), fluids have been documented to reversibly penetrate into crystal structures to form inclusion compounds, especially if the solid structure has channels or large cavities, such as MOFs.[140,347–349,372–376]

Comparing the compression of the same MOF in various hydrostatic fluids can help to understand the relation between intermolecular interactions and the ability of MOF structures to adapt to the changes in experimental conditions (temperature and pressure), as well as to the sorption of various gases or liquids. The structural framework can be preserved to a large extent on compression–decompression, even if it expands or shrinks. Alternatively, it can undergo a displacive (reversible) or a reconstructive (irreversible) rearrangement.[139–141,335–350] The phenomena observed for MOFs and coordination polymers are similar to those previously documented for minerals.

In porous MOFs the hydrostatic medium can penetrate the pores and interact dynamically with the framework. This often causes the framework to expand when pressure is first applied. As pressure on a sample is increased, it can become superfilled (or hyperfilled) with liquid and this behaviour underpins many interesting results that have been observed in MOFs under pressure. The structural response to pressure is highly sensitive to the choice of hydrostatic medium.[162] The selection of a solvent used to apply pressure can be based not only on solubility and hydrostatic limit, but also on molecular size, shape and functionality.[376] The same classification of fluids as "penetrating" and "non-penetrating" as has been introduced earlier for minerals is used also for MOFs. The size of pores in a solid and the chemical nature of fluid determine whether the fluid is "penetrating" or not. For nanoporous (pore diameter <10 Å) MOFs, a small molecule, such as methanol, is a penetrating medium, while the bulkier molecule isopropyl alcohol could be either penetrating or non-penetrating, depending on the exact pore channel dimensions. For larger microporous (10–20 Å) MOFs, however, isopropyl alcohol is a penetrating medium. Long-chain and branched fluorinated hydrocarbons such as Fluorinert FC-70 [perfluoro tri-*N*-pentylamine ($C_{15}F_{33}N$)] or FC-77 [a mixture of perfluorooctane (C_8F_{18}) and perfluoro ox-acyclononane ($C_8F_{16}O$)] tend to be used as non-penetrating media in such cases. As the pore size is increased to mesoporous (>20 Å), however, even the largest Fluorinert liquid could of course be penetrating, although there are currently no high-pressure studies published on mesoporous materials.[347]

The MOF frameworks are usually charge neutral and thus host–guest interactions between the framework and residual cavity-bound solvent are often weak. Solvent removal can give rise to permanent porosity and allow reversible guest uptake, which is a key facet of much interesting host–guest chemistry in MOFs.[377] This is in contrast to zeolites where a cationic skeleton can often collapse on solvent/guest removal due to stronger electrostatic host–guest interactions.[378] Pressure-induced transformations in MOFs generally occur at lower pressures than in zeolites.[347,379–384] This is due to the greater porosities and elaborates connectivity in MOFs, which often result in frameworks with inherent flexibility and a tendency for more extreme and dramatic structural changes in response to pressure. The more moderate pressures required to produce a response in MOFs have allowed them to be studied at much lower, industrially achievable pressures (*ca.* below 1 GPa), and in fact their high-pressure behaviour can be highly relevant to their framework properties at ambient pressure.[347]

The two key properties that account for the structural response of MOFs to external stimuli are the flexibility of the framework and its adsorption characteristics, the latter depending on the size of pores as compared to the size of the guest molecules and on the chemical nature of the functional groups and atoms forming the walls of a cavity. Guest–guest and host–guest interactions affect framework compressibility, the anisotropy of compression, including the negative linear compressibility, pressure-induced amorphization. High-pressure diffraction experiments, which can push the uptake capacity and flexible regions of frameworks to their limits, are very informative for studying all these factors.[347] Guests can be forced into unfavourable environments (*e.g.* hydrophilic molecules into hydrophobic pockets), oversized guests can be modelled in confined cavities and amorphization can be induced in crystalline compounds, causing a breadth of dynamic structural responses to be observed. High-pressure experimental diffraction studies can also benefit from complementary first-principles condensed matter simulations. This allows the potential energy landscape of MOFs to be explored and helps to rationalize their structural behaviour.[347]

An interesting direction of high-pressure research of MOFs is related to the possibility to generate porous structures by applying pressure to non-porous starting samples. The fact that porous materials can be made from non-porous structures is a remarkable discovery. This seemingly counter-intuitive process can be rationalized if one takes into account that compression is accomplished in a fluid medium. The hydrostatic medium molecules included as guests on compression favour different framework topologies as compared to those in the starting sample. The new structures can be stable upon the release of pressure and, in some cases, even on removal of the guest molecules.[344,347]

Hydrostatic compression of the crystal hydrates of coordination polymers can trigger dehydration accompanying the displacive structural rearrangement, when the product phase forms inside the parent phase as crystalline inclusions ("martensitic planes"), and released water crystallizes as solid ice.[334] This solid-state dehydration is not necessarily related to the interaction with the hydrostatic fluid and can be observed on compression in hydrophobic fluids.[334]

The effect of pressure on the photo and thermal reactions in solids is another research direction related to studying intermolecular interactions. Pressure has long been used as a tool for probing the mechanisms of chemical reactions in solution.[385,386] Activation volumes of the reactions derived from variable-pressure experiments, together with activation energies calculated from variable-temperature experiments, shed light on the mechanisms of the transformations.[386] In the solid state, studies of the effect of pressure on the thermal and photochemical reactions in coordination compounds were carried out in the 1990s using linkage isomerization in $[Co(NH_3)_5NO_2]XY$ complexes as a case study.[118,129,387] An unusual effect of pressure on the rates of the photo and thermal reactions was found: pressure induced an opposite effect to what could be expected from changes in the molar volume during the reaction. This could be interpreted within the concept of a reaction cavity[388,389] using the detailed information on the anisotropy of strain on hydrostatic compression and over the course of chemical reactions and the corresponding changes in the intermolecular contacts and interactions.[129] Recently interest in this subject has been revived. Pressure-induced phase transitions have been discovered during which the structural changes are similar to those resulting from the photoisomerization.[390] Lattice strain on hydrostatic compression has been shown to correlate with the structural changes accounting for the photosalient effect.[130] The difference in the strain resulting from variations in pressure and temperature has been shown to have a pronounced effect on the photoisomerization.[391] A very interesting series of studies, also based on the concept of reaction cavity,[388,389] has been published on the effect of hydrostatic compression on the photoinduced dimerization in organic crystals.[392–396] Studies in photo-responsive MOFs at high pressures are another promising direction.[397]

2.8 Conclusions

I hope that this chapter has shown, through several examples, that intermolecular interactions can in fact be probed by varying pressure. The structures formed under selected temperature and pressure conditions and the structural changes resulting from temperature and pressure variation can give us a clue to understanding the structure-forming role of various types of hydrogen bonds, stacking, van der Waals and other non-covalent interactions. This is one of the mainstream directions of high-pressure science, crystal engineering, and solid-state and physical chemistry.

The main directions of studying intermolecular interactions in the crystals of organic and coordination compounds on hydrostatic compression include (i) the analysis of the compressibility and the anisotropy of compression in relation to different types of intermolecular hydrogen bonds and other non-covalent interactions; (ii) the analysis of the thermodynamic and kinetic factors that influence the nucleation and nuclei growth of the high-pressure phases either on crystallization, or on solid-state phase transitions; (iii) the studies of the chemical transformations induced by hydrostatic compression or of the effect of pressure on the (mainly) photochemical and thermal chemical reactions.

High-pressure experiments are difficult and time-consuming. Therefore, even a mere refinement of the cell parameters and volume of a high-pressure phase can be an achievement. If these data are available at multiple pressure points, they can be used for fine-tuning interatomic potentials used for modelling the crystal structures, for example, for computer-assisted structure prediction.[142] Structure solution and refinement at multiple pressure points can shed more light on the structure-forming role of the intermolecular interactions and their properties. Theoretical modelling, such as calculating lattice energies, can be of enormous help for finalizing the choice of a structural model and improving its parameters.[269] The state of the art in both experimental and theoretical approaches today makes it possible to achieve such a quality of structure refinement that a charge-density analysis becomes possible,[108–112] and not only proton migration from one atom to another,[303,304] but also such fine effects as de-aromatization of the aromatic bonds on hydrostatic compression become possible to follow.[312]

Different methods of theoretical analysis of selected interactions and their contributions into the total energy of a crystal structure under variable-pressure conditions are widely used today.[1,4,10,142,150,151,167,172–174,213,219,225,237,238,272,298,303,311,317–320,323,348,352,353] This information is used, in order to rationalize the pressure-induced structural evolution on the anisotropic compression, in the course of phase transitions, or on the interaction of solids with the hydrostatic fluids. In addition to this "microscopic" approach, a "macroscopic" analysis of the reasons of phase transitions and the roles of different intermolecular interactions in structural rearrangement includes, on the one hand, a detailed analysis of the phase diagrams in the system,[31,219,220,222–224] and on the other hand, considering the interplay between the rates of the nucleation and nuclei growth of a new phase.[16,57,178,189–191,193–195,201,206,211,217,219,226] Theoretical modelling is progressing fast. Still, no less understanding of the mechanisms and the role of various interactions can be gained from well-planned model experiments with variable hydrostatic fluids, compression–decompression protocols, including variation of the compression–decompression rate, duration of keeping the sample at a selected pressure, adding temperature as a variable parameter, as well as studying a series of related compounds or different starting polymorphs of the same compound.

While high-pressure studies on organic and coordination compounds have led to many new and interesting discoveries, high-pressure science remains largely in its infancy. With a plethora of new advances in theory and experimentation, and a multitude of phenomena and parameters to explore, high pressure promises to be an exciting and dynamic field for many years to come.

Acknowledgements

The support from the Russian Academy of Sciences (project 0301-2016-0014) is acknowledged. I thank Mr A. Michalchuk for language polishing and useful advise, and Dr Zakharov and Mr Bogdanov for assistance with preparation of the figures.

References

1. *High-pressure Crystallography. From Novel Experimental Approaches to Applications in Cutting-edge Technologies*, ed. E. V. Boldyreva and P. Dera, Springer, Dordrecht, 2010, p. 612.
2. R. M. Hazen, *The Diamond Makers*, Cambridge University Press, Cambridge, UK, 1999, p. 258.
3. P. W. Bridgman, *The Physics of High Pressure*, G Bell and Sons, London, 1931, p. 398.
4. J. Tse and E. Boldyreva, Electron density topology of crystalline solids at high pressure, in *Modern Charge Density Analysis*, ed. C. Gatti and P. Macchi, Springer, 2012, vol. XXIII, pp. 573–623.
5. A. Katrusiak, High-pressure X-ray diffraction studies on organic crystals, *Cryst. Res. Technol.*, 1991, **26**(5), 523–531.
6. A. Katrusiak, Macroscopic and structural effects of hydrogen-bond transformations, *Crystallogr. Rev.*, 1996, **5**(2), 133–175.
7. A. Katrusiak, Macroscopic and structural effects of hydrogen-bond transformations: some recent directions, *Crystallogr. Rev.*, 2003, **9**(2–3), 87–89.
8. E. V. Boldyreva, High-pressure-induced structural changes in molecular crystals preserving the space group symmetry: anisotropic distortion/isosymmetric polymorphism, *Cryst. Eng.*, 2003, **6**(4), 235–254.
9. E. V. Boldyreva, High-pressure studies of the anisotropy of structural distortion of molecular crystals, *J. Mol. Struct.*, 2003, **647**(1), 159–179.
10. *High-pressure Crystallography*, ed. A. Katrusiak and P. McMillan, Springer Science & Business Media, 2004, vol. 140, p. 567.
11. E. V. Boldyreva, High-pressure studies of the hydrogen bond networks in molecular crystals, *J. Mol. Struct.*, 2004, **700**(1), 151–155.
12. E. V. Boldyreva, T. N. Drebushchak, T. P. Shakhtshneider, H. Sowa, H. Ahsbahs, S. V. Goryainov, S. N. Ivashevskaya, E. N. Kolesnik, V. A. Drebushchak and E. B. Burgina, Variable-temperature and variable-pressure studies of small-molecule organic crystals, *ARKIVOC*, 2004, **12**, 128–155.
13. E. V. Boldyreva, High pressure and supramolecular systems, *Russ. Chem. Bull.*, 2004, **53**(7), 1369–1378.
14. F. P. Fabbiani and C. R. Pulham, High-pressure studies of pharmaceutical compounds and energetic materials, *Chem. Soc. Rev.*, 2006, **35**(10), 932–942.
15. P. F. McMillan, Chemistry at high pressure, *Chem. Soc. Rev.*, 2006, **35**(10), 855–857.
16. E. V. Boldyreva, High-pressure polymorphs of molecular solids: when are they formed, and when are they not? Some examples of the role of kinetic control, *Cryst. Growth Des.*, 2007, **7**(9), 1662–1668.
17. E. V. Boldyreva, High-pressure studies of crystalline amino acids and simple peptides, *High Pressure Biosci. Biotechnol.*, 2007, **1**(1), 28–46.

18. E. V. Boldyreva, Crystalline amino acids – a link between chemistry, materials science and biology, in *Models, Mysteries, and Magic of Molecules*, ed. J. C. A. Boeyens and J. F. Ogilvie, Springer Verlag, 2007, pp. 169–194.
19. C.-H. Görbitz, Crystal structures of amino acids: from bond lengths in glycine to metal complexes and high-pressure polymorphs, *Crystallogr. Rev.*, 2015, **21**(3), 160–212.
20. S. K. Sikka, On some hydrogen bond correlations at high pressures, *High Pressure Res.*, 2007, **27**(3), 313–319.
21. A. Katrusiak, High-pressure crystallography, *Acta Crystallogr., Sect. A: Found. Crystallogr.*, 2008, **64**(1), 135–148.
22. S. A. Moggach, S. Parsons and P. A. Wood, High-pressure polymorphism in amino acids, *Crystallogr. Rev.*, 2008, **14**(2), 143–184.
23. E. V. Boldyreva, High-pressure diffraction studies of molecular organic solids. A personal view, *Acta Crystallogr., Sect. A: Found. Crystallogr.*, 2008, **64**(1), 218–231.
24. E. V. Boldyreva, H. Sowa, H. Ahsbahs, S. V. Goryainov, V. V. Chernyshev, V. P. Dmitriev, Y. V. Seryotkin, E. N. Kolesnik, T. P. Shakhtshneider, S. N. Ivashevskaya and T. N. Drebushchak, Pressure-induced phase transitions in organic molecular crystals: a combination of X-ray single-crystal and powder diffraction, Raman and IR-spectroscopy, *J. Phys.: Conf. Ser.*, 2008, **121**(2), 022023.
25. S. K. Sikka and S. M. Sharma, The hydrogen bond under pressure, *Phase Transitions*, 2008, **81**(10), 907–934.
26. S. A. Moggach and S. Parsons, High pressure crystallography of inorganic and organometallic complexes, *Spectrosc. Prop. Inorg. Organomet. Compd.*, 2009, **40**, 324–354.
27. E. V. Boldyreva, Combined X-ray diffraction and Raman spectroscopy studies of phase transitions in crystalline amino acids at low temperatures and high pressures: selected examples, *Phase Transitions*, 2009, **82**(4), 303–321.
28. A. Tiwari, I. S. Butler and J. A. Kozinski, A brief overview of the effect of high pressures on the vibrational spectra of biomaterials, *Appl. Spectrosc. Rev.*, 2009, **44**(6), 552–567.
29. I. D. H. Oswald, I. Chataigner, S. Elphick, F. P. A. Fabbiani, A. R. Lennie, J. Maddaluno, W. G. Marshall, T. J. Prior, C. R. Pulham and R. I. Smith, Putting pressure on elusive polymorphs and solvates, *CrystEngComm*, 2009, **11**(2), 359–366.
30. D. I. Millar, W. G. Marshall, I. D. Oswald and C. R. Pulham, High-pressure structural studies of energetic materials, *Crystallogr. Rev.*, 2010, **16**(2), 115–132.
31. R. Ceolin and I. B. Rietveld, Phenomenology of polymorphism and topological pressure–temperature diagrams, *J. Therm. Anal. Calorim.*, 2010, **102**(1), 357–360.
32. E. V. Boldyreva, Supramolecular systems under extreme conditions, *Herald Russ. Acad. Sci.*, 2012, **82**(6), 423–431.

33. R. Lee, J. A. K. Howard, M. R. Probert and J. W. Steed, Structure of organic solids at low temperature and high pressure, *Chem. Soc. Rev.*, 2014, **43**(13), 4300–4311.
34. E. V. Boldyreva, Multicomponent organic crystals at high pressure, *Z. Kristallogr.*, 2014, **229**(3), 236–245.
35. G. Resnati, E. Boldyreva, P. Bombicz and M. Kawano, Supramolecular interactions in the solid state, *IUCrJ*, 2015, **2**(6), 675–690.
36. E. V. Boldyreva, Non-ambient conditions in the investigation and manufacturing of drug forms, *Curr. Pharm. Des.*, 2016, **22**, 4981–5000.
37. E. V. Boldyreva, Multi-component crystals and non-ambient conditions, in *Multi-component Crystals. Synthesis, Concepts, Function*, ed. E. Tiekink and J. Zukerman-Schpector, DeGruyter, Berlin, 2017.
38. L. Merrill and W. A. Bassett, Miniature diamond anvil pressure cell for single crystal X-ray diffraction studies, *Rev. Sci. Instrum.*, 1974, **45**(2), 290–294.
39. H. Ahsbahs, X-Ray diffraction on single crystals at high pressure, *Prog. Cryst. Growth Charact.*, 1987, **14**, 263–302.
40. R. Letoullec, J. P. Pinceaux and P. Loubeyre, The membrane diamond anvil cell: a new device for generating continuous pressure and temperature variations, *High Pressure Res.*, 1988, **1**, 77–90.
41. W. F. Kuhs, H. Ahsbahs, D. Londono and J. L. Finney, *In situ* crystal growth and neutron four-circle diffractometry under high pressure, *Phys. B*, 1989, **156–157**, 684–687.
42. S. P. Besedin, I. N. Makarenko, S. M. Stishov, V. P. Glazkov, I. N. Goncharenko and V. A. Somenkov, Diamond anvil cells for neutron diffraction, *High Pressure Res.*, 1995, **14**(1–3), 193–197.
43. W. F. Kuhs, F. C. Bauer, R. Hausmann, H. Ahsbahs, R. Dorwarth and K. Hoelzer, Single crystal diffraction with X-rays and neutrons: high quality at high pressure? *High Pressure Res.*, 1996, **14**(4–6), 341–352.
44. M. Eremets, *High Pressure Experimental Methods*, Oxford University Press, New York, 1996, p. 390.
45. D. R. Allan, R. Miletich and R. J. Angel, A diamond-anvil cell for single-crystal X-ray diffraction studies to pressures in excess of 10 GPa, *Rev. Sci. Instrum.*, 1996, **67**(3), 840–842.
46. High pressure techniques in chemistry and physics. A practical approach, *Practical Approach in Chemistry Series*, ed. W. B. Holzapfel and N. S. Isaacs, Oxford Univ. Press, Oxford, 1997, p. 400.
47. R. Miletich, D. R. Allan, W. F. Kuhs, R. Miletich, High-temperature and high-pressure crystal chemistry, in *Mineral. Soc. Amer. Rev. Mineral. Geochem.*, ed. R. M. Hazen and R. T. Downs, 2000, vol. 41, pp. 445–520.
48. R. Fourme, E. Girard and K. Akasaka, High-pressure macromolecular crystallography and NMR: status, achievements and prospects, *Curr. Opin. Struct. Biol.*, 2012, **22**(5), 636–642.
49. H. Ahsbahs, New pressure cell for single-crystal X-ray investigations on diffractometers with area detectors, *Z. Kristallogr.*, 2004, **219**(6), 305–308.

50. R. Boehler and K. de Hantsetters, New anvil designs in diamond-cells, *High Pressure Res.*, 2004, **24**(3), 391–396.
51. R. Boehler, New diamond cell for single-crystal X-ray diffraction, *Rev. Sci. Instrum.*, 2006, **77**(11), 115103.
52. I. N. Goncharenko, New techniques for high-pressure neutron and X-ray studies, *High Pressure Res.*, 2007, **27**(1), 183–188.
53. T. Boffa Ballaran, A. Kurnosov and D. Trots, Single-crystal X-ray diffraction at extreme conditions: a review, *High Pressure Res.*, 2013, **3**(3), 453–465.
54. J. Binns, K. V. Kamenev, G. J. McIntyre, S. A. Moggach and S. Parsons, Use of a miniature diamond-anvil cell in high-pressure single-crystal neutron Laue diffraction, *IUCrJ*, 2016, **3**, 168–179.
55. E. V. Boldyreva, B. A. Zakharov, S. V. Rashchenko, Y. V. Seryotkin and N. A. Tumanov, *Studying Solid-state Transformations Using In situ X-Ray Diffraction Studies at High-pressures*, Publishing House Siber. Branch Russ. Acad. Sci., Novosibirsk, 2016, p. 156.
56. L. Dubrovinsky, N. Dubrovinskaia, V. B. Prakapenka and A. M. Abakumov, Implementation of micro-ball nanodiamond anvils for high-pressure studies above 6 Mbar, *Nat. Commun.*, 2012, **3**, 1163.
57. M. Fisch, A. Lanza, E. V. Boldyreva, P. Macchi and N. Casati, Kinetic control of high-pressure solid-state phase transitions: a case study on L-serine, *J. Phys. Chem. C*, 2015, **119**(32), 18611–18617.
58. G. J. Piermarini, S. Block and J. D. Barnett, Hydrostatic limits in liquids and solids to 100 kbar, *J. Appl. Phys.*, 1973, **44**(12), 5377–5382.
59. R. J. Angel, M. Bujak, J. Zhao, G. D. Gatta and S. D. Jacobsen, Effective hydrostatic limits of pressure media for high-pressure crystallographic studies, *J. Appl. Crystallogr.*, 2007, **40**(1), 26–32.
60. K. Murata, K. Yokogawa, H. Yoshino, S. Klotz, P. Munsch, A. Irizawa, M. Nishiyama, K. Iizuka, T. Nanba, T. Okada, Y. Shiraga and S. Aoyama, Pressure transmitting medium Daphne 7474 solidifying at 3.7 GPa at room temperature, *Rev. Sci. Instrum.*, 2008, **79**(8), 085101.
61. S. Klotz, L. Paumier, G. Le March and P. Munsch, The effect of temperature on the hydrostatic limit of 4:1 methanol-ethanol under pressure, *High Pressure Res.*, 2009, **29**(4), 649–652.
62. S. Klotz, J. C. Chervin, P. Munsch and G. Le Marchand, Hydrostatic limits of 11 pressure transmitting media, *J. Phys. D: Appl. Phys.*, 2009, **42**, 075413.
63. H. Li, L. Chen, X. Zhu, Z. Wang, H. Huang, K. Yang, L. Su, G. Yang and X. Li, Ionic liquid: a good pressure transmitting medium, *J. Solution Chem.*, 2016, 1–8.
64. J. Zhao, R. J. Angel and N. L. Ross, Effects of deviatoric stresses in the diamond-anvil pressure cell on single-crystal samples, *J. Appl. Crystallogr.*, 2010, **43**(4), 743–751.
65. H. K. Mao, P. M. Bell, K. J. Dunn, R. M. Chrenko and R. C. DeVries, Absolute pressure measurements and analysis of diamonds subjected to maximum static pressures of 1.3–1.7 Mbar, *Rev. Sci. Instrum.*, 1979, **50**, 1002–1009.

66. Y. Fei, A. Ricolleau, M. Frank, K. Mibe, G. Shen and V. Prakapenka, Toward an internally consistent pressure scale, *Proc. Natl. Acad. Sci. U. S. A.*, 2007, **104**, 9182–9186.
67. A. L. Ruoff, R. C. Lincoln and Y. C. Chen, A new method of absolute high pressure determination, *J. Phys. D: Appl. Phys.*, 1973, **6**, 1295–1306.
68. R. A. Forman, G. J. Piermarini, J. D. Barnett and S. Block, Pressure measurement by utilization of ruby sharp- line luminescence, *Science*, 1972, **176**, 284–285.
69. G. J. Piermarini, S. Block, J. D. Barnett and R. A. Forman, Calibration of the R1 ruby fluorescence line to 195 kbar, *J. Appl. Phys.*, 1975, **46**, 2774–2780.
70. D. M. Adams, R. Appleby and S. K. Sharma, Spectroscopy at very high pressure: Part X. Use of ruby R-lines in the estimation of pressure at ambient and at low temperatures, *J. Phys. E: Sci. Instrum.*, 1976, **9**, 1140–1144.
71. H. E. King and C. T. Prewitt, Improved pressure calibration system using the ruby R1 fluorescence, *Rev. Sci. Instrum.*, 1979, **51**, 1037–1039.
72. H. K. Mao, J. Xu and P. M. Bell, Calibration of the ruby gauge to 800 kbar under quasihydrostatic conditions, *J. Geophys. Res., B*, 1986, **91**, 4673–4676.
73. P. M. Bell, J. A. Xu and H. K. Mao, Static compression of gold and copper and calibration of the ruby pressure scale to pressures to 1.8 megabares, in *Shock Waves in Condensed Matter*, ed. Y. M. Gupta, Plenum Press, New York, 1986, pp. 125–130.
74. I. F. Silvera, A. D. Chijioke, W. J. Nellis, A. Soldatov and J. Tempere, Calibration of the ruby pressure scale to 150 GPa, *Phys. Status Solidi B*, 2007, **244**(1), 460–467.
75. K. Syassen, Ruby under pressure, *High Pressure Res.*, 2008, **28**, 75–126.
76. H. Jahren, M. Kruger and R. Jeanloz, Alexandrite and ruby as high *P* and *T* calibration standards, *Eos, Trans., Am. Geophys. Union*, 1990, **71**, 1611.
77. N. J. Hess and G. J. Exarhos, Temperature and pressure dependence of laser-induced fluorescence in Sm : YAG—a new pressure calibrant, *High Pressure Res.*, 1989, **2**, 57–64.
78. A. Lacam and C. Chateau, High-pressure measurements at moderate temperatures in a diamond anvil cell with a new optical sensor: $SrB_4O_7:Sm^{2+}$, *J. Appl. Phys.*, 1989, **66**(1), 366–372.
79. N. J. Hess and D. Schiferl, Pressure and temperature dependence of laser-induced fluorescence of Sm:YAG to 100 kbar and 700 °C and an empirical model, *J. Appl. Phys.*, 1990, **68**, 1953–1960.
80. F. Datchi, R. LeToullec and P. Loubeyre, Improved calibration of the $SrB_4O_7:Sm^{2+}$ optical pressure gauge: advantages at very high pressures and temperatures, *J. Appl. Phys.*, 1997, **81**, 3333–3339.
81. Y. Zhao, W. Barvosa-Carter, S. D. Theiss, S. Mitha, M. J. Aziz and D. Schiferl, Pressure measurement at high temperature using ten Sm: YAG fluorescence peaks, *J. Appl. Phys.*, 1998, **84**(8), 4049–4059.

82. Q. Wei, N. Dubrovinskaia, L. Dubrovinsky and Q. Wei, Ruby and Sm:YAG fluorescence pressure gauges up to 120 GPa and 700 K, *J. Appl. Phys.*, 2011, **110**(4), 043513.
83. D. M. Trots, A. Kurnosov, T. Boffa Ballaran, S. Tkachev, K. Zhuravlev, V. Prakapenka, M. Berkowski and D. J. Frost, The Sm:YAG primary fluorescence pressure scale, *J. Geophys. Res.: Solid Earth*, 2013, **118**, 5805–5813.
84. S. V. Rashchenko, A. Kurnosov, L. Dubrovinsky and K. D. Litasov, Revised calibration of the Sm : SrB_4O_7 pressure sensor using the Sm-doped yttrium-aluminum garnet primary pressure scale, *J. Appl. Phys.*, 2015, **117**, 145902.
85. F. Datchi, A. Dewaele, P. Loubeyre, R. Letoullec, Y. Le Godec and B. Canny, Optical pressure sensors for high-pressure – high-temperature studies in a diamond anvil cell, *High Pressure Res.*, 2007, **27**, 447–463.
86. A. Watenphul and C. Schmidt, Calibration of berlinite ($AlPO_4$) as Raman spectroscopic pressure sensor for diamond-anvil cell experiments at elevated temperatures, *J. Raman Spectrosc.*, 2012, **43**(4), 564–570.
87. N. Guo and H.-F. Zheng, Study on the accuracy of pressure determined by Raman spectra of quartz, *Spectrosc. Spect. Anal.*, 2010, **30**(8), 2161–2163.
88. H. Sowa, K. Reithmayer, J. Macavei, W. Rieck, H. Schulz and V. Kupčik, High-pressure single-crystal study on $AlPO_4$ with synchrotron radiation, *J. Appl. Crystallogr.*, 1990, **23**(5), 397–400.
89. D. L. Decker, Equation of state of NaCl and its use as a pressure gauge in high-pressure research, *J. Appl. Phys.*, 1965, **36**, 157–161.
90. D. L. Decker, Equation of state of sodium chloride, *J. Appl. Phys.*, 1966, **37**, 5012–5014.
91. D. L. Decker, High-pressure equation of state for NaCl, KCl and CsCl, *J. Appl. Phys.*, 1971, **42**, 3239–3244.
92. F. Birch, Equation of state and thermodynamic parameters of NaCl to 300 kbar in the high-temperature domain, *J. Geophys. Res.*, 1986, **91**, 4949–4954.
93. J. M. Brown, The NaCl pressure standard, *J. Appl. Phys.*, 1999, **86**, 5801–5808.
94. R. M. Hazen and L. W. Finger, Calcium fluoride as an internal pressure standard in high-pressure/high-temperature crystallography, *J. Appl. Crystallogr.*, 1981, **14**, 234–236.
95. A. Katrusiak and R. J. Nelmes, A test of the accuracy of high-pressure measurements using a Merrill–Bassett diamond-anvil cell, *J. Appl. Crystallogr.*, 1986, **19**(2), 73–76.
96. L. Gerward, J. S. Olsen, S. Steenstrup, M. Malinowski, S. Åsbrink and A. Waskowska, X-ray diffraction investigations of CaF_2 at high pressure, *J. Appl. Crystallogr.*, 1992, **25**, 578–581.
97. R. J. Angel, The high-pressure, high-temperature equation of state of calcium fluoride, CaF_2, *J. Phys.: Condens. Matter*, 1993, **5**, L141–L144.

98. H. J. McSkimin, P. Andreatch and R. N. Thurston, Elastic moduli of quartz *versus* hydrostatic pressure at 25 °C and −195.8°, *J. Appl. Phys.*, 1965, **36**, 1624–1632.
99. R. J. Angel, D. R. Allan, R. Miletich and L. W. Finger, The use of quartz as an internal pressure standard in high-pressure crystallography, *J. Appl. Crystallogr.*, 1997, **30**, 461–466.
100. Z. Wu, R. M. Wentzcovitch, K. Umemoto, B. Li, H. Hirose and J.-C. Zheng, Pressure-volume-temperature relations in MgO: an ultrahigh pressure-temperature scale for planetary sciences applications, *J. Geophys. Res.*, 2008, **113**(6), B06204.
101. I. C. Ming, M. H. Manghnani, J. Balogh, S. B. Qadri, E. F. Skeltonm and J. C. Jamieson, Gold as a reliable internal pressure calibrant at high temperatures, *J. Appl. Phys.*, 1983, **54**, 4390–4397.
102. B. K. Godwal and R. Jeanloz, First-principles equation of state of gold, *Phys. Rev. B: Condens. Matter Mater. Phys.*, 1989, **40**, 7501–7507.
103. A. Dewaele, P. Loubeyre and M. Mezouar, Equations of state of six metals above 94 GPa, *Phys. Rev. B: Condens. Matter Mater. Phys.*, 2004, **70**(9), 094112-1–094112-8.
104. S. Anzellini, A. Dewaele, F. Occelli, P. Loubeyre and M. Mezouar, Equation of state of rhenium and application for ultra high pressure calibration, *J. Appl. Phys.*, 2014, **115**(4), 043511.
105. C. J. Ridley and K. V. Kamenev, High pressure neutron and X-ray diffraction at low temperatures, *Z. Kristallogr.*, 2014, **229**(3), 171–199.
106. X. Wang and K. V. Kamenev, Review of modern instrumentation for magnetic measurements at high pressure and low temperature, *Low Temp. Phys.*, 2014, **40**(8), 735–746.
107. J. S. Tse, D. D. Klug, S. Patchkovskii, Y. Ma and J. K. Dewhurst, Chemical bonding, electron–phonon coupling, and structural transformations in high-pressure phases of Si, *J. Phys. Chem. B*, 2006, **110**, 3721–3726.
108. E. A. Zhurova, V. G. Tsirelson, V. V. Zhurov, A. I. Stash and A. A. Pinkerton, Chemical bonding in pentaerythritol at very low temperature or at high pressure: an experimental and theoretical study, *Acta Crystallogr., Sect. B: Struct. Sci.*, 2006, **62**(3), 513–520.
109. T. Yamanaka, T. Okada and Y. Nakamoto, Electron density distribution and static dipole moment of $KNbO_3$ at high pressure, *Phys. Rev. B: Condens. Matter Mater. Phys.*, 2009, **80**, 094108.
110. F. P. A. Fabbiani, B. Dittrich, C. R. Pulham and J. E. Warren, Towards charge-density analysis of high-pressure molecular crystal structures, *Acta Crystallogr., Sect. A: Found. Crystallogr.*, 2011, **67**, C376.
111. P. Macchi and N. Casati, Strong hydrogen bonds in crystals under high pressure, *Acta Crystallogr., Sect. A: Found. Crystallogr.*, 2011, **67**, C163–C164.
112. N. Casati, A. Genoni, B. Meyer, A. Krawczuk and P. Macchi, Exploring charge density analysis in crystals at high pressure: data collection, data analysis and advanced modeling, *Acta Crystallogr., Sect. B: Struct. Sci., Cryst. Eng. Mater.*, 2017, **73**, 584–597.

113. E. S. Fedorov, in *Symmetry and Crystal Structures. Collection of Basic Papers*, ed. A. V. Shubnikov and I. I. Shafranovski, Publishing House of the USSR Acad. Sci., Moscow, 1949.
114. J. M. Robertson and A. R. Ubbelohde, Structure and thermal properties associated with some hydrogen bonds in crystals. I. The isotope effect, *Proc. R. Soc. London, Ser. A*, 1939, **170**, 222–240.
115. A. R. Ubbelohde, Structure and thermal properties associated with some hydrogen bonds in crystals. III. Further examples of the isotope effect, *Proc. R. Soc. London, Ser. A*, 1939, **173**, 417–427.
116. A. R. Ubbelohde and I. Woodward, Structure and thermal properties of crystals. VI. The role of hydrogen bonds in Rochelle salt, *Proc. R. Soc. London, Ser. A*, 1946, **185**, 448–465.
117. K. Gallagher, A. R. Ubbelohde and I. Woodward, Effects of temperature on some hydrogen-bond networks in crystals, *Acta Crystallogr.*, 1955, **8**, 561–566.
118. E. V. Boldyreva, Intramolecular linkage isomerization in the crystals of some Co (III)-ammine complexes – a link between inorganic and organic solid state chemistry, *Mol. Cryst. Liq. Cryst. Inc. Nonlinear Opt.*, 1994, **242**, 17–52.
119. N. Masciocchi, A. Kolyshev, V. Dulepov, E. V. Boldyreva and A. Sironi Study of the linkage isomerization $[Co(NH_3)_5NO_2]Br_2$ – $[Co(NH_3)_5ONO]Br_2$ in the solid state by X-ray powder diffraction, *Inorg. Chem.*, 1994, **33**, 2579–2585.
120. E. Boldyreva, H. Ahsbahs and H. Uchtmann, Pressure induced lattice distortion in some Co (III)-Ammine complexes, $[Co(NH_3)_5NO_2]XY$ ($XY = Cl_2, Br_2, ClNO_3$), *Ber. Bunsengesell. Phys. Chem.*, 1994, **98**, 738–745.
121. E. V. Boldyreva, J. Kivikoski and J. A. K. Howard, Distortion of crystal structures of some Co^{III} ammine complexes. I. Distortion of crystal structure of $[Co(NH_3)_5NO_2]Cl(NO_3)$ on cooling, *Acta Crystallogr., Sect. B: Struct. Sci.*, 1997, **B53**, 394–404.
122. E. V. Boldyreva, J. Kivikoski and J. A. K. Howard, Distortion of crystal structures of some Co^{III} ammine complexes. II. Distortion of crystal structures of $[Co(NH_3)_5NO_2]X_2$ (X = Cl, Br) on cooling, *Acta Crystallogr., Sect. B: Struct. Sci.*, 1997, **53**, 405–414.
123. E. V. Boldyreva, D. Yu. Naumov and H. Ahsbahs, Distortion of crystal structures of some Co^{III} ammine complexes. III. Distortion of crystal structure of $[Co(NH_3)_5NO_2]Cl_2$ at hydrostatic pressures up to 3.5 GPa, *Acta Crystallogr., Sect. B: Struct. Sci.*, 1998, **54**(6), 798–808.
124. J. F. Nye, *Physical Properties of Crystals: Their Representation by Tensors and Matrices*, Oxford University Press, 1985.
125. Y. Ohashi, R. Hazen and L. Finger, *Comparative Crystal Chemistry. Temperature, Pressure, Composition and Variation of the Crystal Structure*, Wiley, N.Y., 1982.
126. http://www.rossangel.com/text_strain.htm.
127. S. M. Jessen and H. Kueppers, The precision of thermal-expansion tensors of triclinic and monoclinic crystals, *J. Appl. Crystallogr.*, 1991, **24**, 239–242.

128. E. V. Boldyreva, N. V. Podberezskaya, A. V. Virovets, L. P. Burleva and V. E. Dulepov, Powder X-ray diffraction analysis of the products of the solid-state linkage nitro-nitrito isomerization $[Co(NH_3)_5NO_2]XY \rightarrow [Co(NH_3)_5ONO]XY$ ($XY = 2Cl^-$, $2Br^-$, $2I^-$, $Cl^-(NO_3)^-$, $2(NO_3)^-$), *J. Struct. Chem.*, 1993, **34**, 128–138.
129. E. V. Boldyreva, Crystal chemistry aspects of the solid-state intramolecular linkage isomerization in the nitro- (nitrito-) pentammines of cobalt (III), *Russ. J. Coord. Chem.*, 2001, **27**(5), 323–350.
130. P. Naumov, S. C. Sahoo, B. A. Zakharov and E. V. Boldyreva, Dynamic single crystals: kinematic analysis of photoinduced crystal jumping (the photosalient effect), *Angew. Chem., Int. Ed. Engl.*, 2013, **252**(38), 9990–9995.
131. A. C. Holt and M. Ross, Calculations of the Grüneisen parameter for some models of the solid, *Phys. Rev. B: Condens. Matter Mater. Phys.*, 1970, **1**(6), 2700.
132. E. V. Boldyreva, T. P. Shakhtshneider, M. A. Vasilchenko, H. Ahsbahs and H. Uchtmann, Anisotropic crystal structure distortion of the monoclinic polymorph of acetaminophen at high hydrostatic pressures, *Acta Crystallogr., Sect. B: Struct. Sci.*, 2000, **56**(2), 299–309.
133. E. V. Boldyreva, T. P. Shakhtshneider, H. Ahsbahs, H. Sowa and H. Uchtmann, A comparative study of the effect of hydrostatic pressure up to 4 GPa on the monoclinic and the orthorhombic polymorphs of paracetamol, *J. Therm. Anal. Calorim.*, 2001, **68**, 437–452.
134. E. V. Boldyreva, T. N. Drebushchak and E. S. Shutova, Structural distortion of the α, β, and γ-polymorphs of glycine on cooling, *Z. Kristallogr.*, 2003, **218**, 366–376.
135. E. V. Boldyreva, H. Ahsbahs and H.-P. Weber, A comparative study of pressure-induced lattice strain of α- and γ-polymorphs of glycine, *Z. Kristallogr.*, 2003, **218**, 231–236.
136. T. N. Drebushchak and E. V. Boldyreva, Variable temperature (100–360 K) single-crystal X-ray diffraction study of the orthorhombic polymorph of paracetamol (*p*-hydroxyacetanilide), *Z. Kristallogr.*, 2004, **219**(8), 506–512.
137. E. V. Boldyreva, E. N. Kolesnik, T. N. Drebushchak, H. Ahsbahs, J. A. Beukes and H.-P. Weber, A comparative study of the anisotropy of lattice strain induced in the crystals of L-serine by cooling down to 100 K or by increasing pressure up to 4.4 GPa, *Z. Kristallogr.*, 2005, **220**, 58–65.
138. E. V. Boldyreva, E. N. Kolesnik, T. N. Drebushchak, H. Sowa, H. Ahsbahs and Y. V. Seryotkin, A comparative study of the anisotropy of lattice strain induced in the crystals of DL-serine by cooling down to 100 K, or by increasing pressure up to 8.6 GPa, *Z. Kristallogr.*, 2006, **221**, 150–161.
139. G. M. Espallargas, L. Lee Brammer, D. R. Allan, C. R. Pulham, N. Robertson and J. E. Warren, Noncovalent interactions under extreme conditions: high-pressure and low-temperature diffraction studies of the isostructural metal–organic networks (4-chloropyridinium)$_2$[CoX_4] (X = Cl, Br), *J. Am. Chem. Soc.*, 2008, **130**(28), 9058–9071.

140. W. Cai and A. Katrusiak, Giant negative linear compression positively coupled to massive thermal expansion in a metal–organic framework, *Nat. Commun.*, 2014, **5**, 4337.
141. S. R. Madsen, S. A. Moggach, J. Overgaard and B. Brummerstedt Iversen, Anisotropic compressibility of the coordination polymer emim [Mn(btc)], *Acta Crystallogr., Sect. B: Struct. Sci., Cryst. Eng. Mater.*, 2016, **72**(3), 389–394.
142. A. V. Dzyabchenko and E. V. Boldyreva, Modelling the crystal packing and conformation of $[Co^{III}(NH_3)_5NO_2]^{2+}2Cl^-$ in a range of pressures, *Acta Crystallogr., Sect. A: Found. Crystallogr.*, 2000, **56**, s215.
143. K. Adhikari, K. M. Flurchick and L. Valenzano, A hybrid density functional study on the effects of pressure on paracetamol and aspirin polymorphs, *Comput. Theor. Chem.*, 2015, **1062**, 90–98.
144. V. Vaiser and H. Rapaport, Compressibility and elasticity of amphiphilic and acidic β-sheet peptides at the air–water interface, *J. Phys. Chem. B*, 2010, **115**(1), 50–56.
145. I. Orgzall, F. Emmerling, B. Schulz and O. Franco, High-pressure studies on molecular crystals—relations between structure and high-pressure behavior, *J. Phys.: Condens. Matter*, 2008, **20**, 295206.
146. M. Podsiadlo and A. Katrusiak, Competing patterns of weak directional forces in pressure-frozen CH_2ClI and CH_2I_2, *J. Phys. Chem. B*, 2008, **112**, 5355–5362.
147. K. Takizawa, H. Fukudome, Y. Kozaki and S. Ando, Pressure-induced changes in crystalline structures of polyimides analyzed by wide-angle X-ray diffraction at high pressures, *Macromolecules*, 2014, **47**(12), 3951–3958.
148. B. A. Zakharov, B. A. Kolesov and E. V. Boldyreva, Effect of pressure on crystalline L-and DL-serine: revisited by a combined single-crystal X-ray diffraction at a laboratory source and polarized Raman spectroscopy study, *Acta Crystallogr., Sect. B: Struct. Sci.*, 2012, **68**(3), 275–286.
149. E. V. Boldyreva, E. B. Burgina, V. P. Baltakhinov, L. P. Burleva, H. Ahsbahs, H. Uchtmann and V. E. Dulepov, Effect of high pressure on the infra-red spectra of solid nitro- and nitrito-cobalt (III) ammine complexes, *Ber. Bunsengesell. Phys. Chem.*, 1992, **96**(7), 931–937.
150. N. Casati, P. Macchi and A. Sironi, Staggered to eclipsed conformational rearrangement of $[Co_2(CO)_6(PPh_3)_2]$ in the solid state: an X-ray diffraction study at high pressure and low temperature, *Angew. Chem., Int. Ed. Engl.*, 2005, **44**(47), 7736–7739.
151. N. Casati, P. Macchi and A. Sironi, Molecular crystals under high pressure: theoretical and experimental investigations of the solid–solid phase transitions in $[Co_2(CO)_6(XPh_3)_2]$(X = P, As), *Chem.–Eur. J.*, 2009, **15**(17), 4446–4457.
152. G. Molnár, V. Niel, J. A. Real, L. Dubrovinsky, A. Bousseksou and J. J. McGarvey, Raman spectroscopic study of pressure effects on the spin-crossover coordination polymers Fe(Pyrazine)[M(CN)$_4$] $2H_2O$ (M = Ni, Pd, Pt). First observation of a piezo-hysteresis loop at room temperature, *J. Phys. Chem. B*, 2003, **107**(14), 3149–3155.

153. A. Sieber, R. Bircher, O. Waldmann, G. Carver, G. Chaboussant, H. Mutka and H. U. Güdel, Effect of pressure on the magnetic anisotropy in the single-molecule magnet Mn_{12} acetate: an inelastic neutron scattering study, *Angew. Chem., Int. Ed. Engl.*, 2005, **44**(27), 4239–4242.
154. Y. Garcia, G. Bravic, C. Gieck, D. Chasseau, W. Tremel and P. Gütlich, Crystal structure, magnetic properties, and ^{57}Fe Mössbauer spectroscopy of the two-dimensional coordination polymers [M(1, 2-bis(1, 2, 4-triazol-4-yl) ethane)$_2$(NCS)$_2$](M^{II} = Fe, Co), *Inorg. Chem.*, 2005, **44**(26), 9723–9730.
155. P. Gütlich, V. Ksenofontov and A. B. Gaspar, Pressure effect studies on spin crossover systems, *Coord. Chem. Rev.*, 2005, **249**(17), 1811–1829.
156. A. Galet, A. B. Gaspar, M. C. Muñoz, G. V. Bukin, G. Levchenko and J. A. Real, Tunable bistability in a three-dimensional spin-crossover sensory-and memory-functional material, *Adv. Mater.*, 2005, **17**(24), 2949–2953.
157. L. Egan, K. Kamenev, D. Papanikolaou, Y. Takabayashi and S. Margadonna, Pressure-induced sequential magnetic pole inversion and antiferromagnetic–ferromagnetic crossover in a trimetallic Prussian Blue analogue, *J. Am. Chem. Soc.*, 2006, **128**(18), 6034–6035.
158. G. Molnár, T. Guillon, N. O. Moussa, L. Rechignat, T. Kitazawa, M. Nardone and A. Bousseksou, Two-step spin-crossover phenomenon under high pressure in the coordination polymer Fe(3-methylpyridine)$_2$[Ni(CN)$_4$], *Chem. Phys. Lett.*, 2006, **423**(1), 152–156.
159. S. Tancharakorn, F. P. Fabbiani, D. R. Allan, K. V. Kamenev and N. Robertson, Combined magnetic and single-crystal X-ray structural study of the linear chain antiferromagnet [$(CH_3)_4N$][$MnCl_3$] under varying pressure, *J. Am. Chem. Soc.*, 2006, **128**(28), 9205–9210.
160. G. Agustí, A. B. Gaspar, M. C. Muñoz and J. A. Real, Thermal-and pressure-induced cooperative spin transition in the 2D and 3D coordination polymers {Fe(5-Br-pmd)$_z$[M(CN)$_x$]$_y$}(M = Ag^I, Au^I, Ni^{II}, Pd^{II}, Pt^{II}), *Inorg. Chem.*, 2007, **46**(23), 9646–9654.
161. A. Prescimone, C. J. Milios, S. Moggach, J. E. Warren, A. R. Lennie, J. Sanchez-Benitez, K. Kamenev, R. Bircher, M. Murrie, S. Parsons and E. K. Brechin, [Mn_6] under pressure: a combined crystallographic and magnetic study, *Angew. Chem., Int. Ed. Engl.*, 2008, **47**(15), 2828–2831.
162. K. W. Chapman, G. J. Halder and P. J. Chupas, Guest-dependent high pressure phenomena in a nanoporous metal–organic framework material, *J. Am. Chem. Soc.*, 2008, **130**(32), 10524–10526.
163. A. Prescimone, C. J. Milios, J. Sanchez-Benitez, K. V. Kamenev, C. Loose, J. Kortus, S. Moggach, M. Murrie, J. F. Warren, A. R. Lennie, S. Parsons and E. K. Brechin, High pressure induced spin changes and magneto-structural correlations in hexametallic SMMs, *Dalton Trans.*, 2009, **25**, 4858–4867.
164. A. Prescimone, J. Sanchez-Benitez, K. V. Kamenev, S. A. Moggach, J. F. Warren, A. R. Lennie, M. Murrie, S. Parsons and E. K. Brechin, High pressure studies of hydroxo-bridged Cu(II) dimers, *Dalton Trans.*, 2010, **39**(1), 113–123.

165. A. Prescimone, J. Sanchez-Benitez, K. V. Kamenev, J. E. Warren, A. R. Lennie, M. Murrie, S. Parsons and E. K. Brechin, High-pressure study of oxo-bridged mixed-valent Mn^{III}/Mn^{IV} dimers high-pressure study of oxo-bridged mixed-valent Mn^{III}/Mn^{IV} dimers, *Z. Naturforsch. B*, 2010, **65**(3), 221–230.
166. P. Parois, S. A. Moggach, J. Sanchez-Benitez, K. V. Kamenev, A. R. Lennie, J. E. Warren, E. K. Brechin, S. Parsons and M. Murrie, Pressure-induced Jahn–Teller switching in a Mn_{12} nanomagnet, *Chem. Commun.*, 2010, **46**(11), 1881–1883.
167. P. Parois, S. A. Moggach, A. R. Lennie, J. E. Warren, E. K. Brechin, M. Murrie and S. Parsons, The effect of pressure on the crystal structure of $[Gd\,(PhCOO)_3(DMF)]_n$ to 3.7 GPa and the transition to a second phase at 5.0 GPa, *Dalton Trans.*, 2010, **39**(30), 7004–7011.
168. M. Bykov, E. Bykova, S. van Smaalen, L. Dubrovinsky, C. McCammon, V. Prakapenka and H. P. Liermann, High-pressure behavior of FeOCl, *Phys. Rev. B: Condens. Matter Mater. Phys.*, 2013, **88**(1), 014110.
169. P. J. Byrne, P. J. Richardson, J. Chang, A. F. Kusmartseva, D. R. Allan, A. C. Jones, K. V. Kamenev, P. A. Tasker and S. Parsons, Piezochromism in nickel salicylaldoximato complexes: tuning crystal-field splitting with high pressure, *Chem.–Eur. J*, 2012, **18**(25), 7738–7748.
170. K. Tarafder, S. Kanungo, P. M. Oppeneer and T. Saha-Dasgupta, Pressure and temperature control of spin-switchable metal–organic coordination polymers from *ab initio* calculations, *Phys. Rev. Lett.*, 2012, **109**(7), 077203.
171. H. J. Shepherd, P. Rosa, L. Vendier, N. Casati, J. F. Létard, A. Bousseksou, P. Guionneau and G. Molnár, High-pressure spin-crossover in a dinuclear Fe(II) complex, *Phys. Chem. Chem. Phys.*, 2012, **14**(15), 5265–5271.
172. I. F. Bruce-Smith, B. A. Zakharov, J. Stare, E. Boldyreva and C. R. Pulham, Structural properties of nickel dimethylglyoxime at high pressure: single-crystal X-ray diffraction and DFT studies, *J. Phys. Chem. C*, 2014, **118**, 24705–24713.
173. S. G. Patra, N. Mandal, A. Datta and D. Datta, On bonding in bis(dimethylglyoximato)nickel(II), *Comput. Theor. Chem.*, 2017, **1114**, 118–124.
174. K. Liu, Y. Orimoto and Y. Aoki, Theoretical investigation of the pressure-induced insulator-to-metal-to-insulator transitions in one-dimensional bis(dimethylglyoximato)platinum(II), Pt $(dmg)_2$, *Polyhedron*, 2015, **87**, 141–146.
175. E. C. Spencer, N. L. Ross and R. J. Angel, The high pressure behaviour of the 3D copper carbonate framework $\{[Cu(CO_3)_2](CH_6N_3)_2\}_n$, *J. Mater. Chem.*, 2012, **22**(5), 2074–2080.
176. E. C. Spencer, M. S. Kiran, W. Li, U. Ramamurty, N. L. Ross and A. K. Cheetham, Pressure-induced bond rearrangement and reversible phase transformation in a metal–organic framework, *Angew. Chem., Int. Ed. Engl.*, 2014, **53**(22), 5583–5586.
177. D. Pinkowicz, M. Rams, M. Mišek, K. V. Kamenev, H. Tomkowiak, A. Katrusiak and B. Sieklucka, Enforcing multifunctionality: a pressure-induced spin-crossover photomagnet, *J. Am. Chem. Soc.*, 2015, **137**(27), 8795–8802.

178. Y. V. Seryotkin, T. N. Drebushchak and E. V. Boldyreva, Effect of hydrostatic pressure on α-polymorph of chlorpropamide in saturated ethanol solution, *Acta Crystallogr., Sect. B: Struct. Sci.*, 2013, **69**, 77–85.
179. R. Wang, S. Li, K. Wang, K. Wang, D. Duan, L. Tang, T. Cui, B. Liu, Q. Cui, J. Liu, B. Zou and G. Zou, Pressure-induced phase transition in hydrogen-bonded supramolecular structure: guanidinium nitrate, *J. Phys. Chem. B*, 2010, **114**(20), 6765–6769.
180. S. Li, K. Wang, M. Zhou, Q. Li, B. Liu, G. Zou and B. Zou, Pressure-induced phase transitions in ammonium squarate: a supramolecular structure based on hydrogen-bonding and π-stacking interactions, *J. Phys. Chem. B*, 2011, **115**(29), 8981–8988.
181. S. Li, Q. Li, K. Wang, X. Tan, M. Zhou, B. Li, B. Liu, G. Zou and B. Zou, Pressure-induced phase transition in guanidinium perchlorate: a supramolecular structure directed by hydrogen bonding and electrostatic interactions, *J. Phys. Chem. B*, 2011, **115**(41), 11816–11822.
182. S. Li, Q. Li, J. Zhou, R. Wang, Z. Jiang, K. Wang, D. Xu, J. Liu, B. Liu, G. Zou and B. Zou, Effect of high pressure on the typical supramolecular structure of guanidinium methanesulfonate, *J. Phys. Chem. B*, 2012, **116**(10), 3092–3098.
183. S. Li, Q. Li, K. Wang, M. Zhou, X. Huang, J. Liu, J. Liu, K. Yang, B. Liu, T. Cui, G. Zou and B. Zou, Pressure-induced irreversible phase transition in the energetic material urea nitrate: combined Raman scattering and X-ray diffraction study, *J. Phys. Chem. C*, 2012, **117**(1), 152–159.
184. H. L. Wong, D. R. Allan, N. R. Champness, J. McMaster, M. Schröder and A. J. Blake, Bowing to the pressure of π···π interactions: bending of phenyl rings in a palladium (II) thioether crown complex, *Angew. Chem., Int. Ed. Engl.*, 2013, **52**(19), 5093–5095.
185. L. Kang, K. Wang, S. Li, J. Liu, K. Yang, B. Liu and B. Zou, Pressure-induced phase transition in hydrogen-bonded supramolecular structure: ammonium formate, *J. Phys. Chem. C*, 2014, **118**(16), 8521–8530.
186. S. Bergantin, M. Moret, G. Buth and F. P. Fabbiani, Pressure-induced conformational change in organic semiconductors: triggering a reversible phase transition in rubrene, *J. Phys. Chem. C*, 2014, **118**(25), 13476–13483.
187. B. A. Zakharov and E. V. Boldyreva, A high-pressure single-crystal to single-crystal phase transition in DL-alaninium semi-oxalate mono-hydrate with switching-over hydrogen bonds, *Acta Crystallogr., Sect. B: Struct. Sci., Cryst. Eng. Mater.*, 2013, **69**(3), 271–280.
188. E. Marelli, N. Casati, F. Gozzo, P. Macchi, P. Simoncic and A. Sironi, High pressure modification of organic NLO materials: large conformational re-arrangement of 4-aminobenzophenone, *CrystEngComm*, 2011, **13**(22), 6845–6849.
189. N. A. Tumanov, E. V. Boldyreva, B. A. Kolesov, A. V. Kurnosov and R. Q. Cabrera, Pressure-induced phase transitions in L-alanine, revisited, *Acta Crystallogr., Sect. B: Struct. Sci.*, 2010, **66**, 458–471.

190. E. V. Boldyreva, H. Ahsbahs, H. Uchtmann and N. E. Kashcheeva, Effects of pressure on the two polymorphs of $[Co(NH_3)_5NO_2]I_2$: the anisotropy of lattice distortion and a phase transition, *High Pressure Res.*, 2000, **17**, 79–99.
191. E. V. Boldyreva, V. P. Dmitriev and B. C. Hancock, Effect of pressure up to 5.5 GPa on dry powder samples of chlorpropamide form-A, *Int. J. Pharm.*, 2006, **327**, 51–57.
192. E. Eikeland, M. K. Thomsen, S. R. Madsen, J. Overgaard, M. A. Spackman and B. Brummerstedt Iversen, Structural collapse of the hydroquinone–formic acid clathrate: a pressure-medium-dependent phase transition, *Chem.–Eur. J.*, 2016, **22**(12), 4061–4069.
193. B. A. Zakharov, N. A. Tumanov and E. V. Boldyreva, β-Alanine under pressure: towards understanding the nature of phase transitions, *CrystEngComm*, 2015, **17**, 2074–2079.
194. B. A. Zakharov, S. V. Goryainov and E. V. Boldyreva, Unusual seeding effect in the liquid-assisted high-pressure polymorphism of chlorpropamide, *CrystEngComm*, 2016, **18**(29), 5423–5428.
195. B. A. Zakharov, Y. V. Seryotkin, N. A. Tumanov, D. Paliwoda, M. Hanfland, A. V. Kurnosov and E. V. Boldyreva, The role of fluids in high-pressure polymorphism of drugs: different behaviour of β-chlorpropamide in different inert gas and liquid media, *RSC Adv.*, 2016, **6**(95), 92629–92637.
196. R. Gajda, A. Katrusiak and J. Crassous, Pressure-controlled aggregation in carboxylic acids. A case study on the polymorphism of bromochlorofluoroacetic acid, *CrystEngComm*, 2009, **11**(12), 2668–2676.
197. K. Roszak, A. Katrusiak and A. Katrusiak, High-pressure preference for the low Z′ polymorph of a molecular crystal, *Cryst. Growth Des.*, 2016, **16**(7), 3947–3953.
198. D. Paliwoda, K. F. Dziubek and A. Katrusiak, Imidazole hidden polar phase, *Cryst. Growth Des.*, 2012, **12**(9), 4302–4305.
199. M. Anioła and A. Katrusiak, Pressure-preferred symmetric reactions of 4,4′-bipyridine hydrobromide, *CrystEngComm*, 2016, **18**(18), 3223–3228.
200. E. V. Boldyreva, S. N. Ivashevskaya, H. Sowa, H. Ahsbahs and H. P. Weber, Effect of high pressure on crystalline glycine: a new high-pressure polymorph, *Dokl. Phys. Chem.*, 2004, **396**, 358–361.
201. E. V. Boldyreva, S. N. Ivashevskaya, H. Sowa, H. Ahsbahs and H. P. Weber, Effect of hydrostatic pressure on the γ-polymorph of glycine. 1. A polymorphic transition into a new δ-form, *Z. Kristallogr.*, 2005, **220**, 50–57.
202. S. V. Goryainov, E. N. Kolesnik and E. V. Boldyreva, A pressure-induced phase transition in β-glycine at 0.76 GPa, *Phys. B*, 2005, **357**, 340–347.
203. N. A. Tumanov, E. V. Boldyreva and H. Ahsbahs, Structure solution and refinement from powder or single-crystal diffraction data? Pros and cons: An example of the high-pressure β′-polymorph of glycine, *Powder Diffr.*, 2008, **23**, 307–316.

204. S. V. Goryainov, E. V. Boldyreva and E. N. Kolesnik, Raman evidence of a new (ζ) polymorph of glycine? *Chem. Phys. Lett.*, 2006, **419**, 496–500.
205. A. K. Mishra, C. Murli and S. M. Sharma, High pressure Raman spectroscopic study of deuterated γ-glycine, *J. Phys. Chem. B*, 2008, **112**(49), 15867–15874.
206. H. N. Bordallo, E. V. Boldyreva, A. Buchsteiner, M. M. Koza and S. Landsgesell, Dynamic properties of the polymorphs of glycine: an incoherent inelastic neutron scattering study, *J. Phys. Chem. B*, 2008, **112**, 8748–8759.
207. A. Dawson, D. R. Allan, S. A. Belmonte, S. J. Clark, W. I. F. David, P. A. McGregor, S. Parsons, C. R. Pulham and L. Sawyer, Effect of high pressure on the crystal structures of polymorphs of glycine, *Cryst. Growth Des.*, 2005, **5**, 1415–1427.
208. E. N. Kolesnik, S. V. Goryainov and E. V. Boldyreva, Different behavior of L-and DL-serine crystals at high pressures: phase transitions in L-serine and stability of the DL-serine structure, *Dokl. Phys. Chem.*, 2005, **404**, 169–172.
209. S. A. Moggach, D. R. Allan, C. A. Morrison, S. Parsons and L. Sawyer, Effect of pressure on the crystal structure of L-serine-I and the crystal structure of L-serine-II at 5.4 GPa, *Acta Crystallogr., Sect. B: Struct. Sci.*, 2005, **61**, 58–68.
210. T. N. Drebushchak, H. Sowa, Y. V. Seryotkin, E. V. Boldyreva and H. Ahsbahs, L-Serine III at 8.0 GPa, *Acta Crystallogr., Sect. E: Struct. Rep. Online*, 2006, **62**, 4052–4054.
211. E. V. Boldyreva, H. Sowa, Y. V. Seryotkin, T. N. Drebushchak, H. Ahsbhas, V. V. Chernyshev and V. P. Dmitriev, Pressure-induced phase transitions in crystalline L-serine studied by single-crystal and high-resolution powder X-ray diffraction, *Chem. Phys. Lett.*, 2006, **429**, 474–478.
212. S. A. Moggach, W. G. Marshall and S. Parsons, High-pressure neutron diffraction study of L-serine-I and L-serine-II, and the structure of L-serine-III at 8.1 GPa, *Acta Crystallogr., Sect. B: Struct. Sci.*, 2006, **62**(5), 815–825.
213. D. A. Rychkov, J. Stare and E. V. Boldyreva, Pressure-driven phase transition mechanisms revealed by quantum chemistry: L-serine polymorphs, *Phys. Chem. Chem. Phys.*, 2017, **19**, 6671–6676.
214. A. Olejniczak, K. Ostrowska and A. Katrusiak, H-bond breaking in high-pressure urea, *J. Phys. Chem. C*, 2009, **113**(35), 15761–15767.
215. C. Murli, S. M. Sharma, S. Karmakar and S. K. Sikka, α-Glycine under high pressures: a Raman scattering study, *Phys. B*, 2003, **339**(1), 23–30.
216. B. B. Sharma, C. Murli, R. Chitra and S. M. Sharma, Hydrogen-bonding interactions in fully deuterated α-glycine at high pressures, *J. Raman Spectrosc.*, 2012, **43**, 138–145.
217. J. Ridout, L. S. Price, J. A. K. Howard and M. R. Probert, Polymorphism arising from differing rates of compression of liquids, *Cryst. Growth Des.*, 2014, **14**, 3384–3391.

218. R. Gajda and A. Katrusiak, Pressure-promoted CH···O hydrogen bonds in formamide aggregates, *Cryst. Growth Des.*, 2011, **11**(11), 4768–4774.
219. C. L. Bull, G. Flowitt-Hill, S. de Gironcoli, E. Küçükbenli, S. Parsons, C. H. Pham, H. Y. Playford and M. G. Tucker, ζ-Glycine: insight into the mechanism of a polymorphic phase transition, *IUCrJ*, 2017, **4**(5), 569–574.
220. P. Espeau, R. Céolin, J. L. Tamarit, M. A. Perrin, J. P. Gauchi and F. Leveiller, Polymorphism of paracetamol: Relative stabilities of the monoclinic and orthorhombic phases inferred from topological pressure-temperature and temperature-volume phase diagrams, *J. Pharm. Sci.*, 2005, **94**(3), 524–539.
221. J. Ledru, C. T. Imrie, C. R. Pulham, R. Céolin and J. M. Hutchinson, High pressure differential scanning calorimetry investigations on the pressure dependence of the melting of paracetamol polymorphs I and II, *J. Pharm. Sci.*, 2007, **96**(10), 2784–2794.
222. L. McGregor, D. A. Rychkov, P. L. Coster, S. Day, V. A. Drebushchak, A. F. Achkasov, G. S. Nichol, C. R. Pulham and E. V. Boldyreva, A new polymorph of metacetamol, *CrystEngComm*, 2015, **17**(32), 6183–6192.
223. M. Barrio, J. Huguet, I. B. Rietveld, B. Robert, R. Céolin and J. L. Tamarit, The pressure-temperature phase diagram of metacetamol and its comparison to the phase diagram of paracetamol, *J. Pharm. Sci.*, 2017, **106**(6), 1538–1544.
224. I. Gana, R. Céolin and I. B. Rietveld, Bicalutamide polymorphs I and II, *J. Therm. Anal. Calorim.*, 2013, **112**(1), 223–228.
225. R. Céolin and I. B. Rietveld, The topological pressure-temperature phase diagram of fluoxetine nitrate: monotropy unexpectedly turning into enantiotropy, *Eur. Phys. J.: Spec. Top.*, 2017, **226**(5), 881–888.
226. A. Yu. Fedorov, D. A. Rychkov, E. A. Losev, B. A. Zakharov, J. Stare and E. V. Boldyreva, Effect of pressure on two polymorphs of tolazamide. Why no interconversion? *CrystEngComm*, 2017, **19**, 2243–2252.
227. A. J. Cruz-Cabeza and J. Bernstein, Conformational polymorphism, *Chem. Rev.*, 2013, **114**, 2170–2190.
228. M. Podsiadło, K. Jakóbek and A. Katrusiak, Density, freezing and molecular aggregation in pyridazine, pyridine and benzene, *CrystEngComm*, 2010, **12**, 2561–2567.
229. A. Katrusiak, M. Szafrański and M. Podsiadło, Pressure-induced collapse of guanidinium nitrate N-H···O bonded honeycomb layers into a 3D pattern with varied H-acceptor capacity, *Chem. Commun.*, 2011, **47**, 2107–2109.
230. B. A. Zakharov and A. F. Achkasov, A compact device for loading diamond anvil cells with low-boiling pressure-transmitting media, *J. Appl. Crystallogr.*, 2013, **46**, 267–269.
231. F. P. A. Fabbiani, G. Buth, D. C. Levendis and A. J. Cruz-Cabeza, Pharmaceutical hydrates under ambient conditions from high-pressure seeds: a case study of GABA monohydrate, *Chem. Commun.*, 2014, **50**, 1817–1819.

232. F. P. A. Fabbiani, D. R. Allan, A. Dawson, W. I. F. David, P. A. McGregor, I. D. H. Oswald, S. Parsons and C. R. Pulham, Pressure-induced formation of a solvate of paracetamol, *Chem. Commun.*, 2003, **9**, 3004–3005.
233. A. Olejniczak and A. Katrusiak, Pressure-induced hydration of 1,4-diazabicyclo[2.2.2]octane hydroiodide (dabcoHI), *Cryst. Growth Des.*, 2011, **11**, 2250–2256.
234. M. Anioła, A. Olejniczak and A. Katrusiak, Pressure-induced solvate crystallization of 1,4-diazabicyclo[2.2.2]octane perchlorate with methanol, *Cryst. Growth Des.*, 2014, **14**, 2187–2191.
235. H. Tomkowiak, A. Olejniczak and A. Katrusiak, Pressure-dependent formation and decomposition of thiourea hydrates, *Cryst. Growth Des.*, 2013, **13**, 121–125.
236. W. Zielinski and A. Katrusiak, Pressure-induced preference for solvation of 5,6-dimethylbenzimidazole, *CrystEngComm*, 2016, **18**(18), 3211–3215.
237. A. Olejniczak, M. Podsiadło and A. Katrusiak, High pressure used for producing a new solvate of 1,4-diazabicyclo[2.2.2]octane hydroiodide, *New J. Chem.*, 2016, **40**, 2014–2020.
238. F. P. A. Fabbiani, S. Bergantin, A. Gavezzotti, S. Rizzato and M. Moret, X-ray diffraction and computational studies of the pressure-dependent tetrachloroethane solvation of diphenylanthracene, *CrystEngComm*, 2016, **18**, 2173–2181.
239. F. P. A. Fabbiani, D. C. Levendis, G. Buth, W. F. Kuhs, N. Shankland and H. Sowa, Searching for novel crystal forms by *in situ* high-pressure crystallisation: the example of gabapentin heptahydrate, *CrystEngComm*, 2010, **12**, 2354–2360.
240. F. P. A. Fabbiani, B. Dittrich, A. J. Florence, T. Gelbrich, M. B. Hursthouse, W. F. Kuhs, N. Shankland and H. Sowa, Crystal structures with a challenge: High-pressure crystallisation of ciprofloxacin sodium salts and their recovery to ambient pressure, *CrystEngComm*, 2009, **11**, 1396–1406.
241. Y. Sakata, Studies on the polymorphism of L-glutamic acid. Part I. Effect of co-existing substances on polymorphic crystallization, *Agric. Biol. Chem.*, 1964, **25**, 829–834.
242. Y. Sakata, Studies on the polymorphism of L-glutamic acid. Part II. Measurement of solubilities, *Agric. Biol. Chem.*, 1964, **25**, 835–837.
243. J. Schöll, D. Bonalumi, L. Vicum and M. Mazzotti, *In situ* monitoring and modeling of the solvent-mediated polymorphic transformation of L-glutamic acid, *Cryst. Growth Des.*, 2006, **6**, 881–891.
244. A. Gavezzotti, Molecular aggregation, *International Union of Crystallography Monographs on Crystallography*, Oxford University Press, Oxford, 2006.
245. A. Gavezzotti, Crystal formation and stability: physical principles and molecular simulation, *Cryst. Res. Technol.*, 2013, **48**, 793–810.
246. N. B. Leonidov, Historic aspects of the concept of polymorphism of chemical compounds, *Russ. Chem. J.*, 1997, **41**, 10–21.

247. N. B. Leonidov, P. M. Zorky and A. E. Masunov, Structure and non-eqiuvalence of the polymorphs of methyluracil, *Acta Crystallogr., Sect. A: Found. Crystallogr.*, 1993, **49**, c185.
248. D. Gidalevitz, R. Feidenhans'l, S. Matlis, D.-M. Smilgies, M. J. Christensen and L. Leiserowitz, Monitoring *in situ* growth and dissolution of molecular crystals: towards determination of the growth units, *Angew. Chem., Int. Ed. Engl.*, 1997, **36**, 955–959.
249. S. Block, C. E. Weir and G. J. Piermarini, Polymorphism in benzene, naphthalene, and anthracene at high pressure, *Science*, 1970, **169**, 586–587.
250. A. Budzianowski and A. Katrusiak, Pressure-frozen benzene I revisited, *Acta Crystallogr., Sect. B: Struct. Sci.*, 2006, **62**, 94–101.
251. R. Fourme, D. André and M. Renaud, A redetermination and group-refinement of the molecular packing of benzene II at 25 kilobars, *Acta Crystallogr., Sect. B: Struct. Crystallogr. Cryst. Chem.*, 1971, **27**, 1275–1276.
252. G. J. Piermarini, A. D. Mighell, C. E. Weir and S. Block, Crystal structure of benzene II at 25 kilobars, *Science*, 1969, **165**, 1250–1255.
253. P. Raiteri, R. Martoňák and M. Parrinello, Exploring polymorphism: the case of benzene, *Angew. Chem., Int. Ed.*, 2005, **44**, 3769–3773.
254. M. M. Thiéry and J. M. Léger, High pressure solid phases of benzene. I. Raman and X-ray studies of C_6H_6 at 294 K up to 25 GPa, *J. Chem. Phys.*, 1988, **89**, 4255–4271.
255. D. R. Allan and S. J. Clark, Comparison of the high-pressure and low-temperature structures of ethanol and acetic acid, *Phys. Rev. B: Condens. Matter Mater. Phys.*, 1999, **60**, 6328–6334.
256. *Physics and Chemistry of Ice*, ed. W. Kuhs, RSC Publishing, Cambridge, 2007.
257. E. R. Pounder, *The Physics of Ice*, Pergamon Press, Oxford, 1965.
258. D. R. Allan, S. J. Clark, R. M. Ibberson, S. Parsons, C. R. Pulham and L. Sawyer, The influence of pressure and temperature on the crystal structure of acetone, *Chem. Commun.*, 1999, 751–752.
259. J. Ridout and M. R. Probert, Low-temperature and high-pressure polymorphs of isopropyl alcohol, *CrystEngComm*, 2014, **16**(32), 7397–7400.
260. E. Patyk, M. Podsiadło and A. Katrusiak, CH···N bonds and dynamics in isostructural pyrimidine polymorphs, *Cryst. Growth Des.*, 2015, **15**(8), 4039–4044.
261. R. Lee, A. J. Firbank, M. R. Probert and J. W. Steed, Expanding the pyridine–formic acid cocrystal landscape under extreme conditions, *Cryst. Growth Des.*, 2016, **16**(7), 4005–4011.
262. R. Lee, D. S. Yufit, M. R. Probert and J. W. Steed, High pressure/low temperature polymorphism in 2,6-dimethylpyridine–formic acid cocrystals, *Cryst. Growth Des.*, 2017, **17**(4), 1647–1653.
263. M. Bujak and A. Katrusiak, Molecular association in low-temperature and high-pressure polymorphs of 1,1,1,2-tetrachloroethane, *CrystEngComm*, 2010, **12**, 1263–1268.

264. K. Dziubek, M. Podsiadło and A. Katrusiak, Nearly isostructural polymorphs of ethynylbenzene: resolution of ⋮CH···π(arene) and cooperative ⋮CH···π(C⋮C) interactions by pressure freezing, *J. Am. Chem. Soc.*, 2007, **129**, 12620–12621.
265. A. Olejniczak and A. Katrusiak, Pressure induced transformations of 1, 4-diazabicyclo [2.2.2] octane (dabco) hydroiodide: diprotonation of dabco, its *N*-methylation and co-crystallization with methanol, *CrystEngComm*, 2010, **12**, 2528–2532.
266. A. Olejniczak and A. Katrusiak, Pressure-induced hydration of 1, 4-diazabicyclo [2.2.2] octane hydroiodide (dabcoHI), *Cryst. Growth Des.*, 2011, **11**, 2250–2256.
267. A. Olejniczak, A. Katrusiak, M. Szafrański, A. Olejniczak, A. Katrusiak and M. Szafrański, Ten polymorphs of NH^+···N hydrogen-bonded 1,4-diazabicyclo [2.2.2] octane complexes: Supramolecular origin of giant anisotropic dielectric response in polymorph V, *Cryst. Growth Des.*, 2010, **10**, 3537–3546.
268. F. P. A. Fabbiani, D. R. Allan, W. I. F. David, S. A. Moggach, S. Parsons and C. R. Pulham, High-pressure recrystallisation – a route to new polymorphs and solvates, *CrystEngComm*, 2004, **6**, 504–511.
269. M. A. Neumann, J. Van De Streek, F. P. A. Fabbiani, P. Hidber and O. Grassmann, Combined crystal structure prediction and high-pressure crystallization in rational pharmaceutical polymorph screening, *Nat. Commun.*, 2015, **6**, 7793.
270. N. Abbas, I. D. Oswald and C. R. Pulham, Accessing mefenamic acid form II through high-pressure recrystallization, *Pharmaceutics*, 2017, **9**(2), 16.
271. F. P. A. Fabbiani, D. R. Allan, W. I. F. David, A. J. Davidson, A. R. Lennie, S. Parsons, C. R. Pulham and J. E. Warren, High-pressure studies of pharmaceuticals: an exploration of the behavior of piracetam, *Cryst. Growth Des.*, 2007, **7**, 1115–1124.
272. F. P. A. Fabbiani, D. R. Allan, S. Parsons and C. R. Pulham, An exploration of the polymorphism of piracetam using high pressure, *CrystEngComm*, 2005, **7**, 179–186.
273. D. A. Rychkov, S. Hunter, V. Y. Kovalskii, A. A. Lomzov, C. R. Pulham and E. V. Boldyreva, Towards an understanding of crystallization from solution. DFT studies of multi-component serotonin crystals, *Comput. Theor. Chem.*, 2016, **1088**, 52–61.
274. T. Q. Luong, S. Kapoor and R. Winter, Pressure – a gateway to fundamental insights into protein solvation, dynamics, and function, *ChemPhysChem*, 2015, **16**(17), 3555–4357.
275. S. G. Arkhipov, D. A. Rychkov, A. M. Pugachev and E. V. Boldyreva, New hydrophobic L-amino acid salts: maleates of L-leucine, L-isoleucine and L-norvaline, *Acta Crystallogr., Sect. C: Struct. Chem.*, 2015, **71**(7), 584–592.
276. S. G. Arkhipov, B. A. Zakharov and E. V. Boldyreva, Semi-maleate salts of L- and DL-serinium: the first example of chiral and racemic serinium

salts with the same composition and stoichiometry, *Acta Crystallogr., Sect. C: Cryst. Struct. Commun.*, 2013, **69**(5), 517–521.

277. D. A. Rychkov, S. G. Arkhipov and E. V. Boldyreva, Structure-forming units of amino acid maleates. Case study of L-valinium hydrogen maleate, *Acta Crystallogr., Sect. B: Struct. Sci., Cryst. Eng. Mater.*, 2016, **72**(1), 160–163.
278. S. G. Arkhipov, E. A. Losev and E. V. Boldyreva, New 1 : 1 and 2 : 1 salts in the "DL-norvaline–maleic acid" system as an example of assembling various crystal structures from similar supramolecular building blocks, *Acta Crystallogr., Sect. C: Struct. Chem.*, 2017, **73**(1), 13–19.
279. B. A. Zakharov and E. V. Boldyreva, DL-alaninium semi-oxalate monohydrate, *Acta Crystallogr., Sect. C: Cryst. Struct. Commun.*, 2011, **67**, o47–o51.
280. B. A. Zakharov, B. A. Kolesov and E. V. Boldyreva, Monitoring selected hydrogen bonds in crystal hydrates of amino acid salts: combining variable-temperature single-crystal X-ray diffraction and polarized Raman spectroscopy, *Phys. Chem. Chem. Phys.*, 2011, **13**(28), 13106–13116.
281. K. A. Kerr and J. P. Ashmore, Structure and conformation of orthorhombic L-cysteine, *Acta Crystallogr., Sect. B: Struct. Crystallogr. Cryst. Chem.*, 1973, **29**, 2124–2127.
282. K. A. Kerr, J. P. Ashmore and F. Koetzle, A neutron diffraction study of L-cysteine, *Acta Crystallogr., Sect. B: Struct. Crystallogr. Cryst. Chem.*, 1975, **31**, 2022–2026.
283. C. H. Görbitz and B. Dalhus, L-cysteine, monoclinic form, redetermination at 120 K, *Acta Crystallogr., Sect. C: Cryst. Struct. Commun.*, 1996, **52**, 1756–1759.
284. P. Luger and M. Weber, DL-Cysteine at 298 K, *Acta Crystallogr., Sect. C: Cryst. Struct. Commun.*, 1999, **55**, 1882–1885.
285. S. A. Moggach, D. R. Allan, S. J. Clark, M. J. Gutmann, S. Parsons, C. R. Pulham and L. Sawyer, High-pressure polymorphism in L-cysteine: the crystal structures of L-cysteine-III and L-cysteine-IV, *Acta Crystallogr., Sect. B: Struct. Sci.*, 2006, **62**, 296–309.
286. V. S. Minkov, S. V. Goryainov, E. V. Boldyreva and C.-H. Görbitz, Raman study of pressure-induced phase transitions in the crystals of orthorhombic and monoclinic polymorphs of L-cysteine: dynamics of the side-chain, *J. Raman Spectrosc.*, 2010, **41**(12), 1458–1468.
287. V. S. Minkov, N. A. Tumanov, R. Q. Cabrera and E. V. Boldyreva, Low temperature/high pressure polymorphism in DL-cysteine, *CrystEngComm*, 2010, **12**, 2551–2560.
288. V. S. Minkov, E. V. Boldyreva, T. N. Drebushchak and C.-H. Görbitz, Stabilizing structures of cysteine-containing crystals with respect to variations of temperature and pressure by immobilizing amino acid side chains, *CrystEngComm*, 2012, **14**, 5943–5954.
289. A. K. Mishra, C. Murli, N. Garg, R. Chitra and S. M. Sharma, Pressure-induced structural transformations in bis(glycinium) oxalate, *J. Phys. Chem. B*, 2010, **114**(51), 17084–17091.

290. H. Bhatt, C. Murli, N. Garg, M. N. Deo, R. Chitra, R. R. Choudhury and S. M. Sharma, High pressure phase transformations in bis(glycinium) oxalate – an infrared absorption study, *Chem. Phys. Lett.*, 2012, **532**, 57–62.
291. H. Bhatt, M. N. Deo, C. Murli, S. R. Vishwakarma, R. Chitra and S. M. Sharma, A temperature-dependent infrared absorption study of strong hydrogen bonds in bis(glycinium) oxalate, in *The Solid State Physics Symposium*, 2015, vol. 1731, p. 040018.
292. H. Bhatt, C. Murli, A. K. Mishra, A. K. Verma, N. Garg, M. N. Deo, R. Chitra and S. M. Sharm, Hydrogen bond symmetrization in glycinium oxalate under pressure, *J. Phys. Chem. B*, 2016, **120**(4), 851–859.
293. B. A. Zakharov, E. A. Losev and E. V. Boldyreva, A new high-pressure polymorph of a glycine-glutaric acid co-crystal: the same as the low-temperature form, *CrystEngComm*, 2013, **15**, 1693–1697.
294. B. A. Zakharov, E. A. Losev, B. A. Kolesov, V. A. Drebushchak and E. V. Boldyreva, Low-temperature phase transition in glycine–glutaric acid co-crystals studied by single-crystal X-ray diffraction, Raman spectroscopy and differential scanning calorimetry, *Acta Crystallogr., Sect. B: Struct. Sci.*, 2012, **68**, 287–296.
295. E. A. Losev, B. A. Zakharov and E. V. Boldyreva, Polymorphic transformations in glycine co-crystals at low temperature and high pressure: two new examples as a follow-up to a glycine-glutaric acid study, *CrystEngComm*, 2016, **18**(31), 5869–5875.
296. A. K. Mishra, C. Murli, A. K. Verma, Y. Song, M. S. Kumar and S. M. Sharma, Conformation and hydrogen-bond-assisted polymerization in glycine lithium sulfate at high pressures, *J. Phys. Chem. A*, 2013, **117**(28), 5734–5741.
297. N. A. Tumanov and E. V. Boldyreva, X-ray diffraction and Raman study of DL-alanine at high pressure: revision of phase transitions, *Acta Crystallogr., Sect. B: Struct. Sci.*, 2012, **68**(4), 412–423.
298. N. P. Funnell, A. Dawson, D. Francis, A. R. Lennie, W. G. Marshall, S. A. Moggach, J. E. Warren and S. Parsons, The effect of pressure on the crystal structure of L-alanine, *CrystEngComm*, 2010, **12**, 2573–2583.
299. R. D. Johnstone, D. Francis, A. R. Lennie, W. G. Marshall, S. A. Moggach, S. Parsons, E. Pidcock and J. E. Warren, High-pressure polymorphism in L-serine monohydrate: identification of driving forces in high pressure phase transitions and possible implications for pressure-induced protein denaturation, *CrystEngComm*, 2008, **10**(12), 1758–1769.
300. B. A. Zakharov and E. V. Boldyreva, Reversible pressure-induced disordering in bis(DL-serinium) oxalate dihydrate, *J. Mol. Struct.*, 2014, **1078**, 151–157.
301. S. G. Arkhipov, B. A. Zakharov and E. V. Boldyreva, High pressure studies of L-serine-L-ascorbic acid co-crystal, *Acta Crystallogr., Sect. A: Found. Crystallogr.*, 2014, **70**, C988.
302. M. Sikora and A. Katrusiak, Pressure-controlled neutral–ionic transition and disordering of NH···N hydrogen bonds in pyrazole, *J. Phys. Chem. C*, 2013, **117**(20), 10661–10668.

303. N. Casati, P. Macchi and A. Sironi, Hydrogen migration in oxalic acid di-hydrate at high pressure? *Chem. Commun.*, 2009, **19**, 2679–2681.
304. P. Macchi, N. Casati, W. G. Marshall and A. Sironi, The α and β forms of oxalic acid di-hydrate at high pressure: a theoretical simulation and a neutron diffraction study, *CrystEngComm*, 2010, **12**(9), 2596–2603.
305. R. Kumai, S. Horiuchi, J. Fujioka and Y. Tokura, Ferroelectricity and pressure-induced phenomena driven by neutral ionic valence instability of acid–base supramolecules, *J. Am. Chem. Soc.*, 2012, **134**, 1036–1046.
306. S. Horiuchi, R. Kumai and Y. Tokura, High-temperature and pressure-induced ferroelectricity in hydrogen-bonded supramolecular crystals of anilic acids and 2,3-di(2-pyridinyl)pyrazine, *J. Am. Chem. Soc.*, 2013, **135**, 4492–4500.
307. S. Scheins, J. Overgaard, G. A. Timco, A. Stash, Y. S. Chen, F. K. Larsen, M. Christensen, M. R. V. Jørgensen, S. R. Madsen, M. S. Schmøkel and B. Brummerstedt Iversen, Pressure *versus* temperature effects on intramolecular electron transfer in mixed-valence complexes, *Chem.–Eur. J.*, 2013, **19**(1), 195–205.
308. J. Wu, J. Tang, H. Wang, Q. Qi, X. Fang, Y. Liu, S. Xu, Z. Xiao-An, H. Zhang and W. Xu, Reversible piezofluorochromic property and intrinsic structure changes of tetra(4-methoxyphenyl) ethylene under high pressure, *J. Phys. Chem. A*, 2015, **119**(35), 9218–9224.
309. R. D. Johnstone, D. Allan, A. Lennie, E. Pidcock, R. Valiente, F. Rodríguez, J. Gonzalez, J. Warren and S. Parsons, The effect of pressure on the crystal structure of bianthrone, *Acta Crystallogr., Sect. B: Struct. Sci.*, 2011, **67**(3), 226–237.
310. D. R. Allan, D. Bailey, N. Bird, A. J. Blake, N. R. Champness, D. Huang, C. P. Keane, J. McMaster, T. J. Prior, J. P. Tidey and M. Schroeder, High-pressure studies of palladium and platinum thioether macrocyclic dihalide complexes, *Acta Crystallogr., Sect. B: Struct. Sci., Cryst. Eng. Mater.*, 2014, **70**(3), 469–486.
311. D. Kurzydłowski, P. Zaleski-Ejgierd, W. Grochala and R. Hoffmann, Freezing in resonance structures for better packing: XeF_2 becomes $(XeF^+)(F^-)$ at large compression, *Inorg. Chem.*, 2011, **50**(8), 3832–3840.
312. N. Casati, A. Kleppe, A. P. Jephcoat and P. Macchi, Putting pressure on aromaticity along with *in situ* experimental electron density of a molecular crystal, *Nat. Commun.*, 2016, **7**, 10901.
313. I. Hernández and W. P. Gillin, Influence of high hydrostatic pressure on Alq_3, Gaq_3, and Inq_3 (q = 8-hydroxyquinoline), *J. Phys. Chem. B*, 2009, **113**(43), 14079–14086.
314. E. C. Spencer, R. J. Angel, N. L. Ross, B. E. Hanson and J. A. K. Howard, Pressure-induced cooperative bond rearrangement in a zinc imidazolate framework: a high-pressure single-crystal X-ray diffraction study, *J. Am. Chem. Soc.*, 2009, **131**(11), 4022–4026.
315. K. W. Galloway, S. A. Moggach, P. Parois, A. R. Lennie, J. E. Warren, E. K. Brechin, R. D. Peacock, R. Valiente, J. González, F. Rodríguez, S. Parsons and M. Murrie, Pressure-induced switching in a copper (II) citrate dimer, *CrystEngComm*, 2010, **12**(9), 2516–2519.

316. R. Engelke and N. C. Blais, Chemical dimerization of crystalline anthracene produced by transient high pressure, *J. Chem. Phys.*, 1994, **101**(12), 10961–10972.
317. F. P. Fabbiani, D. R. Allan, S. Parsons and C. R. Pulham, Exploration of the high-pressure behaviour of polycyclic aromatic hydrocarbons: naphthalene, phenanthrene and pyrene, *Acta Crystallogr., Sect. B: Struct. Sci.*, 2006, **62**(5), 826–842.
318. V. M. Tapilin, N. N. Bulgakov, A. P. Chupakhin and A. A. Politov, On the mechanism of mechanochemical dimerization of anthracene. Quantum-chemical calculation of the electronic structure of anthracene and its dimer, *J. Struct. Chem.*, 2008, **49**(4), 581–586.
319. V. M. Tapilin, N. N. Bulgakov, A. P. Chupakhin, A. A. Politov and A. G. Druganov, On mechanochemical dimerization of anthracene. Different possible reaction pathways, *J. Struct. Chem.*, 2010, **51**(4), 635–641.
320. A. A. Politov, A. P. Chupakhin, V. M. Tapilin, N. N. Bulgakov and A. G. Druganov, To mechanochemical dimerization of anthracene. Crystalline phenanthrene under high pressure and shear conditions, *J. Struct. Chem.*, 2010, **51**(6), 1064–1069.
321. J. B. Conant and W. R. Peterson, Polymerization reactions under high pressure. II. The mechanism of the reaction, *J. Am. Chem. Soc.*, 1932, **54**(2), 628–635.
322. C. Walling and J. Pellon, Organic reactions under high pressure. I. The polymerization of styrene, *J. Am. Chem. Soc.*, 1957, **79**(17), 4776–4782.
323. K. Aoki, S. Usuba, M. Yoshida, Y. Kakudate, K. Tanaka and S. Fujiwara, Raman study of the solid-state polymerization of acetylene at high pressure, *J. Chem. Phys.*, 1988, **89**(1), 529–534.
324. M. Bernasconi, G. L. Chiarotti, P. Focher, M. Parrinello and E. Tosatti, Solid-state polymerization of acetylene under pressure: *ab initio* simulation, *Phys. Rev. Lett.*, 1997, **78**(10), 2008.
325. A. F. Goncharov, M. R. Manaa, J. M. Zaug, R. H. Gee, L. E. Fried and W. B. Montgomery, Polymerization of formic acid under high pressure, *Phys. Rev. Lett.*, 2005, **94**(6), 065505.
326. C. Wilhelm, S. A. Boyd, S. Chawda, F. W. Fowler, N. S. Goroff, G. P. Halada, C. P. Grey, J. W. Lauher, L. Luo, D. Martin, J. B. Parise, C. Tarabrella and J. A. Webb, Pressure-induced polymerization of diiodobutadiyne in assembled cocrystals, *J. Am. Chem. Soc.*, 2008, **130**(13), 4415–4420.
327. C. Murli and Y. Song, Pressure-induced polymerization of acrylic acid: a Raman spectroscopic study, *J. Phys. Chem. B*, 2010, **114**(30), 9744–9750.
328. B. B. Sharma, C. Murli and S. M. Sharma, Hydrogen bonds and polymerization in acrylamide under pressure, *J. Raman Spectrosc.*, 2013, **44**(5), 785–790.
329. S. A. Moggach, K. W. Galloway, A. R. Lennie, P. Parois, N. Rowantree, E. K. Brechin, J. E. Warren, M. Murrie and S. Parsons, Polymerisation of a Cu (II) dimer into 1D chains using high pressure, *CrystEngComm*, 2009, **11**(12), 2601–2604.

330. P. Ray, J. L. Gray, J. V. Badding and A. D. Lueking, High-pressure reactivity of triptycene probed by Raman spectroscopy, *J. Phys. Chem. B*, 2016, **120**(42), 11035–11042.
331. W. J. Evans, M. J. Lipp, C. S. Yoo, H. Cynn, J. L. Herberg, R. S. Maxwell and M. F. Nicol, Pressure-induced polymerization of carbon monoxide: disproportionation and synthesis of an energetic lactonic polymer, *Chem. Mater.*, 2006, **18**(10), 2520–2531.
332. A. Y. Likhacheva, S. V. Rashchenko and K. D. Litasov, High-pressure structural properties of naphthalene up to 6 GPa, *J. Appl. Crystallogr.*, 2014, **47**(3), 984–991.
333. C. Fujimoto, A. Shinozaki, K. Mimura, T. Nishida, H. Gotou, K. Komatsu and H. Kagi, Pressure-induced oligomerization of alanine at 25 °C, *Chem. Commun.*, 2015, **51**(69), 13358–13361.
334. B. A. Zakharov, P. A. Gribov, A. A. Matvienko and E. V. Boldyreva, Isostructural crystal hydrates of rare-earth metal oxalates at high pressure: from strain anisotropy to dehydration, *Z. Kristallogr. - Cryst. Mater.*, 2017, **232**(11), 751–757.
335. S. A. Moggach, T. D. Bennett and A. K. Cheetham, The effect of pressure on ZIF-8: increasing pore size with pressure and the formation of a high-pressure phase at 1.47 GPa, *Angew. Chem., Int. Ed. Engl.*, 2009, **121**(38), 7221–7223.
336. T. D. Bennett, J. C. Tan, S. A. Moggach, R. Galvelis, C. Mellot-Draznieks, B. A. Reisner, A. Thirumurugan, D. R. Allan and A. K. Cheetham, Mechanical properties of dense zeolitic imidazolate frameworks (ZIFs): a high-pressure X-ray diffraction, nanoindentation and computational study of the zinc framework $Zn(Im)_2$, and its lithium-boron analogue, $LiB(Im)_4$, *Chem.-Eur. J.*, 2010, **16**(35), 10684–10690.
337. C. H. Woodall, J. Christensen, J. M. Skelton, L. E. Hatcher, A. Parlett, P. R. Raithby, A. Walsh, S. C. Parker, C. M. Beavers, S. J. Teat, M. Intissar and C. Reber, Observation of a re-entrant phase transition in the molecular complex tris (μ2-3,5-diisopropyl-1,2,4-triazolato-$\kappa_2N_1:N_2$)trigold (I) under high pressure, *IUCrJ*, 2016, **3**(5), 367–376.
338. A. J. Graham, D. R. Allan, A. Muszkiewicz, C. A. Morrison and S. A. Moggach, The effect of high pressure on MOF-5: guest-induced modification of pore size and content at high pressure, *Angew. Chem., Int. Ed. Engl.*, 2011, **123**(47), 11334–11337.
339. A. J. Graham, J. C. Tan, D. R. Allan and S. A. Moggach, The effect of pressure on Cu-btc: framework compression *vs.* guest inclusion, *Chem. Commun.*, 2012, **48**(10), 1535–1537.
340. D. Fairen-Jimenez, S. A. Moggach, M. T. Wharmby, P. A. Wright, S. Parsons and T. Duren, Opening the gate: framework flexibility in ZIF-8 explored by experiments and simulations, *J. Am. Chem. Soc.*, 2011, **133**(23), 8900–8902.
341. Y. Hu, H. Kazemian, S. Rohani, Y. Huang and Y. Song, *In situ* high pressure study of ZIF-8 by FTIR spectroscopy, *Chem. Commun.*, 2011, **47**(47), 12694–12696.

342. T. D. Bennett, P. Simoncic, S. A. Moggach, F. Gozzo, P. Macchi, D. A. Keen, J.-C. Tan and A. K. Cheetham, Reversible pressure-induced amorphization of a zeolitic imidazolate framework (ZIF-4), *Chem. Commun.*, 2011, **47**(28), 7983–7985.
343. K. J. Gagnon, C. M. Beavers and A. Clearfield, MOFs under pressure: the reversible compression of a single crystal, *J. Am. Chem. Soc.*, 2013, **135**(4), 1252–1255.
344. S. H. Lapidus, G. J. Halder, P. J. Chupas and K. W. Chapman, Exploiting high pressures to generate porosity, polymorphism, and lattice expansion in the nonporous molecular framework $Zn(CN)_2$, *J. Am. Chem. Soc.*, 2013, **135**(20), 7621–7628.
345. E. C. Spencer, M. S. Kiran, W. Li, U. Ramamurty, N. L. Ross and A. K. Cheetham, Pressure-induced bond rearrangement and reversible phase transformation in a metal–organic framework, *Angew. Chem., Int. Ed. Engl.*, 2014, **53**(22), 5583–5586.
346. A. A. Yakovenko, K. W. Chapman and G. J. Halder, Pressure-induced structural phase transformation in cobalt (II) dicyanamide, *Acta Crystallogr., Sect. B: Struct. Sci., Cryst. Eng. Mater.*, 2015, **71**(3), 252–257.
347. S. C. McKellar and S. A. Moggach, Structural studies of metal–organic frameworks under high pressure, *Acta Crystallogr., Sect. B: Struct. Sci., Cryst. Eng. Mater.*, 2015, **71**(6), 587–607.
348. F. X. Coudert, Responsive metal–organic frameworks and framework materials: under pressure, taking the heat, in the spotlight, with friends, *Chem. Mater.*, 2015, **27**(6), 1905–1916.
349. C. L. Hobday, R. J. Marshall, C. F. Murphie, J. Sotelo, T. Richards, D. R. Allan, T. Düren, F.-X. Coudert, R. S. Forgan, C. A. Morrison, S. A. Moggach and T. D. Bennett, A computational and experimental approach linking disorder, high-pressure behavior, and mechanical properties in UiO frameworks, *Angew. Chem., Int. Ed. Engl.*, 2016, **55**(7), 2401–2405.
350. J. Im, D. Seoung, G. C. Hwang, J. W. Jun, S. H. Jhung, C. C. Kao, T. Vogt and Y. Lee, Pressure-dependent structural and chemical changes in a metal–organic framework with one-dimensional pore structure, *Chem. Mater.*, 2016, **28**(15), 5336–5341.
351. E. A. Kapustin, S. Lee, A. S. Alshammari and O. M. Yaghi, Molecular retrofitting adapts a metal–organic framework to extreme pressure, *ACS Cent. Sci.*, 2017, **3**(6), 662–667.
352. W. Grochala, R. Hoffmann, J. Feng and N. W. Ashcroft, The chemical imagination at work in very tight places, *Angew. Chem., Int. Ed. Engl.*, 2007, **46**(20), 3620–3642.
353. W. Grochala, Atypical compounds of gases, which have been called 'noble', *Chem. Soc. Rev.*, 2007, **36**(10), 1632–1655.
354. E. Zurek and W. Grochala, Predicting crystal structures and properties of matter under extreme conditions *via* quantum mechanics: the pressure is on, *Phys. Chem. Chem. Phys.*, 2015, **17**(5), 2917–2934.
355. W. Grochala, L. Khriachtchev and M. Räsänen, Noble-gas chemistry, *Phys. Chem. Low Temp.*, 2011, **m13**, 419.

356. G. D. Gatta and Y. Lee, Zeolites at high pressure: a review, *Mineral. Mag.*, 2014, **78**(2), 267–291.
357. G. Gottardi and E. Galli, *Natural Zeolites*, Springer-Verlag, Berlin, 1985.
358. I. A. Belitsky, B. A. Fursenko, S. P. Gabuda, O. V. Kholdeev and Y. V. Seryotkin, Structural transformations in natrolite and edingtonite, *Phys. Chem. Miner.*, 1992, **18**(8), 497–505.
359. Y. Lee, J. A. Hriljac, T. Vogt, J. B. Parise and G. Artioli, First structural investigation of a super-hydrated zeolite, *J. Am. Chem. Soc.*, 2001, **123**, 12732–12733.
360. Y. V. Seryotkin, V. V. Bakakin, B. A. Fursenko, I. A. Belitsky, W. Joswig and P. G. Radaelli, Structural evolution of natrolite during over-hydration, *Eur. J. Mineral.*, 2005, **17**, 305–314.
361. S. V. Rashchenko, Y. V. Seryotkin and V. V. Bakakin, An X-ray single crystal study of alkaline cations influence on laumontite hydration ability: II. Pressure-induced hydration of Na, K-rich laumontite, *Microporous Mesoporous Mater.*, 2012, **159**, 126–131.
362. S. Quartieri, G. Montagna, R. Arletti and G. Vezzalini, Elastic behavior of MFI-type zeolites: compressibility of H-ZSM-5 in penetrating and non-penetrating media, *J. Solid State Chem.*, 2011, **184**, 1505–1516.
363. Y. V. Seryotkin, Influence of content of pressure-transmitting medium on structural evolution of heulandite: single-crystal X-ray diffraction study, *Microporous Mesoporous Mater.*, 2015, **214**, 127–135.
364. G. D. Gatta and Y. Lee, Anisotropic elastic behaviour and structural evolution of zeolite phillipsite at high pressure: a synchrotron powder diffraction study, *Microporous Mesoporous Mater.*, 2007, **105**, 239–250.
365. Y. V. Seryotkin, Evolution of the bikitaite structure at high pressure: a single-crystal X-ray diffraction study, *Microporous Mesoporous Mater.*, 2016, **226**, 415–423.
366. G. Shen, Q. Mei, V. B. Prakapenka, P. Lazor, S. Sinogeikin, Y. Meng and C. Park, Effect of helium on structure and compression behavior of SiO_2 glass, *Proc. Natl. Acad. Sci. U. S. A.*, 2011, **108**, 6004–6007.
367. B. Coasne, C. Weigel, A. Polian, M. Kint, J. Rouquette, J. Haines, M. Foret, R. Vacher and B. Rufflé, Poroelastic theory applied to the adsorption-induced deformation of vitreous silica, *J. Phys. Chem. B*, 2014, **118**, 14519–14525.
368. K. Niwa, T. Tanaka, M. Hasegawa, T. Okada, T. Yagi and T. Kikegawa, Pressure-induced noble gas insertion into Linde-type A zeolite and its incompressible behaviors at high pressure, *Microporous Mesoporous Mater.*, 2013, **182**, 191–197.
369. T. Sato, N. Funamori and T. Yagi, Helium penetrates into silica glass and reduces its compressibility, *Nat. Commun.*, 2011, **2**, 345.
370. T. Sato, H. Takada, T. Yagi, H. Gotou, T. Okada, D. Wakabayashi and N. Funamori, Anomalous behavior of cristobalite in helium under high pressure, *Phys. Chem. Miner.*, 2013, **40**, 3–10.
371. C. Weigel, A. Polian, M. Kint, B. Rufflé, M. Foret and R. Vacher, Vitreous silica distends in helium gas: acoustic *versus* static compressibilities, *Phys. Rev. Lett.*, 2012, **109**, 245504.

372. W. Cai, A. Gładysiak, M. Anioła, V. J. Smith, L. J. Barbour and A. Katrusiak, Giant negative area compressibility tunable in a soft porous framework material, *J. Am. Chem. Soc.*, 2015, **137**, 9296–9301.
373. J. Joo, H. Kim and S. S. Han, Volume shrinkage of a metal–organic framework host induced by the dispersive attraction of guest gas molecules, *Phys. Chem. Chem. Phys.*, 2013, **15**, 18822–18826.
374. A. J. Graham, A.-M. Banu, T. Düren, A. Greenaway, S. C. McKellar, J. P. S. Mowat, K. Ward, P. A. Wright and S. A. Moggach, Stabilization of scandium terephthalate MOFs against reversible amorphization and structural phase transition by guest uptake at extreme pressure, *J. Am. Chem. Soc.*, 2014, **136**, 8606–8613.
375. S. C. McKellar, J. Sotelo, J. P. S. Mowat, P. A. Wright and S. A. Moggach, Perfluorocarbon liquid under pressure: a medium for gas delivery, *CrystEngComm*, 2016, **18**, 1273–1276.
376. S. C. McKellar, A. J. Graham, D. R. Allan, M. I. H. Mohideen, R. E. Morris and S. A. Moggach, The effect of pressure on the post-synthetic modification of a nanoporous metal–organic framework, *Nanoscale*, 2014, **6**, 4163–4173.
377. S. Kitagawa, R. Kitaura and S. I. Noro, Functional porous coordination polymers, *Angew. Chem., Int. Ed.*, 2004, **43**(18), 2334–2375.
378. G. Férey, Hybrid porous solids: past, present, future, *Chem. Soc. Rev.*, 2008, **37**, 191–214.
379. K. W. Chapman and P. J. Chupas, Pressure enhancement of negative thermal expansion behavior and induced framework softening in zinc cyanide, *J. Am. Chem. Soc.*, 2007, **129**, 10090–10091.
380. R. M. Hazen, Zeolite molecular sieve 4A: anomalous compressibility and volume discontinuities at high pressure, *Science*, 1983, **219**, 1065–1067.
381. R. M. Hazen and L. W. Finger, Compressibility of zeolite 4A is dependent on the molecular size of the hydrostatic pressure medium, *J. Appl. Phys.*, 1984, **56**, 1838–1840.
382. R. M. Hazen and C. T. Prewitt, Effects of temperature and pressure on interatomic distances in oxygen-based minerals, *Am. Mineral.*, 1977, **62**, 309–315.
383. M. Colligan, P. M. Forster, A. K. Cheetham, Y. Lee, T. Vogt and J. A. Hriljac, Synchrotron X-ray powder diffraction and computational investigation of purely siliceous zeolite Y under pressure, *J. Am. Chem. Soc.*, 2004, **126**, 12015–12022.
384. J. Haines, O. Cambon, C. Levelut, M. Santoro, F. Gorelli and G. J. Garbarino, Deactivation of pressure-induced amorphization in silicalite SiO_2 by insertion of guest species, *J. Am. Chem. Soc.*, 2010, **132**, 8860–8861.
385. *Vibrational Spectroscopy at High External Pressures. The Diamond Anvil Cell*, ed. J. R. Ferraro, Academic Press, 1984, eBook ISBN: 9780323158107.
386. *High Pressure Chemistry*, ed. R. van Eldik and F.-G. Klarner, Wiley, 2008, p. 474, ISBN: 978-3-527-61263-5.

387. E. V. Boldyreva, S. L. Kuzmina and H. Ahsbahs, Kinetics of the solid-state nitrito–nitro linkage isomerization in $[Co(NH_3)_5ONO]Br_2$ at high hydrostatic pressures, *J. Struct. Chem.*, 1998, **39**(5), 934–946.
388. E. V. Boldyreva, The concept of the 'reaction cavity': a link between solution and solid-state chemistry, *Solid State Ionics*, 1997, **101**, 843–849.
389. T. Luty, Lattice mediation in thermo- and photo-induced reactions; co-operative activation, *Mol. Cryst. Liq. Cryst.*, 2001, **356**(1), 539–548.
390. B. A. Zakharov, A. S. Marchuk and E. V. Boldyreva, A single-crystal to single-crystal phase transition in $[Co(NH_3)_5NO_2]Br_2$ at high pressure: a step towards understanding linkage photo-isomerisation, *CrystEngComm*, 2015, **17**(46), 8812–8816.
391. A. A. Sidelnikov, S. A. Chizhik, B. A. Zakharov, A. P. Chupakhin and E. V. Boldyreva, The effect of thermal expansion on photoisomerisation in the crystals of $[Co(NH_3)_5NO_2]Cl(NO_3)$: different strain origins–different outcomes, *CrystEngComm*, 2016, **18**(38), 7276–7283.
392. J. Bąkowicz and I. Turowska-Tyrk, Structural transformations in crystals induced by radiation and pressure. Part 1. How pressure influences the intramolecular photochemical reactions in crystals, *CrystEngComm*, 2014, **16**(27), 6039–6048.
393. J. Bąkowicz and I. Turowska-Tyrk, Photo-induced structural transformations in crystals at high pressure. Part 1. The crystallographic studies of the photochemical reaction at high pressure, *J. Photochem. Photobiol., A*, 2012, **232**, 41–43.
394. K. Konieczny, J. Bąkowicz and I. Turowska-Tyrk, Structural transformations in crystals induced by radiation and pressure. Part 3. The pressure-induced structural changes *versus* the rate of the Norrish–Yang reaction in crystals, *J. Photochem. Photobiol., A*, 2016, **325**, 111–115.
395. J. Bąkowicz and I. Turowska-Tyrk, Structural transformations in crystals induced by radiation and pressure. Part 4. The complex influence of high pressure on the path and kinetics of the [2 + 2] photodimerization, *CrystEngComm*, 2016, **18**(46), 8898–8905.
396. J. Bąkowicz, T. Galica and I. Turowska-Tyrk, Influence of pressure on molecular packing and photochemical properties in three chalcone analogs, *Z. Kristallogr.*, 2015, **230**(2), 131–137.
397. C. L. Jones, A. J. Tansell and T. L. Easun, The lighter side of MOFs: structurally photoresponsive metal–organic frameworks, *J. Mater. Chem. A*, 2016, **4**(18), 6714–6723.

CHAPTER 3

Intermolecular Interactions in In situ *Cryocrystallized Compounds*

DHANANJAY DEY[a], SUSANTA K. NAYAK[b] AND DEEPAK CHOPRA*[a]

[a]Department of Chemistry, Indian Institute of Science Education and Research, Bhopal 462066, Madhya Pradesh, India; [b]Department of Chemistry, Visvesvaraya National Institute of Technology, Nagpur, 440010, Maharashtra, India
*E-mail: dchopra@iiserb.ac.in

3.1 Introduction

In situ cryocrystallization[1–3] is a unique technique to determine the crystal structure of compounds that are liquids at room temperature or low melting in nature. This method has also been applicable to ionic liquids and gaseous hydrates, and a comprehensive examination of the work in this area by different research groups, up to 2007, has been given in a review article.[2] The current chapter is an attempt to unravel the application of this procedure in the understanding of intermolecular interactions in the crystal structures of *in situ* cryocrystallized compounds in the past decade. This crystallization technique allows for the unequivocal characterization of the compound, that is a liquid at ambient conditions, in the solid state, and subsequent analysis

Monographs in Supramolecular Chemistry No. 26
Understanding Intermolecular Interactions in the Solid State: Approaches and Techniques
Edited by Deepak Chopra

Published by the Royal Society of Chemistry, www.rsc.org

of the molecular conformation and various intermolecular interactions that characterize the crystal packing. According to Dunitz, "Weak interactions are not only important in modern supramolecular chemistry, they hold the organic world together and are responsible for the very existence of liquids and solids; a characteristic feature of weak interactions is the phenomenon of enthalpy-entropy compensation".[4] In this chapter, we investigate the utility of this technique for the crystallization of different ionic liquids, solvents, gas hydrates, low-melting solids, co-crystal formation through hydrogen bonding (HB) and halogen bonding (XB). In addition, a detailed analysis, depicting the role of different computational methods, adopted for obtaining quantitative insights into the "*nature*", "*energetics*" and "*topological characteristics*" associated with these interactions is also highlighted. Not only in last 20 years, but also in the earlier days of developments in crystallography, scientists have used this technique of *in situ* crystallization to crystallize liquids or a mixture of liquids or low melting solids. In 1935, the group of Keesom and Taconis developed a method to determine the crystal structures of solidified gases.[5] From India, the work reported in this area, in 1962, was on the determination of the crystal structure of dimethyl sulphoxide (DMSO) at −60 °C, which was crystallized at low temperature using a Weissenberg camera by Viswamitra and Kannan.[6] The study of the Nobel Laureate, Odd Hassel, on the structural characterization of halogen bonded adducts presents an overview of the cryocrystallization technique before 1970.[7] The group of Bond was actively involved in the determination of the crystal structures of heterocyclic amines,[8a,b] terpenes[8c,d] and hydrocarbons, which are liquids at low temperature or low-melting solids.

The single crystal X-ray diffraction technique is extremely useful for mapping the 3D atomic contents of different compounds that are 'solids' under normal conditions of temperature and pressure. However, for the compounds that are liquids at normal temperature and pressure, different procedures have to be adopted for obtaining their single crystal and 3D structure. In principle, the liquid state is a complex dynamic system characterized by the presence of intermolecular interactions, which play an important role in molecular association. For example, the solubility and miscibility properties of a liquid strongly depend on the strength and directionality of these interactions. *In situ* cryocrystallization in crystallography allows for the study of molecules without any influence of solvent participation either as promoters for crystal growth or as incorporated solvatomorphs.[2] This is one of the cutting-edge research techniques used for structural investigation of single crystals as was highlighted recently by Howard.[9]

3.2 Methodology, Equipment and Instrumentation

The main aspect of this methodology is to start from a liquid and to convert into the solid state by the lowering of the temperature of the system. The nature of the crystallized solid may vary from being amorphous to

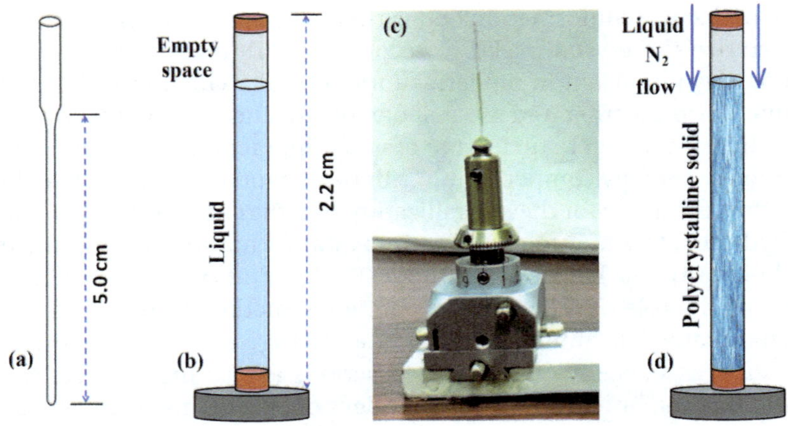

Figure 3.1 (a) Lindemann glass capillary, (b) the target liquid inside the capillary, (c) the capillary containing the liquid mounted on the goniometer, (d) solidification of the liquid under the LT device.

polycrystalline. We delineate below the detailed strategy adopted by several research groups to perform *in situ* cryocrystallization.

The research group of Row at Bangalore has successfully demonstrated this technique wherein the process of crystallization is controlled *via* the manual adjustment of the nitrogen cryostream.[2] The technique underwent a major renaissance when Nussbaumer and Boese, described that the method of *in situ* cryocrystallization could be efficiently controlled with the help of an optical heating and crystallization device (OHCD),[10] the details of which are provided later. The working principles and limitations of both these methods are discussed below.

Prior information about the nature of the liquid from differential scanning calorimetry (DSC)[11] is useful for obtaining the crystallization window for the samples. The DSC data provide input towards the design of the experimental strategy using the freezing point and the melting point of the sample under investigation. A 2.5 cm length Lindemann glass capillary (Figure 3.1a) with a diameter of 0.3 mm is used to fill up the target liquid and then both sides of the capillary are sealed using a glue. A small portion above the top of the liquid meniscus has to be kept empty (Figure 3.1b). The capillary is mounted very safely on the goniometer (Figure 3.1c) of an X-ray diffractometer equipped with the detector (Figure 3.2a). The goniometer, along with the different ramp rates, are applied and after repeated trials and experimentation, in general, 70 K h^{-1} is suggested as an optimized cooling rate to achieve the solidification of the liquid inside the capillary. The ramp rate is another important physical parameter that has been standardized to overcome the glass transition temperature and avoid glass formation inside the capillary. The liquid nitrogen stream has to be kept parallel to the capillary and the nozzle of the low-temperature device is positioned above the centre of the goniometer. Under a liquid nitrogen stream, the liquid sample inside

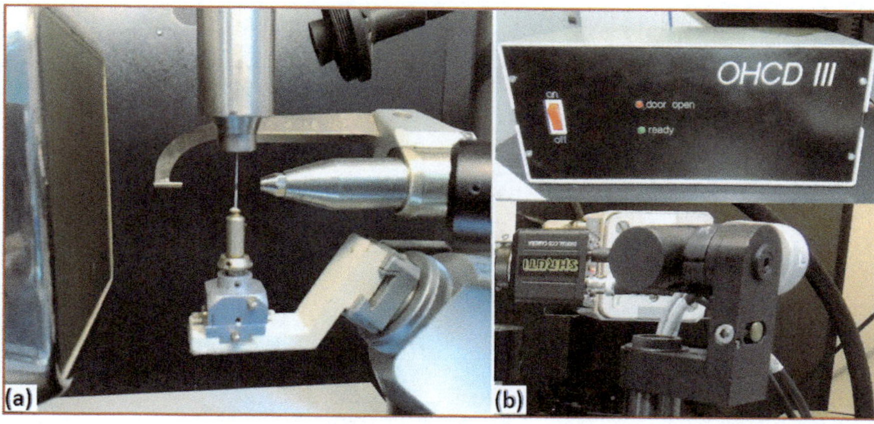

Figure 3.2 (a) Bruker SMART APEX CCD area detector, Oxford nitrogen cryosystem, and goniometer attachment with liquid mounted in a capillary and (b) OHCD attached to an Oxford Apex II XRD.

the capillary converts into a polycrystalline solid (Figure 3.1d) at a certain temperature.

3.2.1 OHCD (Optical Heating and Crystallization Device)

The OHCD is a system for local heating and *in situ* crystallization by IR laser. In general, crystallization by cooling results in the formation of a polycrystalline material because nucleation occurs at lower temperatures than the melting point of the neat sample or the saturation point of a solution. Therefore local heating is applied to produce a molten zone or a temperature gradient within a solution. The focused IR laser is applied as a heat source and the molten zone (or the clear solvent) is kept very small (up to 0.1 mm) and then there exists an opportunity to grow crystals moving against the direction of the cold gas stream, which runs usually from top to bottom. The fine focus of the laser is also used for local heating to destroy unwanted satellites with an exact tuning of its position. In general crystals are obtained at far below its crystallization temperature by repeated melting and solidification using an IR laser.[10] One of the advantages of the OHCD system is the use of the autocrystallization program (Figure 3.2b). For example, it can be used for 10 min for the first cycle heating and then 20 min for another heating cycle during the process of autocrystallization. Sometimes several attempts are repeated to get better quality diffraction data. The RLATT program is being frequently used to identify manually, by visual inspection, groups of reflections that belong to different reciprocal lattices. Figure 3.3 describes the procedure followed step-by-step using the OHCD.

In the absence of any OHCD, it is also possible to crystallize the liquid inside the capillary using a manually adjusted cryostream. A temperature gradient is

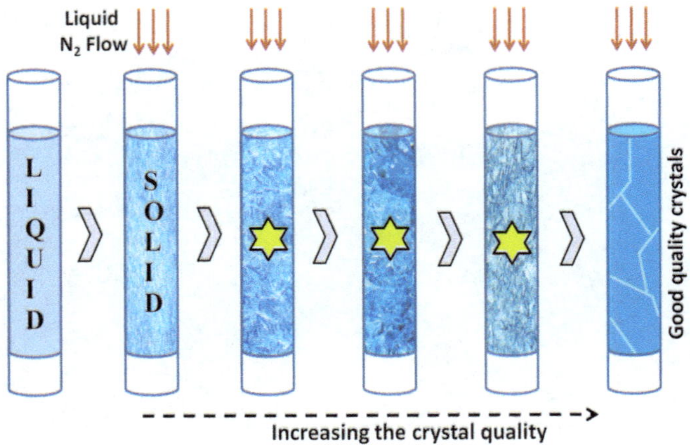

Figure 3.3 Crystallization process inside the capillary and the formation of good quality crystals obtained after several scans.

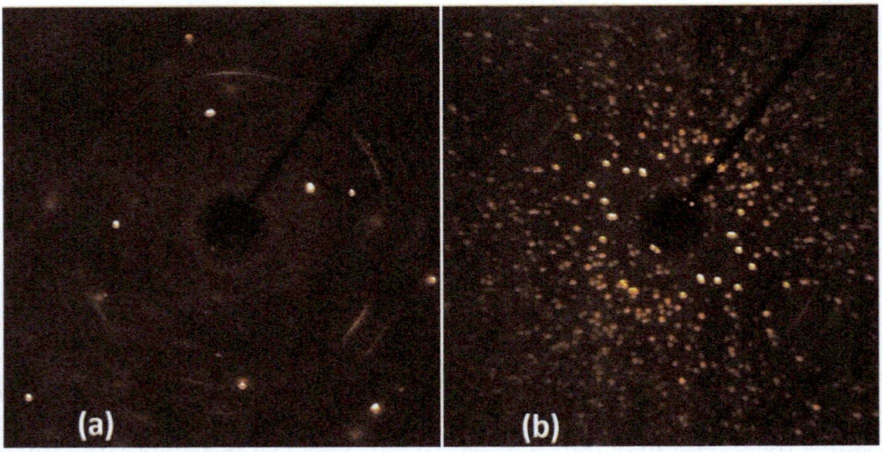

Figure 3.4 (a) Still photograph and (b) rotation photograph of a cryocrystallized sample before the data collection.

created by initially moving the nitrogen stream along the capillary length away from the goniometer. After the melting of the polycrystalline solid, the stream is realigned with a time delay to obtain a single crystal inside the capillary. This annealing process is repeated until a well-defined crystal is obtained. The quality of the obtained single crystals is ascertained by collection of the still images and rotation photographs (Figure 3.4). The obtained frames are then processed and the spots are analysed using a reciprocal lattice viewer to determine the unit cell dimensions. The procedure for obtaining crystals through this process is indeed physically and mentally challenging, requires long hours of manual work and the chances of success are low.

3.2.2 Problems and Concerns During the OHCD Experiment

Such experiments are very challenging and there exist several concerns at the time of crystallization.

(i) If the solidification of the liquid sample happens at a very low temperature (<150 K), then the glassy state can be formed. For this purpose, we should try to crystallize that liquid with a slow ramp rate (20–30 K h^{-1}) and with low laser intensity. It is not necessarily the case that this procedure will work for any liquid sample or that good quality single crystals can be obtained.
(ii) If the viscosity of the sample is very high, then there is a possibility of conversion into the glassy state during the experiment.
(iii) There is a limit on the intensity of the operation of the CO_2 laser. If we use high laser intensity to create the molten zone, the capillary can break or the liquid sample can also decompose (if the target liquid is sensitive to higher temperature, *i.e.* the liquid sample is not stable at high temperature). In such cases, we should try to crystallize the liquid in the absence of laser.
(iv) There are additional considerations that should be kept in mind during the process of crystallization, such as scan rate, laser intensity and the time for the scan. We have to optimize these parameters before the OHCD scans are performed in such a manner that we can get good quality single crystals.
(v) The formation of a twinned crystal is also a problem during this experiment. On certain occasions, due to the presence of multiple overlapping domains, we are not able to select the required domain.
(vi) The target sample has to be pure. The impurities present in the liquid can affect the nucleation process.
(vii) In the case of more flexible molecules, there exists a chance to obtain multiple phases together inside the cylindrical crystals.
(viii) The formation of air bubbles is also a very serious problem.
(ix) As the capillary is fixed parallel to the cryosystem during the OHCD experiment, ω scans are only possible and not the ϕ scan during the process of data collection. Hence, the data completeness is considerably reduced (~60–80%) for such a procedure.

3.3 Applications of *In situ* Cryocrystallization

The *in situ* cryocrystallization technique is primarily used for the understanding of intermolecular interactions in the solid state. This work paves the way for the utilization of this technique to evaluate various non-covalent interactions, including HB and XB interactions for polymorphism, co-crystal formation, gas hydrates, pharmaceutical drug molecules and technologically important materials. The detailed understanding of intermolecular interactions through the *in situ* cryocrystallization technique in different systems is described below.

3.3.1 Investigation of Strong and Weak Hydrogen Bonds (HBs) in *In situ* Cryocrystallized Liquids

The utilization of the *in situ* cryocrystallization technique has been well known for a long time for gaining understanding of both strong and weak hydrogen bonds in the solid state. In 1966, an *in situ* cryocrystallization study reported the crystal structure of dimethyl sulfoxide.[6] This liquid crystallizes in the monoclinic centrosymmetric $P2_1/c$ space group with $Z = 4$, which shows the formation of bifurcated C–H···O HB infinite chains along the *a*-axis (Figure 3.5).

The study of the electron density distribution of ethane-1,2-diol[12] shows the preference of the gauche conformation with the crystal structure stabilized by a network of intermolecular O–H···O and C–H···O HBs (Figure 3.6). It is indeed noteworthy that the topological evidence accounts for the non-existence of an intramolecular O–H···O hydrogen bond.

Similarly, hydroxyacetone (HA) is studied using low temperature FTIR, supersonic jet expansion, and *in situ* cryocrystallization to unravel the conformational preferences of HA. It confirms that the crystal structure prefers a *trans*-HA conformation[13] with intermolecular O–H···O HBs in the crystal structure (Figure 3.7) in comparison to the expected *cis*-HA with intramolecular O–H···O HBs.

In situ cryocrystallization of monofluoro- and monochloro-phenols[14] at low temperature and high pressure are achieved through O–H···O hydrogen bond interactions. It is observed that phenol, 2-fluorophenol and 2-chlorophenol form different polymorphs obtained utilizing the cryocrystallization technique as well as *via* the application of high pressure. In the case of polymorphic form I (space group $P3_2$) of 2-chlorophenol, which is obtained at ambient pressure and 100 K temperature, there was a formation of helices of OH···OH···OH hydrogen-bonded chains around 3_2 axes (Figure 3.8a). The three symmetry independent molecules are represented with different colour codes. Interestingly, on application of 0.12 GPa pressure, the

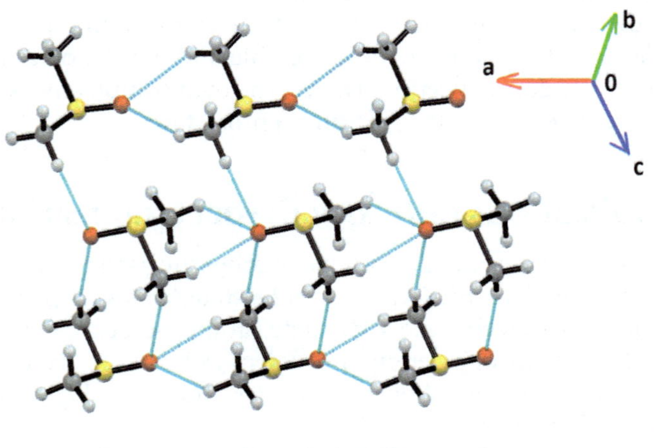

Figure 3.5 DMSO forms C–H···O HB chains along *a*-axis.

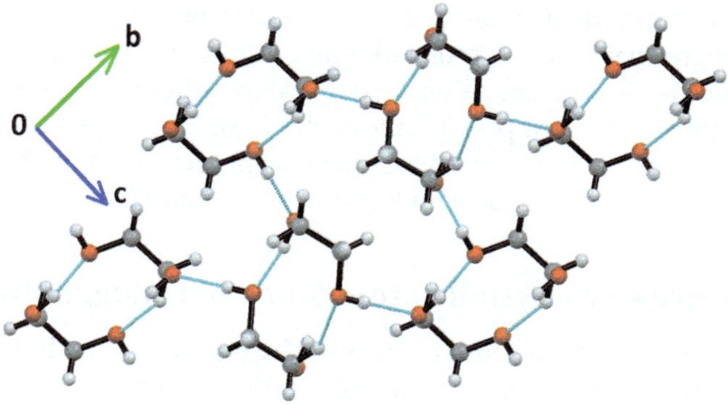

Figure 3.6 Packing network of ethane-1,2-diol shows the formation of a C–H···O dimer down the *bc* plane.

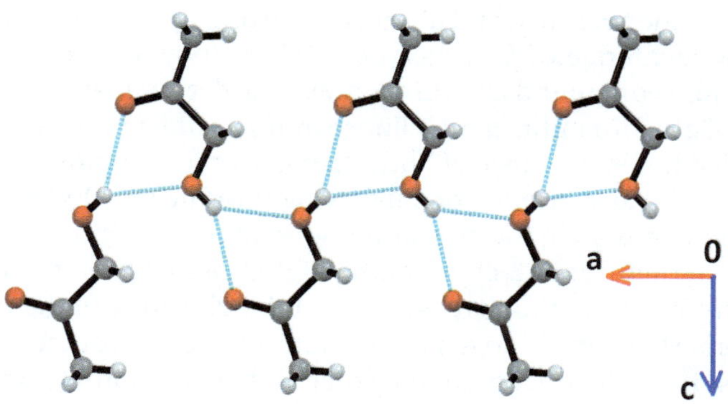

Figure 3.7 Molecular assembly in hydroxyl acetone shows bifurcated O–H···O hydrogen bonds.

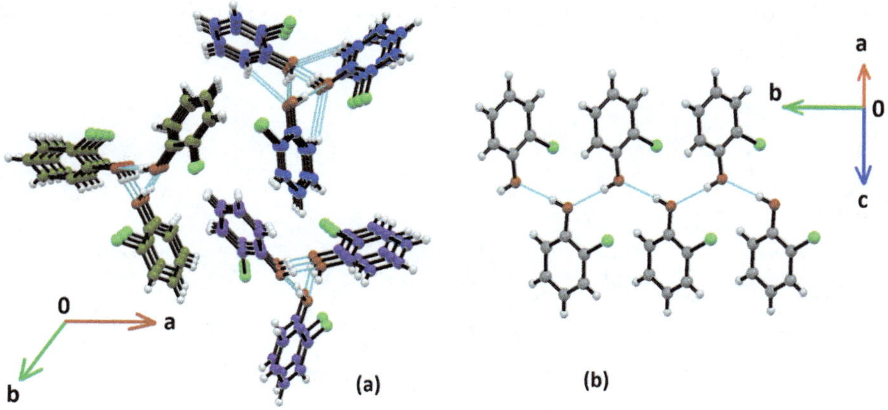

Figure 3.8 Molecular network in (a) polymorph I of 2-chlorophenol which forms a helical OH···OH···OH hydrogen bond chain, (b) polymorph II depicting O–H···O–H-bonded chains.

polymorphic form II of 2-chlorophenol is obtained, which crystallizes in $P2_1/n$ space group where the molecular assembly is stabilized *via* the formation of an O–H···O–H chain along the *b*-direction (Figure 3.8b). Similarly, the high-pressure structure of phenol (monoclinic, space group $P2_1$, which is the same as the ambient pressure phase) is also characterized by the formation of helical OH···OH···OH hydrogen bonded molecular chains.

3.3.2 *In situ* Cryocrystallization Study of Halogen Bonding

According to an International Union of Pure and Applied Chemistry (IUPAC) recommendation, the halogen bonds definition states that "A halogen bond occurs when there is evidence of a net attractive interaction between an electrophilic region associated with a halogen atom in a molecular entity and a nucleophilic region in another, or the same, molecular entity".[15] Our quest for the understanding of halogen driven interactions both in the presence and absence of strong hydrogen bonds led us to employ the technique of *in situ* cryocrystallography for the study of some selected liquid compounds. In this context, the determination of the crystal structure of 4-fluoroaniline and 2-fluoroaniline using *in situ* cryocrystallization revealed the importance of both strong hydrogen bonds (N–H···N) and weak intermolecular interactions involving fluorine (H···F) generating different packing motifs in the crystalline solid in these isomeric molecules.[16] The crystal structures of 2-chloroaniline (2ClA) and 2-bromoaniline (2BrA) obtained from this method are isostructural and belong to the trigonal $P3_1$ space group. Halogen···halogen (Cl/Br) intermolecular contacts along with stronger N–H···N hydrogen bonds generate helical motifs along the crystallographic *c*-axis. Br···Br (3.637(1) Å) and Cl···Cl (3.665(3) Å) interactions (type-II contacts) are observed in the case of 2BrA and 2ClA respectively, and the helical nature of these interactions build the crystal structure along the 3_1-screw axis. The helices form pseudo triangular contacts with Br···Br (or Cl···Cl) homo halogen···halogen interactions and N–H···N HBs along the *c*-axis (Figure 3.9).[16,17]

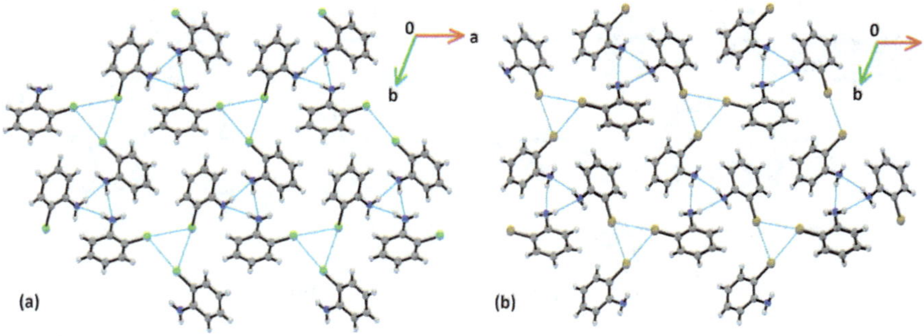

Figure 3.9 Formation of helices consisting of pseudo triangular Type-II directional (a) Cl···Cl and (b) Br···Br contacts along with N–H···N hydrogen bonds in 2ClA and 2BrA, respectively.

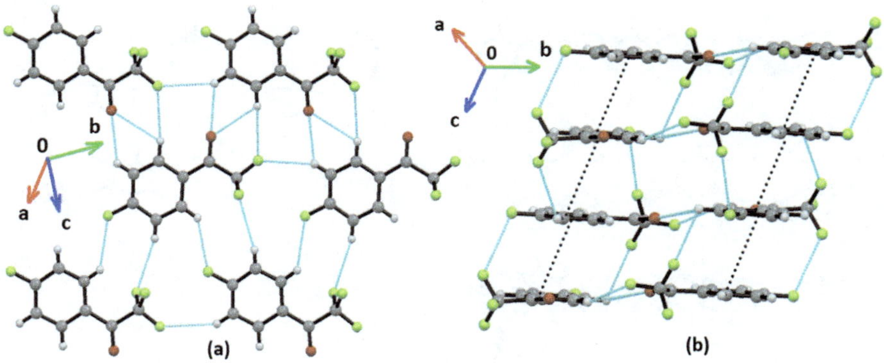

Figure 3.10 Crystal packing of 2,2,2-trifluoro-1-(4-fluorophenyl)ethanone shows (a) the molecular sheet down the (101) plane associated with intermolecular C–H···O and C–H···F interactions and (b) molecular stacking of the planar sheets.

Structural investigation of the four low-melting halogenated trifluoroacetophenones using this technique, along with the corresponding nitro derivative, have been carried out to understand the nature of short halogen···halogen and halogen···oxygen intermolecular contacts. The basic building block in the crystal lattice is invariant because of an intramolecular C–H···F contact, which locks the molecular conformation. The supramolecular assembly is stabilized with additional C–H···O, C–H···F, and aromatic π···π intermolecular interactions. This study demonstrates that halogens, depending on their size, generate a variety of interactions, given the conformationally rigid molecular architecture, wherein fluorine prefers C–H···F, occasionally F···F, and rarely F···X (X = Br, O) in these five compounds. Bromine derivatives among the compounds show priorities towards the formation of Br···O and Br···F along with C–H···F and C–H···O interactions. Short F···F contacts display both features similar to the classification of type I and II XB interactions. The representative crystal structure of 2,2-trifluoro-1-(4-fluorophenyl) ethanone is presented in Figure 3.10.[18]

In situ cryocrystallization of mixtures of 1,3-dibromotetrafluorobenzene and various oxygenated solvents are achieved through halogen bond (XB) interactions. A Br···O XB synthon plays the major role in driving the self-assembly process.[19] There are numerous structures available in the Cambridge Structure Database (CSD) that show the existence of C–Br···O interactions. The role of the C–Br···O supramolecular synthon could not be evaluated as these were obtained serendipitously and/or involve quite complex and random molecules. In this context, the structural study of 1,3-dibromotetrafluorobenzene (DBrTFB) and its halogen bonded adducts with DMSO, hexamethylphosphoramide (HMPA) and *N*,*N*-dimethyl formamide were performed. These adducts are liquids at room temperature and the OHCD allowed us to obtain single crystals of DBrTFB and its adducts, followed by determination of the crystal structures. Single crystal analysis clearly confirmed the formation of C–Br···O XBs, which partly

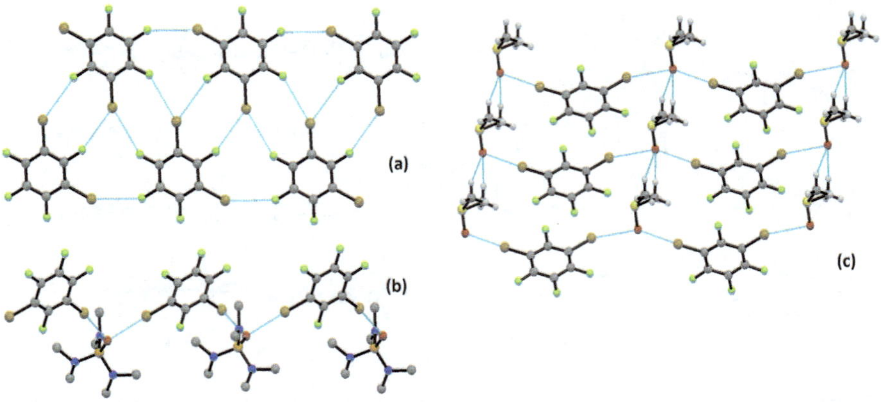

Figure 3.11 (a) Br···F XB in DBrTFB and (b and c) Br···O XB in adducts with solvents of HMPA and DMSO.

substitute for the presence of C–Br···F and C–H···O interactions that are present in the homomeric starting crystals of DBrTFB or solvents, as shown in Figure 3.11.[19]

Further, the *in situ* cryocrystallization technique has been used to obtain four adducts between HMPA and 1,2-dihalotetrafluoroethanes having iodine, bromine and chlorine as halogen-bonding donor atoms.[20] These systems allowed for a precise comparison of different C–X···O synthons. This work evidences how the C–X···O XB synthons are able to completely substitute for the C–H···O HB synthon present in the crystal structure of pure HMPA and to drive the formation of isostructural architectures. It also clearly validates that not only in iodine, but also in bromine and chlorine atoms of 1,2-dihalotetrafluoroethanes, the halogens can act as a donor atom. Among the four adducts, the diiodotetrafluoroethane : HMPA co-crystal exhibits the highest melting point of the series (90 °C), whereas bromoiodotetrafluoroethane : HMPA and dibromotetrafluoroethane : HMPA melt at 28 °C and 148 °C, respectively, and chloroiodotetrafluoroethane : HMPA melts at 212 °C.

The trends in the melting point of co-crystals parallel the expected XB donor ability of the heavy halogen atoms with the stronger XB formation. All the four co-crystals are isostructural and crystallize in the $P2_1/c$ space group. The supramolecular arrangement of the four co-crystals are mainly controlled by XBs between the oxygen atom of HMPA and the activated I, Br and Cl atoms of dihalotetrafluoroethane. Both the XB donor and acceptor tectons of dihalotetrafluoroethane and HMPA, respectively, behave as bidentate modules forming an infinite chain, propagating along the crystallographic *b*-axis (Figure 3.12).[19]

Again, it is demonstrated that the occurrence of orthogonal HBs and XBs in the peptide bond model *N*-methylacetamide self-assembles with a range of dihalotetrafluorobenzenes forming co-crystals using the technique of *in situ* cryocrystallization.[21] In this context of orthogonal XB and HB interactions, *N*-methylacetamide (NMA) functions as the smallest molecule

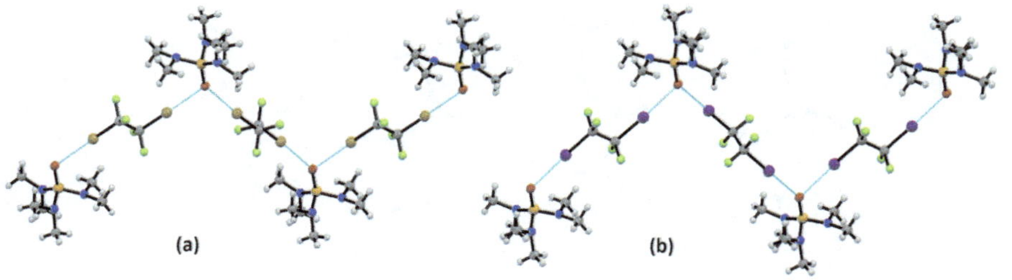

Figure 3.12 Trimeric motif in the co-crystals of (a) dibromotetrafluoroethane : HMPA and (b) diiodotetrafluoroethane : HMPA.

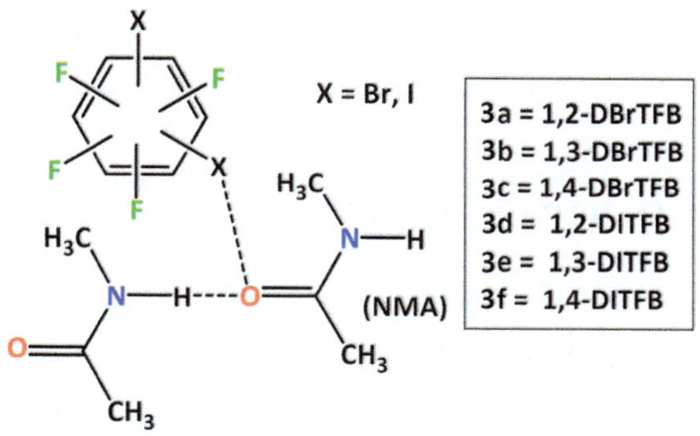

Figure 3.13 Synthesis of *N*-methylacetamide (NMA) co-crystals with various dihalotetrafluorobenzenes (DXTFB, X = Br or I).

that mimics the peptide bond (–NH–C=O) and thus the protein backbone undergoes self-assembly. To rationalize the design and verify the robustness of the formation of the orthogonal HB/XB supramolecular synthon in crystal engineering, NMA, affording the carbonyl oxygen acceptor site, was co-crystallized with several dihalotetrafluorobenzenes (DXTFBs), the XB donors, for example 1,2-dibromotetrafluorobenzene (12DBrTFB, 2a) and others, as shown in Figure 3.13, and the corresponding co-crystals 3a–f were obtained.

Six co-crystals were obtained and all the crystal structures depict the presence of geometrically orthogonal hydrogen and halogen bonds involving simultaneously the carbonyl oxygen atom, which demonstrates the great robustness of the orthogonal XB and HB interactions. Classical N–H···O HBs form infinite 1D chains running along the crystallographic *a*-axis (Figure 3.14). The two interacting groups N–H and C=O are in the *trans*-conformation, mimicking the molecular arrangement of the polypeptide backbone observed in β-sheet structures.[22]

It has been observed that the C–X···O angles in the crystal structures of 3a–f are in the range 168.5–177.7°. X···O interaction distances are in the range 2.800(2)–2.871(4) Å for X = Br (3a–c) and 2.706(2)–3.001(1) Å for X = I (3d–f), which correlate well with the calculated interaction energy curves of

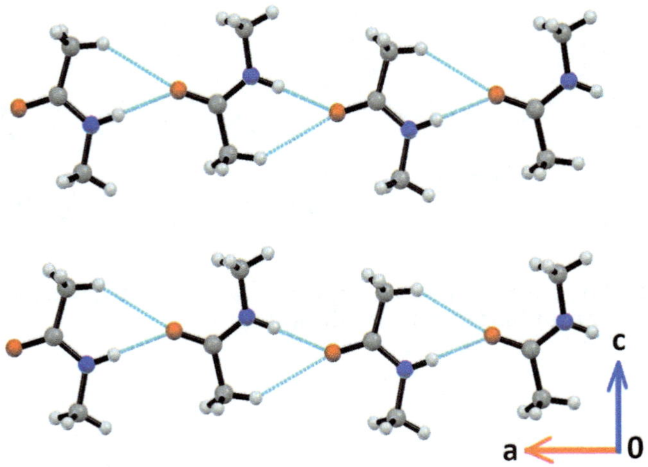

Figure 3.14 N–H···O HBs drive the formation of 1D infinite chains in the crystal packing of the homocrystal.

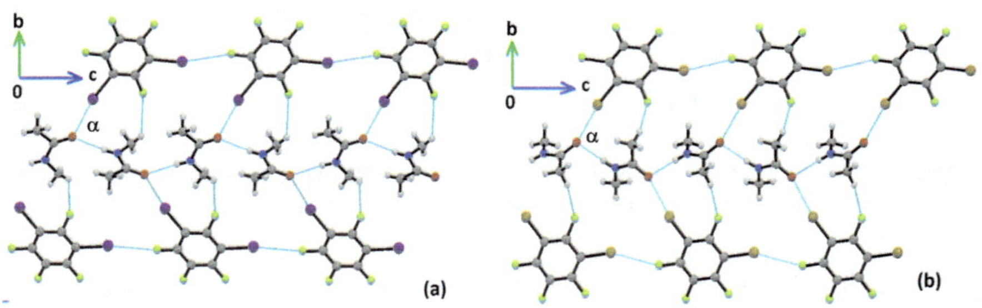

Figure 3.15 Partial crystal packing showing how infinite 1D chains (horizontally positioned) formed by NMA *via* N–H···O HBs further interact with 14DITFB *via* (a) Br···O XBs in 3b and (b) I···O XBs in 3e. The disorder on the NMA molecule is omitted for clarity.

different complexes between halophenyl derivatives and NMA. Only one of the XB donor sites of the DXTFBs in 3a–e is involved in the XB with the carbonyl oxygen atom. Differently, in 3e both of the iodine atoms of 13DITFB are XB to the carbonyl oxygen of two NMA molecules belonging to two different HB chains. In all of the six co-crystals, the oxygen atom of NMA is simultaneously involved in short HBs. Special attention during the analysis of the crystal structures has been given to the angle α, which has been defined as the angle ∠X···O···H–N, between XB and HB, as shown in Figure 3.13. These angles in the series vary in the range 77.9–98.5°, thus demonstrating the intrinsic tendency of HBs and XBs to occur orthogonal to each other when sharing the same sp^2 hybridized oxygen atom (Figure 3.15).[21]

Vasylyeva and Merz investigated different supramolecular motifs based on short F···F, C–F···N, C–F···C_{sp2}, N···N and N···C_{sp2} contacts in the absence of

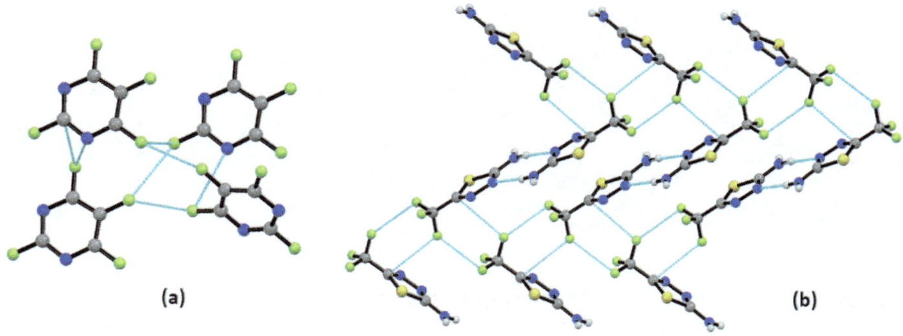

Figure 3.16 Intermolecular interactions and short contacts in tetrafluoropyrimidine (a) F···F, C–F···N and C–F···C_{sp2} contacts. (b) Intermolecular interactions and short contacts in trifluorotriazine F···F, C–F···C_{sp2} and N···C_{sp2} contacts with rosette-shaped motifs.

the hydrogen atoms.[23] They used the method of *in situ* cryocrystallization to obtain single crystals of the low-melting compounds tetrafluoropyrimidine and trifluorotriazine. Quasi type I short F···F contacts with several edge-to-face arrangements of C–F···C_{sp2} and N···C_{sp2} interactions (Figure 3.16) were identified.

3.3.3 Investigation of Other Weak Interactions in *In situ* Cryocrystallized Liquids

It is of interest to examine other weak interactions in cryocrystallized solids, for example, C–H···π and π···π interactions. The importance of the former has been demonstrated in the formation of polymorphs obtained *via* cryocrystallizing diphenyl ether wherein the crystal packing depicts the relevance of C–H···π mediated polymorphic forms that control the packing of the molecules in the unit cell. In form I, the intermolecular C–H···π interactions form a 3D network of molecules (Figure 3.17).[24] Further, a weak intermolecular C–H···O hydrogen bond links neighbouring tetramers in form II. This additional C–H···O interaction provides thermodynamic stability to form II and is probably responsible for the monoclinic form being a "disappearing polymorph",[25] as several attempts under identical conditions have resulted in access to only the orthorhombic form.

Liquids of benzyl derivatives were crystallized through a manually adjusted cryostream in the X-ray diffractometer, which reveals that the molecular packing in these molecules is through the formation of C(sp^3/sp^2)–H···π interactions, which depends on the intrinsic acidity of the benzyl proton. Further, it brings out the subtle features of such weak interactions and points to the directional preferences depending on the nature (electron withdrawing and polarizability) of the neighbouring functional groups (Figure 3.18).[19]

Low melting isomers of nitrotoluene (*ortho*- and *meta*- derivatives) have been grown using the OHCD technique. These compounds are related in

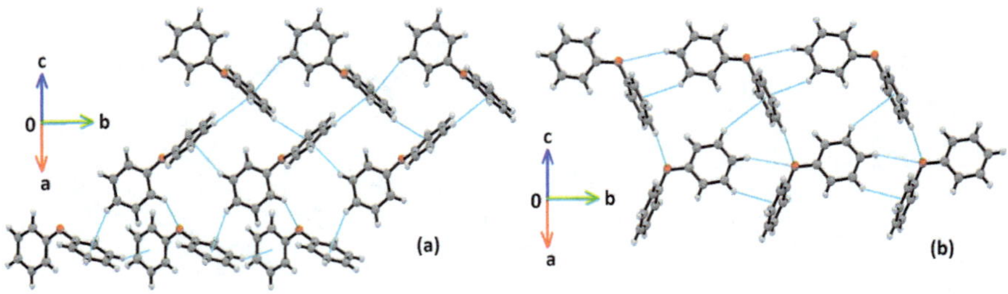

Figure 3.17 Packing of molecules of diphenyl ether in (a) form I and (b) form II associated with intermolecular C–H···O and C–H···π interactions.

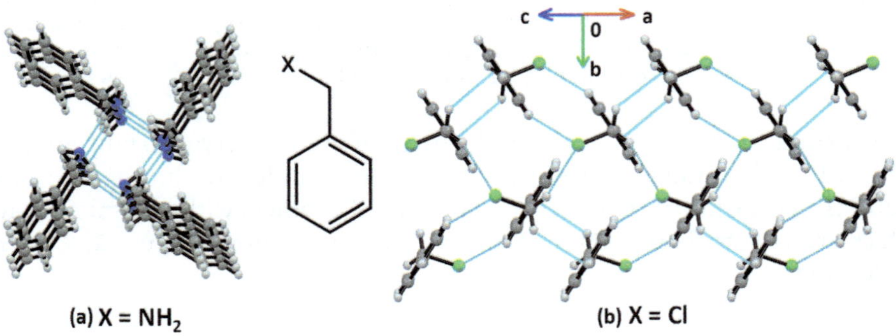

Figure 3.18 (a) N–H···N hydrogen bonds forming a helical motif across the fourfold symmetry axis in benzyl amine. (b) C(sp^3)–H···π and C–H···Cl interactions forming dimers viewed down the c-axis in benzyl chloride.

part to their properties as high energetic materials. From structural analysis, it is found that *m*-nitrotoluene undergoes, on cooling, an order–disorder phase transition from monoclinic to triclinic phase at about 200 K. The crystal structure of the nitro toluene showed the presence of C–H···O and π···π interactions in the molecular assembly.[26]

This method has also been used successfully to determine the crystal structure of gas hydrates. When a gas is cooled in the presence of a solvent under elevated pressure, co-crystals may form in which the gas molecules act as guests in the lattice of the host solvent molecules. In most instances, the lattice of the host is modified markedly and will be unstable without the guest.[27] The structures of the hydrates depend on the size of the guest molecules so that the regular cages will only accommodate molecules up to a maximum size, whereas if the appropriate size is not present in enough guests, the cage structure will be unstable and hexagonal ice may form. It means that water will form a solid through good interactions (ice) or good space filling (clathrate hydrate) and cages are filled with guest molecules, which stabilize the host network. Clathrate hydrates have been characterized into three main structural types by using neutron or

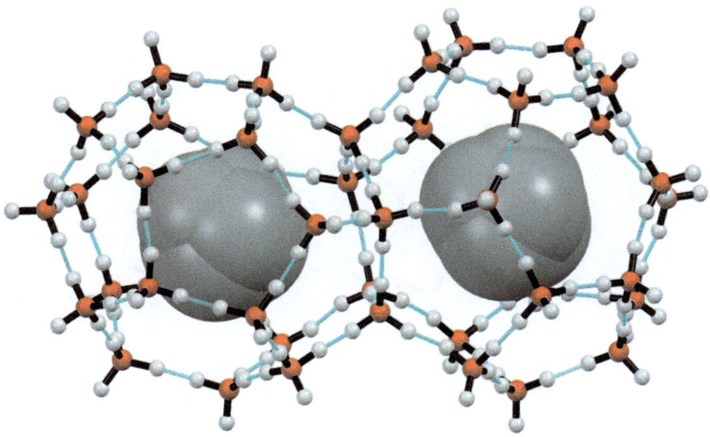

Figure 3.19 A tetrakaidecahedron and a dodecahedron in structure I clathrate hydrate with acetylene as guest; each water and ethylene molecule is disordered.

X-ray diffraction techniques performed on their single crystals: structure I, 6X·2Y·46H2O, structure II, 8X·16Y·136H2O, and the hexagonal structure H, 1X·3Y·2Z·34H2O, in which X, Y and Z represent the different types of cages in the clathrate hydrates.[27]

When acetylene is co-condensed with water, *in situ* crystallization shows the formation of structure I clathrate hydrates (*Pm*-3*n* space group, $a = 11.877(3)$ Å at 23(2) K).[27] The dodecahedral cages are formed by oxygen atoms (Figure 3.19) and adopt a nearly body-centred arrangement with the central dodecahedron turned by 90°. The surrounding tetrakaidecahedra, which form columns along their hexagonal faces, are arranged vertically, horizontally and out of the plane.[28]

Low-melting organic salts ("ionic liquids") are known for possible technical applications. Of interest, the crystal structures of ionic liquids have been determined *via* the zone melting method. An *in situ* crystallization study performed on (1-ethyl-3-methylimidazolium) BF_4 confirmed the presence of imidazolium cations and BF_4^- anions, which are linked *via* weak C–H···F interactions. The cations are oriented such that they display intermolecular C–H···π interactions and π···π stacking with a horizontal displacement that possibly contributes to the stability of the crystal packing.[29]

1-ethyl-3-methylimidazolium-based ionic liquids, [emim]OTf, OTf = trifluoromethanesulfonate (m.p. = −25.7 °C), and [emim]NTf2, NTf2 = bis(trifluoromethanesulfonyl)amide (m.p. = −25.7 °C) were crystallized by *in situ* cryocrystallization, which confirms the packing *via* interionic C–H···O and F···F interactions (Figure 3.20).[30]

The crystals of several low-melting molecular complexes of chloroform and dichloromethane with organic solvents have been grown *in situ* at low temperature. The results demonstrated that not only chloroform but also

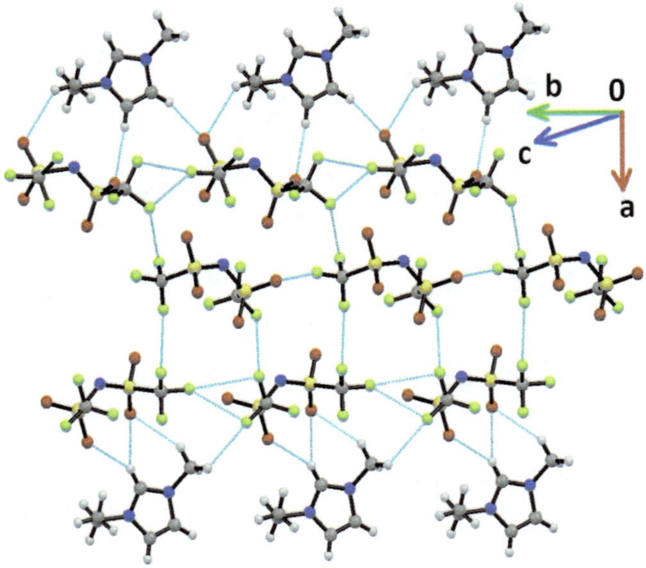

Figure 3.20 Packing diagram of [emim]OTf associated with intermolecular C–H···O and F···F interactions.

dichloromethane form low-melting molecular complexes with ketones. These ketones could be either cyclic or acyclic and similar complexes also formed with 1,4-dioxane. (Cl)CH···O contacts are the main driving force for the complex formation and the important role of Cl···O halogen bonds observed in some cases. Interestingly, it is pointed out that the change in co-crystallization "partner" changes the selected motif.[31] Chloroform forms low-melting co-crystals with various organic solvents and their 3D structures have been determined. In these co-crystals, the Cl_3C–H···X (X= O, N) interactions primarily direct the packing of molecules. Chloroform molecules tend to form corrugated layers *via* Cl···Cl interactions (Figure 3.21).[32]

Hydrazoic acid (HN_3) is known for being potentially explosive, highly toxic, and very hygroscopic in nature and it is the simplest covalent azide and contains 97.7 wt% nitrogen. The crystal structure was solved by X-ray diffraction in 2011. Single crystals of HN_3 were grown *in situ*. The molecules are connected to each other by N–H···N bifurcated hydrogen bonds (~2.0 Å and weak ones with ~2.6 Å) in nearly planar layers parallel to (001) by forming a unique tetramer (Figure 3.22).[33]

Further, *in situ* cryocrystallization methods provided polymorphs and two-component crystals (co-crystals) of pentafluorophenol and pentafluoroaniline systems, which clarified the inconsistencies in the literature and defined a new structural landscape (Figure 3.23).[34]

In co-crystals of acetylene and acetone in 1 : 2 (α phase), the structure indicates a three-component molecular complex with a dumbbell shape and two almost linear C–H···O hydrogen bonds.[35] In the crystal, these complexes are

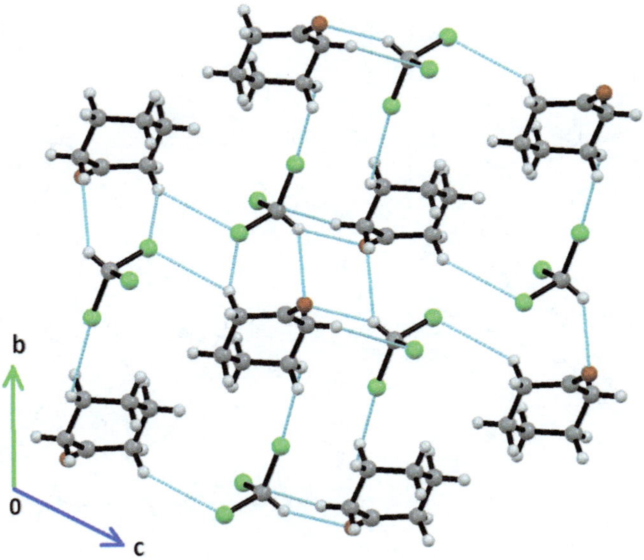

Figure 3.21 Centrosymmetric dimeric C–H···O HBs between dichloro methane and cyclic hexanone.

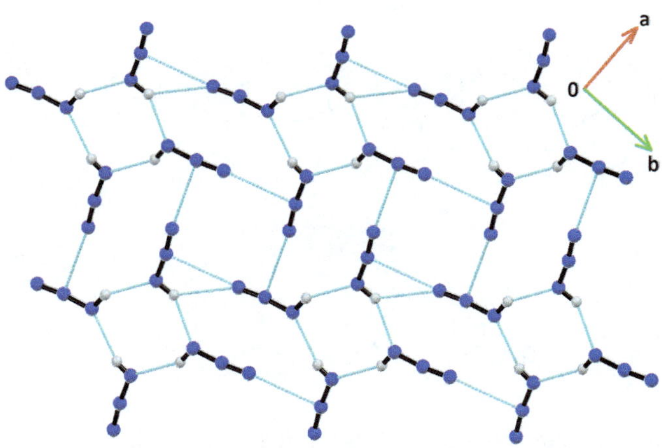

Figure 3.22 Hydrazoic acid (HN_3) forms a network of tetramers through N–H···N HBs.

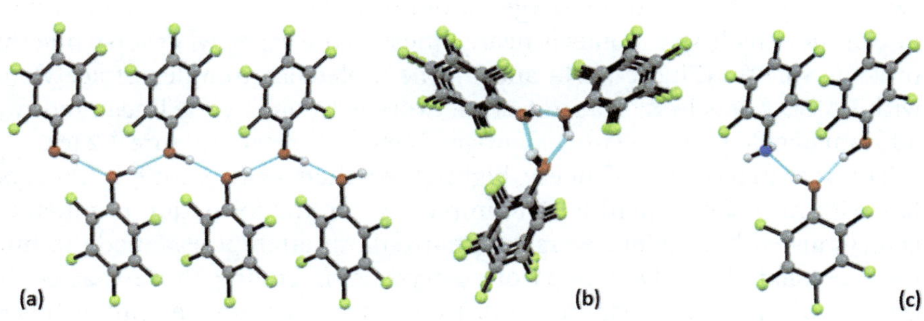

Figure 3.23 Formation of (a) a O–H···O chain along the *b*-direction in form I. (b) A helical O–H···O chain long the *a*-direction in form II of petafluorophenol. (c) 2 : 1 co-crystal of petafluorophenol and petafluoroaniline.

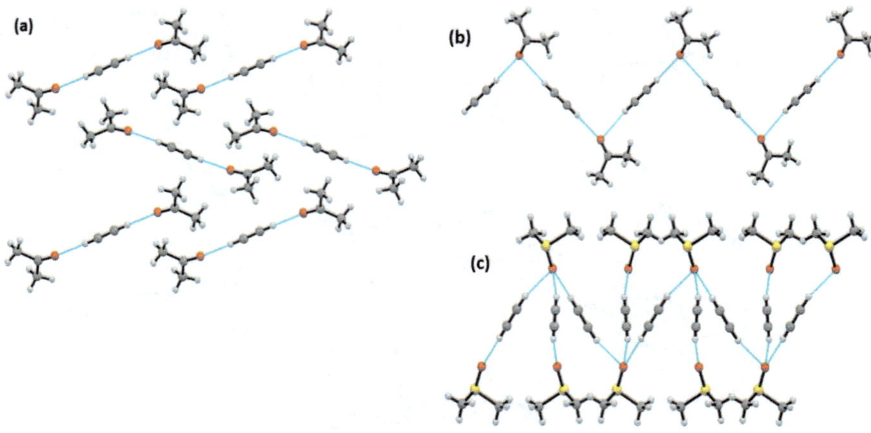

Figure 3.24 Co-crystals of (a) acetylene : acetone (1 : 2) α phase, (b) acetylene : acetone (1 : 1) β phase view of the (100) plane and (c) co-crystal of acetylene : dimethyl sulfoxide (2 : 1) view of the (101) plane with C≡C–H···O hydrogen bonds.

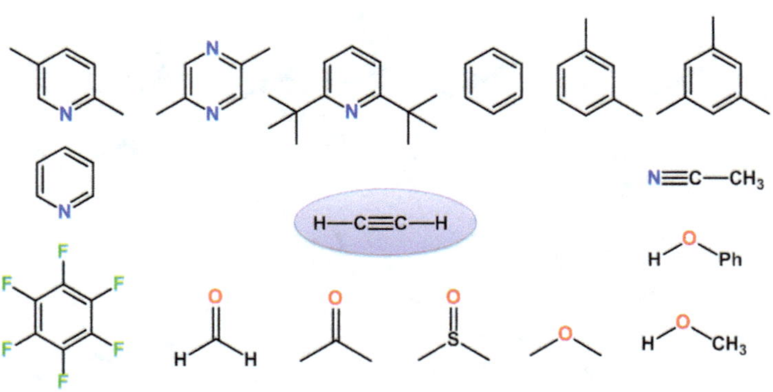

Figure 3.25 Scheme of the co-crystals formed with acetylene.

arranged in a herringbone fashion. However, in co-crystals of acetylene and acetone in 1 : 1 (β phase), the molecules are arranged in a zigzag chain with two hydrogen bonds to the oxygen atom of each acetone molecule, where one of these hydrogen bonds is nearly linear to the carbonyl group while the other is side-on (Figure 3.24a and b) The molecular complex of acetylene with DMSO exists in 2 : 1 ratio. A 2D network with one nearly linear and two additional side-on C≡C–H···O hydrogen bonds is present (Figure 3.24c).

The research group of Boese highlighted their work entitled "Co-crystals with Acetylene: Small Is Not Simple", where the formation of nineteen co-crystals with acetylene were summarized. An attempt was made to find features shared by the groups of co-crystals, including those that could not be co-crystallized (Figure 3.25). However, their intuitive approach was

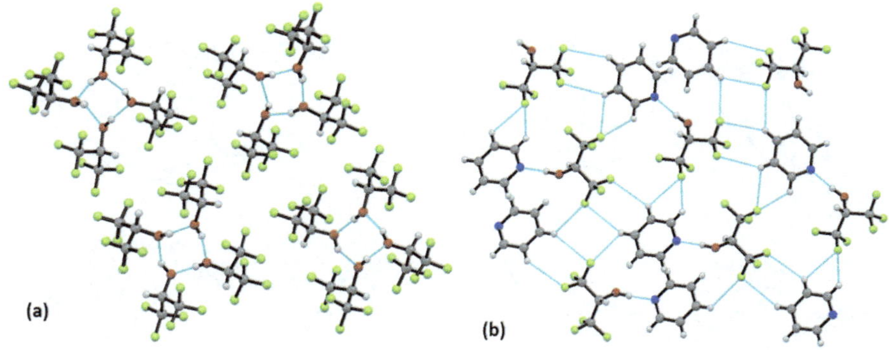

Figure 3.26 Crystal packing of (a) HFIP *via* a O–H⋯O chain, (b) HFIP-PY *via* O–H⋯N hydrogen bonds and C–HF interactions.

not fulfilled, which demonstrates that the prediction of the composition of multi-component crystals for even small molecules still remains a great challenge.[27,28]

Merz *et al.* demonstrated the *in situ* cryocrystallization of 1,1,1,3,3,3-hexafluoro-2-propanol (HFIP).[36] They highlighted the formation of the new polymorphic form II of HFIP, which is stabilized *via* O–H⋯O hydrogen bonds (Figure 3.26a). The two polymorphic forms of HFIP differ in terms of the number of molecules in the asymmetric unit (form I; $Z' = 8$, form II; $Z' = 4$) and hydrogen bonds. Figure 3.26(b) shows the formation of the co-crystal of HFIP with pyridine *via* O–H⋯N hydrogen bond. It was proposed that repulsive interactions between the $-CF_3$ groups and the neighbouring pyridine molecules separate each other. However the attractive interactions between HFIP and the pyridine molecules *via* O–H⋯N, indicate hydrogen bonds with a lower energy value than in forms I and II of HFIP. In HFIP : pyridine co-crystals, the observed intermolecular interactions are in the form of short O–H⋯N hydrogen bonds. In addition, repulsion between the $-CF_3$ groups, and the presence of π⋯π interactions between the pyridine rings, contribute to a staggered arrangement of the zigzag motif in the crystal packing.

This technique has been also utilized to study crystal structures of 1 : 1 co-crystals of *o*-cresol with aniline, *o*-fluoroaniline and *m*-fluoroaniline, and also *m*-cresol with *p*-fluoroaniline, which are liquid at ambient conditions.[37] Co-crystals depict the formation of different intermolecular interactions involving strong O–H⋯N, N–H⋯O HBs along with weak C–H⋯O, C–H⋯π, and C–H⋯F interactions (Figure 3.27).

3.3.4 Computational Analysis

The method of *in situ* cryocrystallography of liquids can be a useful technique to determine the molecular structure as well as to study the arrangement of molecules in the crystal. Once the crystal structure of a liquid compound is

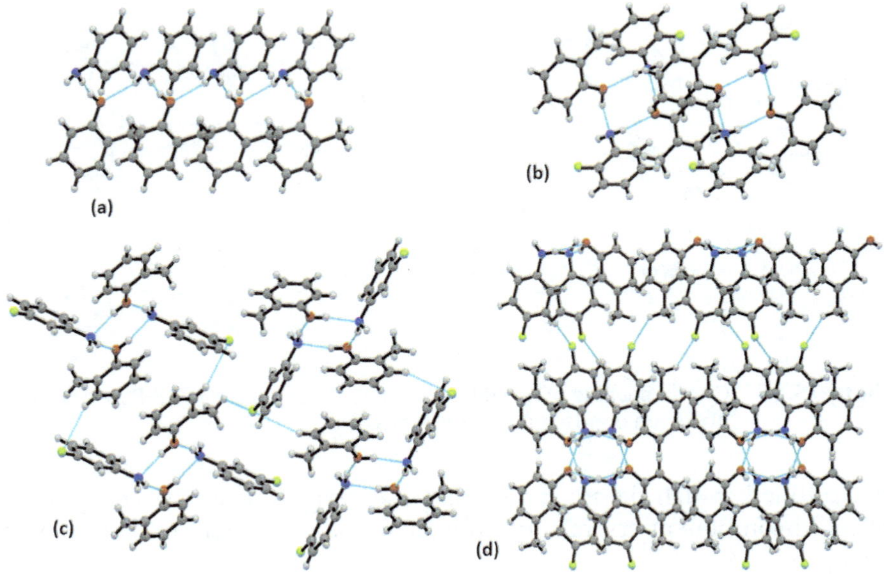

Figure 3.27 (a) Packing of the 1:1 co-crystal of *o*-cresol and aniline *via* H⋯H contact and strong N–H⋯O chains. (b), (c) The molecules of *o*-cresol with *o*-fluoroaniline and *m*-fluoroaniline form tetramer units involving O–H⋯N and N–H⋯O hydrogen bonds. (d) C–H⋯F hydrogen bonds of *p*-fluoroaniline and *m*-cresol co-crystal.

obtained, there are many computational tools that can be utilized to understand the nature and energetics of the various strong/weak intermolecular interactions involving different types of donors and acceptors. The applicability of these computational methods is as follows.

3.3.4.1 *Determination of Intermolecular Interaction Energy Using the* PIXEL *Method*

To understand the process of aggregation of molecules in cryocrystallized liquids, intermolecular interactions energies of the molecular dimers are determined using the *PIXEL* method (version 12.5.2014).[38,39] The *PIXEL* method is based on numerical integration over a large number of quantum chemical electron densities pixels. It is a semi-empirical method where a molecule is represented by *ab initio* free-molecular electron density. The electron densities are calculated using the Gaussian 09[40] program at MP2/6-31G** level to generate the *PIXEL* input file. The method partitions the total crystal energy into molecule⋯molecule contributions over the molecular pairs in crystals. The total interaction energy as well as the total lattice energy is separated into four energy components; Coulombic, polarization, dispersion and repulsion energies between the molecules.

$$E_{\text{total}}\,(\text{lattice energy}) = E_{\text{coul}} + E_{\text{pol}} + E_{\text{disp}} + E_{\text{rep}}.$$

3.3.4.2 *Topological Analysis for Intermolecular Interactions Using QTAIM*

To get more insight into the intra- and interatomic contacts, a detailed QTAIM analysis[41] is performed using AIMALL (version 13.05.06)[42] software to obtain the topological parameters at the BCP. The topological features and energetic parameters such as the electron densities (ρ), Laplacian ($\nabla^2\rho$), local potential energy (V_b), and kinetic potential energy (G_b) at the bond critical points have been evaluated. This methodology is developed by Bader[41] and permits the examination of interatomic interactions through a topological analysis of the electron density distribution $\rho(r)$. The Laplacian of the electron density determines whether the electron density is concentrated ($\nabla^2\rho_{BCP} < 0$) or depleted ($\nabla^2\rho_{BCP} > 0$). For a shared interaction, electron density is accumulated between the nuclei and concentrated along the bond path so that ρ_{BCP} is large and $\nabla^2\rho_{BCP} < 0$.

3.3.4.3 *Non-covalent Interaction (NCI) Index and RDG Isosurface*

The non-covalent interaction (NCI) index,[43,44] introduced by Johnson and co-workers, is a method to study the nature of two interacting atoms or molecular fragments based on properties of the electron density. This analysis has been performed using the program NCImilano.[45] The NCI index provides a real-space graphical visualization around the critical point (BCP, RCP or CCP) where non-covalent interactions occur for both attractive (van der Waals and hydrogen bonding) and repulsive (steric) interactions based on the properties of the electron density. This method is based on the electron density distribution, in particular electron density (ρ_{BCP}), reduced density gradient (RDG) of the electron density (s) and the Laplacian of the electron density ($\nabla^2\rho_{BCP}$). Out of three eigenvalues, the second derivative λ_2 provides interesting information as to whether the interaction is bonding ($\lambda_2 < 0$) or non-bonding ($\lambda_2 > 0$). In the 2D plot, the reduced density gradient (RDG) is plotted with respect to the electron density as well as the quantity, "sign (λ_2) ρ", which shows the change in the reduced density gradient between the two interacting atoms. Generally the RDG isosurface appears as a trough between two interacting atoms. In certain cases, the attractive and the repulsive region can both appear simultaneously in the 3D isosurface. The magnitude of the electron density as well as the colour of the trough indicates the strength of the interaction.

3.3.4.4 *Analysis of Distributed Atomic Polarizability Tensor Using* Polaber

The distributed atomic polarizabilities for a given system are calculated using the program *PolaBer* developed by Macchi and co-workers, where the changes in polarizability ellipsoids can be determined.[46] The significance of the different inter/intramolecular interactions can be obtained *via* detailed analysis of the distributed atomic polarizability and a comparison with their isolated counterparts.[47] From this study, it is possible to understand how the atomic polarizability of the hydrogen atom is influenced when in contact with electron rich species in a strong hydrogen bond.[48] In *PolaBer*, the atomic polarizability tensors are visualized as an ellipsoid with the main axis having units of volume. In addition, the axes are a measure of the anisotropy of the atomic electron density, therefore, polarizability tensors are visualized in the space as the molecule assuming 1 Å^3 ≡ 1 Å, and in addition a scaling factor is used for better visualization of the polarizability ellipsoid.[49] The distributed atomic polarizability is calculated using the QTAIM based topological partitioning of the electron density at the isolated geometry, extracted from the crystal structure.

3.3.5 *In situ* Cryocrystallization Study in Fluorinated Benzoyl Chlorides

Recently, the effects of fluorination on intermolecular interactions and the different interaction motifs have been analysed in *in situ* cryocrystallized benzoyl chloride and their fluorinated analogues (Figure 3.28), which are liquids at ambient conditions.[50,51] The target was to determine the crystal structures of these liquids and see the effect of chemical modification (replacement of F/H on the phenyl ring) on the subsequent molecular assembly in the solid state.

For this purpose, differential scanning calorimetry on these liquids was performed to obtain the freezing point and melting point of the crystallized samples. The nature of the curve also reflects the crystal quality of the obtained cryocrystallized liquids. From DSC experiments, it was observed that the freezing points and the melting points of these liquids are in the range of −37.5 °C to −6.3 °C and −23.4 °C to 10.9 °C respectively. The DSC experimental data provide inputs towards the design of the experimental strategy, as regards the temperature during which the crystallization occurs. The process of initial solidification will result in the formation of many small crystals that are not suitable for single-crystal X-ray diffraction. For further improvements in the quality of the obtained crystals (Figure 3.29), the OHCD technique was performed.

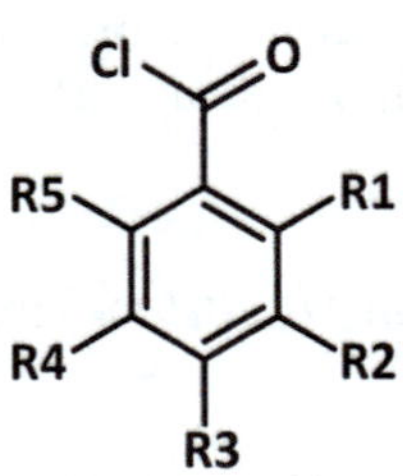

BC: R1 = R2 = R3 = R4 = R5 = H
2FBC: R1 = F, R2 = R3 = R4 = R5 = H
3FBC: R2 = F, R1 = R3 = R4 = R5 = H
4FBC: R3 = F, R1 = R2 = R4 = R5 = H
23DFBC: R1 = R2 = F, R3 = R4 = R5 = H
24DFBC: R1 = R3 = F, R2 = R4 = R5 = H
25DFBC: R1 = R4 = F, R2 = R3 = R5 = H
26DFBC: R1 = R5 = F, R2 = R3 = R4 = H
35DFBC: R2 = R4 = F, R1 = R3 = R5 = H
345TFBC: R2 = R3 = R4 = F, R1 = R5 = H

Figure 3.28 Molecular scheme of fluorinated benzoyl chlorides.

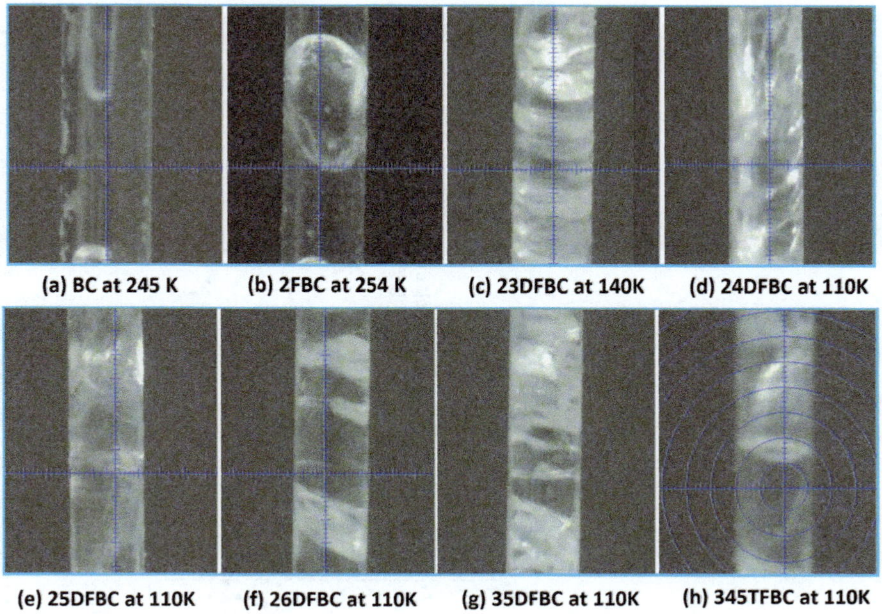

Figure 3.29 (a)–(h) Crystal images of *in situ* cryocrystallized liquids at the corresponding temperatures.

In this study, we observed the parallel alignment of C–F and O=C; C–Cl and C–H bond dipoles in the crystal structures of fluorinated benzoyl chlorides, primarily when the fluorine atom is at the *ortho* position, except in case of 26DFBC. The crystal structure of the parent compound (BC) also enables us to compare the electronic features in the absence of the C–F bond. The most notable feature was the observation of a short intramolecular contact between the fluorine and the oxygen atom, which was observed in the case of *ortho* fluoro substituted benzoyl chloride [$d_{F\cdots O}$ = 2.637 (2)Å in 2FBC] and related compounds in the presence of an intramolecular short C–H···Cl contact. To investigate the nature of these contacts, a detailed investigation of the planar geometry was performed (Figure 3.30). Between the crystal structures of BC (no fluorine substitution), 2FBC (fluorine substitution at *ortho* position) and 26DFBC (fluorine substitution at both the *ortho* positions), it was observed that the crystal geometry of BC (3.5°) and 2FBC (5.0°) is almost planar, whereas in case of 26DFBC, the dihedral angle between the –COCl group and the plane of the phenyl ring was 44.1°. After geometry optimization, BC and 2FBC becomes planar but in 26DFBC the –COCl group is more relaxed from the ring plane (Figure 3.30a). Furthermore, the plot of relative potential energy *vs.* dihedral angle around the C–C single bond supports the role of intramolecular F···O contact, which facilitates the conformation lock in the planar conformation (Figure 3.30b).

The existence of the intramolecular F···O contact in cryocrystallized 2FBC was further supported *via* detailed inputs from the QTAIM, NCI index and

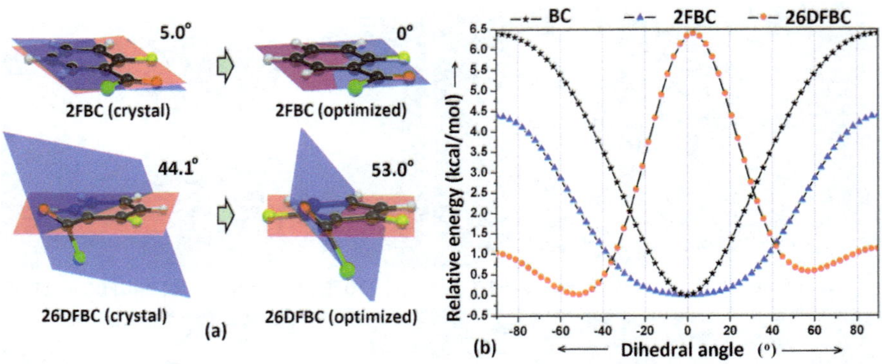

Figure 3.30 (a) Molecular conformation in crystal geometry and gas phase. (b) Plot of relative potential energy *vs.* dihedral angle in BC, 2FBC and 26DFBC. Reproduced from ref. 50 with permission from the Royal Society of Chemistry.

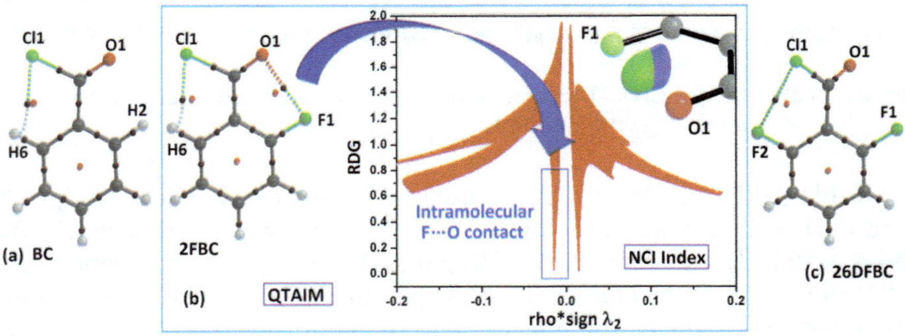

Figure 3.31 Molecular graphs for (a) BC, (b) 2FBC and (c) 26DFBC where deep red and red colours depict the bond critical points and ring critical points respectively. The RDG isosurface and the plot of RDG *vs.* ρ*sign (λ_2) for the intramolecular F···O contact obtained from NCI analysis in 2FBC. Reproduced from ref. 50 with permission from the Royal Society of Chemistry.

NICS[52,53] methods. Figure 3.31 shows the molecular graphs for BC, 2FBC and 26DFBC obtained from AIMALL analysis. The bond critical points for intramolecular F···O contact were observed only for 2FBC, 23DFBC, 24DFBC and 25DFBC. The topological parameters at the BCP for intramolecular F···O in 2FBC are R_{ij} = 2.643 Å, ρ_{BCP} = 0.101 e Å^{-3} and $\nabla^2\rho_{BCP}$ = 0.1.554 e Å^{-5}. As the accumulation or depletion of the electron density in the plane perpendicular to the interaction is characterized by the second eigenvalue λ_2, the RDG isosurface obtained between two interacting atoms F and O is a combination of the green region, which indicates stabilized F···O contact, and the blue region, which specifies the region (destabilized) around the RCP with positive λ_2 value. From the RDG plot, it was understood that the dispersive nature of the

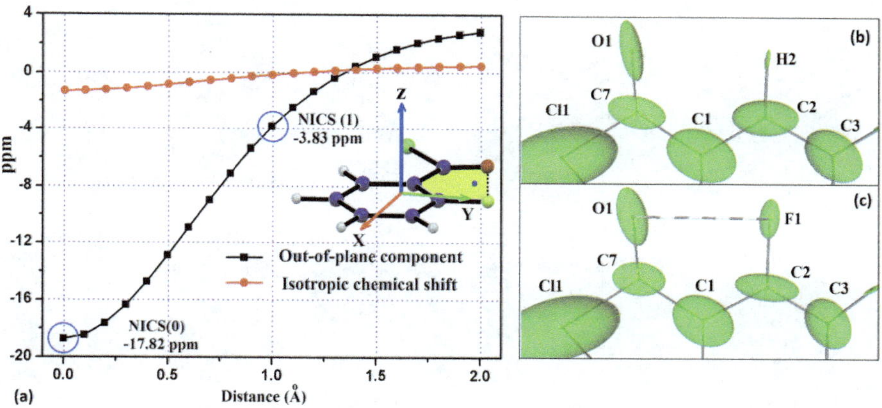

Figure 3.32 (a) NICS values as a function of distance (Å) of the five-membered ring associated with intramolecular F···O contact in 2FBC. Distributed atomic polarizability for (b) BC and (c) 2FBC in which the dashed line indicates the intramolecular F···O contact. Reproduced from ref. 50 with permission from the Royal Society of Chemistry.

intramolecular F···O contact is responsible for the stability of the molecular conformation. Furthermore, the NICS analysis also supports the diatropic ring current present in the five-membered ring associated with intramolecular F···O contact (Figure 3.32a). The negative value (−3.83 ppm) of $NICS(1)_{zz}$ at 1 Å above from the RCP indicates the π-electron delocalization, which confirms the presence of aromaticity due to the formation of intramolecular F···O contact. The distributed atomic polarizability analysis for benzoyl chloride and 2-fluorobenzoyl chloride gives important information regarding the formation of intramolecular F···O contact. The inclusion of fluorine atom at the *ortho* position of the phenyl ring shows the substantial change (increment of polarizability tensor towards the direction of fluorine atom) in the polarizability tensors (Figure 3.32b and c) as well as the volume of the polarizability ellipsoids for the oxygen atom in comparison of BC with 2FBC. This is due to the involvement of the oxygen atom of 2FBC in the formation of an intramolecular F···O contact.

The role of weak intermolecular interactions, which play a crucial role in stabilizing the overall crystal packing in the absence of any strong interaction, was also analysed. The supramolecular construct associated with different intermolecular interactions in isomeric benzoyl chlorides depicts the mechanism by which the molecules assemble the formation of the final crystal structure for these molecules. From the overall crystal packing analysis (Figure 3.33) it is observed that there are two different sets of molecular arrangements. First one was due to formation alternate parallel column where zig-zag layers are stacked on each other with π···π stacking in the case of BC, 2FBC, 3FBC, 24DFBC, 26DFBC and 345TFBC. The second one is due to the formation of planar molecular sheets involving weak intermolecular C–H···O/F/Cl and halogen···halogen interactions (like Cl···F, F···F, Cl···Cl) in

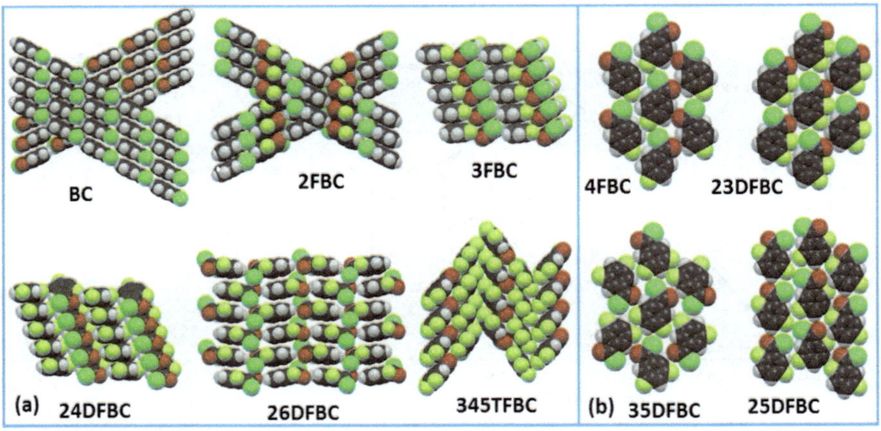

Figure 3.33 (a) Formation of zig-zag molecular sheets *via* weak C–H···O/F/Cl interactions, stacked on each other. (b) Orientation of the molecules in the planar molecular sheets. Reproduced from ref. 51 with permission from the PCCP Owner Societies.

the case of 4FBC, 23DFBC, 35DFBC and 25DFBC. Finally, such types of sheets that are parallel to each other are stacked *via* π···π stacking with a perpendicular interlayer distance of ~3.2 Å.

From the crystal packing of benzoyl chloride (BC) and fluorinated BCs, it was observed that there exist common building blocks, as shown in Figure 3.33. The intermolecular interaction energies obtained from the *PIXEL* method for these motifs are represented as a range of energy with the modification of the fluorine position on the phenyl ring. Some motifs, namely M1, M6, M9 and M10 are present in most of the structures (Figure 3.34). For example, in the case of M1, which is associated with C–H···O and C–H···Cl interactions, the electrostatic and dispersion contributions are ~55% and ~45% respectively for all the equivalent molecular pairs.

3.3.6 *In situ* Cryocrystallization in Organometallic Liquids

Recently, *in situ* cryocrystallization was also performed on organometallic liquid compounds,[54] namely (η^4-1, 3 cyclohexadiene) iron tricarbonyl (complex I) and methyl cyclopentadienyl manganese tricarbonyl (complex II) using the OHCD technique. After 7–8 h of CO_2 laser scans at a temperature of 258 K for complex I and 240 K for complex II under liquid nitrogen flow, good quality single crystals were obtained. The diffraction data were collected at 110 K temperature. Figure 3.35 depicts the molecular scheme as well as the still images of the cryocrystallized organometallic liquids.

The electronic features associated with the presence of both covalent bonds (between the metal and the ligand) and non-covalent interactions were analysed with quantitative inputs from charge density analysis. The packing networks of both the metal carbonyl complexes are shown in

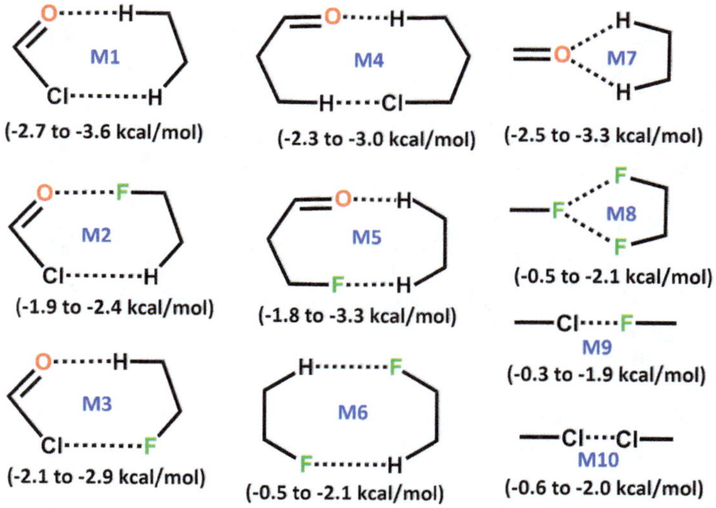

Figure 3.34 Energy of the equivalent molecular motifs (M) present in fluorinated benzoyl chlorides. Reproduced from ref. 51 with permission from the PCCP Owner Societies.

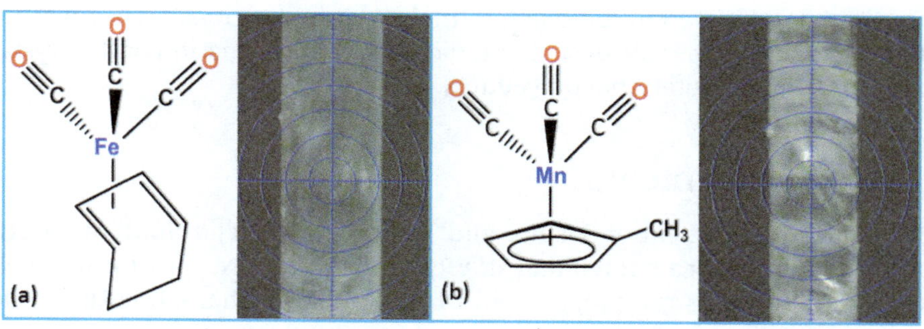

Figure 3.35 Crystal images for the *in situ* cryocrystallized (a) complex I and (b) complex II at 110 K.

Figure 3.36. In case of complex I, there is a formation of two different types of centrosymmetric C–H···O dimers, which form together a molecular chain. Such two parallel molecular chains are interlinked *via* weak C–H···π interactions, whereas in complex II, similar types of C–H···O dimeric chains are connected *via* C–H···O and lp(O=C)···π(C) interactions.

3.4 Overview

In future, the well-established method of *in situ* cryocrystallization can be applied for the crystallization of different two-component systems (non-reacting liquid compounds of organic or organometallic origin), which can form co-crystals, including application to potent drugs that are liquids at

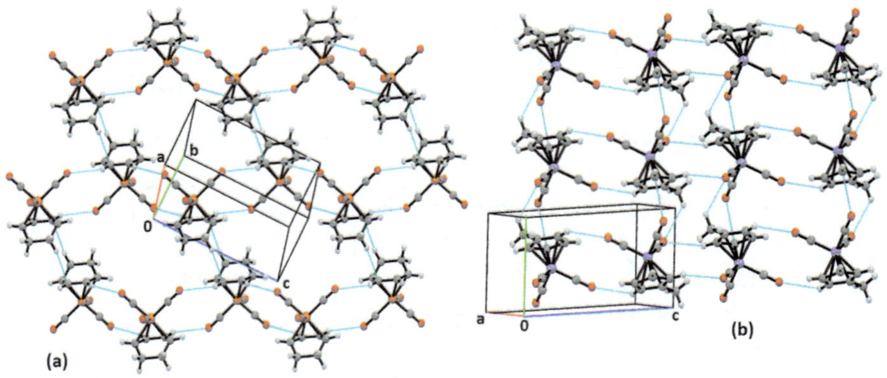

Figure 3.36 Molecular network in (a) complex I down the (2-22) plane (b) in complex II down the (101) plane.

room temperature. There also exists the possibility to study the formation of polymorphs and solvatomorphs in *in situ* cryocrystallized liquids *via* alteration of the conditions of crystallization. Thus, this technique, in addition to the determination of the 3D molecular structure in the solid state, also enables us to understand the process of molecular aggregation, *via* the involvement of different, unique intermolecular interactions, with quantitative insights into the role of such interactions in the formation of the crystal *via* different computational procedures.

Acknowledgements

SK thanks Department of Science and Technology (DST) & SERB and CSIR for funding for our research and IISER, Bhopal, VNIT, Nagpur for research facilities and infrastructure. DC and DD would like to thank IISERB for the research facilities and infrastructure, and DST-SERB for research funding.

References

1. R. Boese, *Z. Kristallogr.*, 2014, **229**, 595–601.
2. D. Chopra and T. N. G. Row, *J. Indian Inst. Sci.*, 2007, **87**, 167–211 and all references therein.
3. D. Dey and D. Chopra, *Resonance*, 2014, **19**, 1104–1114.
4. J. D. Dunitz, *Mol. Cryst. Liq. Cryst.*, 1996, **279**, 209.
5. W. H. Keesom and K. W. Taconis, *Physica*, 1935, **2**, 463–471.
6. (a) M. A. Viswamitra, *J. Sci. Inst.*, 1962, **39**, 381; (b) M. A. Viswamitra and K. K. Kannan, *J. Sci. Inst.*, 1962, **39**, 318; (c) M. A. Viswamitra and K. K. Kannan, *Nature*, 1966, **209**, 1016; (d) R. Thomas, C. Brink-Shoemaker and K. Eriks, *Acta Crystallogr.*, 1966, **21**, 12.
7. (a) O. Hassel and J. Hvoslef, *Acta Chem. Scand.*, 1954, **8**, 873; (b) O. Hassel and K. O. Stromme, *Acta Chem. Scand.*, 1958, **12**, 1146; (c) O. Hassel and

K. O. Stromme, *Acta Chem. Scand.*, 1959, **13**, 1781–1786; (d) T. Bjorvatten and O. Hassel, *Acta Chem. Scand.*, 1962, **16**, 249–255; (e) O. Hassel and K. O. Stromme, *Acta Chem. Scand.*, 1959, **13**, 1775–1780.

8. (a) J. E. Davis and A. D. Bond, *Acta Crystallogr.*, 2002, **E58**, o961; (b) J. E. Davis and A. D. Bond, *Acta Crystallogr.*, 2002, **E58**, o326; (c) J. E. Davis and A. D. Bond, *Acta Crystallogr.*, 2002, **E58**, o331; (d) J. E. Davis and A. D. Bond, *Acta Crystallogr.*, 2001, **E57**, o1039.
9. J. A. K. Howard and M. R. Probert, *Science*, 2014, **343**, 1098.
10. (a) R. Boese and M. Nussbaumer, in *Organic Crystal Chemistry*, ed. J. B. Garbarczyk and D. W. Jones, Oxford University Press, Oxford, 1994, pp. 20–37; (b) http://www.sci-ohcd.eu/.
11. (a) D. Giron, *Thermochim. Acta*, 1995, **248**, 1–59; (b) P. Atkins and J. d. Paula, *Physical Chemistry*, Oxford University Press, 9th edn, 2010.
12. D. Chopra, T. N. Guru Row, E. Arunan and R. A. Klein, *J. Mol. Struct.*, 2010, **964**, 126–133.
13. A. Sharma, I. Reva, R. Fausto, S. Hesse, Z. Xue, M. A Suhm, S. K. Nayak, R. Sathishkumar, R. Pal and T. N. Guru Row, *J. Am. Chem. Soc.*, 2011, **133**, 20194–20207.
14. (a) D. R. Allan, S. J. Clark, A. Dawson, P. A. McGregor and S. Parsons, *Acta Crystallogr.*, 2002, **B58**, 1018–1024; (b) I. D. H. Oswald, D. R. Allan, W. D. S. Motherwell and S. Parsons, *Acta Crystallogr.*, 2005, **B61**, 69–79; (c) I. D. H. Oswald, D. R. Allan, G. M. Day, W. D. S. Motherwell and S. Parsons, *Cryst. Growth Des.*, 2005, **5**, 1055–1071.
15. G. R. Desiraju, P. S. Ho, L. Kloo, A. C. Legon, R. Marquardt, P. Metrangolo, P. Politzer, G. Resnati and K. Rissanen, *Pure Appl. Chem.*, 2013, **85**, 1711–1713.
16. D. Chopra, V. Thiruvenkatam and T. N. Guru Row, *Cryst. Growth Des.*, 2006, **6**, 843–845.
17. S. K. Nayak, M. K. Reddy, T. N. Guru Row and D. Chopra, *Cryst. Growth Des.*, 2011, **11**, 1578–1596.
18. D. Chopra, V. Thiruvenkatam, S. G. Manjunath and T. N. Guru Row, *Cryst. Growth Des.*, 2007, **7**, 868–874.
19. S. K. Nayak, G. Terraneo, A. Forni, P. Metrangolo and G. Resnati, *CrystEngComm*, 2012, **14**, 4259–4261.
20. F. Bertolotti, G. Cavallo, P. Metrangolo, S. K. Nayak, G. Resnati and G. Terraneo, *Supramol. Chem.*, 2013, **25**, 718–727.
21. V. Vasylyeva, S. K. Nayak, G. Terraneo, G. Cavallo, P. Metrangolo and G. Resnati, *CrystEngComm*, 2014, **16**, 8102–8105.
22. T. Clark, M. Hennemann, J. S. Murray and P. Politzer, *J. Mol. Model.*, 2007, **13**, 291–296.
23. V. Vasylyeva and K. Merz, *Z. Kristallogr.*, 2014, **229**, 603–608.
24. A. R. Choudhury, K. Islam, M. T. Kirchner, G. Mehta and T. N. Guru Row, *J. Am. Chem. Soc.*, 2004, **126**, 12274–12275.
25. J. D. Dunitz and J. Bernstein, *Acc. Chem. Res.*, 1995, **28**, 193–200.
26. A. A. Yakovenko, M. Y. Antipin and T. V. Timofeeva, *Cryst. Growth Des.*, 2009, **9**, 57–65.

27. (a) M. T. Kirchner, R. Boese, W. E. Billups and L. R. Norman, *J. Am. Chem. Soc.*, 2004, **126**, 9407–9412; (b) E. D. Sloan, *Clathrate Hydrates of Natural Gases*, Dekker, New York, 1997; (c) G. A. Jeffrey and R. K. McMullan, *Prog. Inorg. Chem.*, 1967, **8**, 43–108.
28. M. T. Kirchner, D. Bläser and R. Boese, *Chem.-Eur. J.*, 2010, **16**, 2131–2146.
29. A. R. Choudhury, N. Winterton, A. Steiner, A. I. Cooper and K. A. Johnson, *J. Am. Chem. Soc.*, 2005, **127**, 16792–16793.
30. A. R. Choudhury, N. Winterton, A. Steiner, A. I. Cooper and K. A. Johnson, *CrystEngComm*, 2006, **8**, 742–745.
31. D. S. Yufit and J. A. K. Howard, *CrystEngComm*, 2012, **14**, 2003–2008.
32. D. S. Yufit and J. A. K. Howard, *CrystEngComm*, 2010, **12**, 737–741.
33. J. Evers, M. Göbel, B. Krumm, F. Martin, S. Medvedyev, G. Oehlinger, F. X. Steemann, I. Troyan, T. M. Klapötke and M. I. Eremets, *J. Am. Chem. Soc.*, 2011, **133**, 12100–12105.
34. (a) D. Das, R. Banerjee, R. Mondal, J. A. K. Howard, R. Boese and G. R. Desiraju, *Chem. Commun.*, 2006, 555–557; (b) M. Gdaniec, *CrystEngComm*, 2007, **9**, 286–288; (c) M. T. Kirchner, D. Bläser, R. Boese and G. R. Desiraju, *CrystEngComm*, 2009, **11**, 229–231.
35. R. Boese, M. T. Kirchner, W. E. Billups and L. R. Norman, *Angew. Chem., Int. Ed.*, 2003, **42**, 1961–1963.
36. G. Lipinski, B. Mallik and K. Merz, *Z. Anorg. Allg. Chem.*, 2017, **643**, 683–688.
37. A. R. Choudhury, D. S. Yufit and J. A. K. Howard, *Z. Kristallogr.*, 2014, **229**, 625–634.
38. J. D. Dunitz and A. Gavezzotti, *Chem. Soc. Rev.*, 2009, **38**, 2622–2633.
39. L. Maschio, B. Civalleri, P. Ugliengo and A. Gavezzotti, *J. Phys. Chem. A*, 2011, **115**, 11179–11186.
40. M. J. Frisch, G. W. Trucks, H. B. Schlegel, G. E. Scuseria, M. A. Robb, J. R. Cheeseman, G. Scalmani, V. Barone, B. Mennucci, G. A. Petersson, H. Nakatsuji, M. Caricato, X. Li, H. P. Hratchian, A. F. Izmaylov, J. Bloino, G. Zheng, J. L. Sonnenberg, M. Hada, M. Ehara, K. Toyota, R. Fukuda, J. M. HasegawaIshida, T. Nakajima, Y. Honda, O. Kitao, H. Nakai, T. Vreven, J. A. Montgomery Jr, J. E. Peralta, F. Ogliaro, M. Bearpark, J. J. Heyd, E. Brothers, K. N. Kudin, V. N. Staroverov, R. Kobayashi, J. Normand, K. Raghavachari, A. Rendell, J. C. Burant, S. S. Iyengar, J. Tomasi, M. Cossi, N. Rega, J. M. Millam, M. Klene, J. E. Knox, J. B. Cross, V. Bakken, C. Adamo, J. Jaramillo, R. Gomperts, R. E. Stratmann, O. Yazyev, A. J. Austin, R. Cammi, C. Pomelli, J. W. Ochterski, R. L. Martin, K. Morokuma, V. G. Zakrzewski, G. A. Voth, P. Salvador, J. J. Dannenberg, S. Dapprich, A. D. Daniels, Ö. Farkas, J. B. Foresman, J. V. Ortiz, J. Cioslowski and D. J. Fox, *Gaussian 09, Revision D.01*, Gaussian, Inc., Wallingford, CT, 2009.
41. R. F. W. Bader, *Atoms in Molecules: A Quantum Theory*, Oxford University Press, Oxford, U.K., 1990.
42. T. A. Keith, *AIMALL, Version 13.05.06*, TK Gristmill Software, Overland Park KS, 2013, aim.tkgristmill.com.

43. E. R. Johnson, S. Keinan, P. Mori-Sanchez, J. Contreras-Garcia, A. J. Cohen and W. Yang, *J. Am. Chem. Soc.*, 2010, **132**, 6498–6506.
44. J. Contreras-Garcia, E. R. Johnson, S. Keinan, R. Chaudret, J.-P. Piquemal, D. N. Beratan and W. Yang, *J. Chem. Theory Comput.*, 2011, **7**, 625–632.
45. G. Saleh, L. L. Presti, C. Gatti and D. Ceresoli, *J. Appl. Crystallogr.*, 2013, **46**, 1513–1517.
46. A. Krawczuk, D. Perez and P. Macchi, *J. Appl. Crystallogr.*, 2014, **47**, 1452–1458.
47. A. Krawczuk and P. Macchi, *Chem. Cent. J.*, 2014, **8**, 68.
48. A. S. Chimpri, M. Gryl, L. H. R. Santos, A. Krawczuk and P. Macchi, *Cryst. Growth Des.*, 2013, **13**, 2995–3010.
49. A. Krawczuk-Pantula, D. Perez, K. Stadnicka and P. Macchi, *Trans. Am. Crystallogr. Assoc.*, 2011, **42**, 1–25.
50. D. Dey, S. Bhandary, A. Sirohiwal, V. R. Hathwar and D. Chopra, *Chem. Commun.*, 2016, **52**, 7225–7228.
51. D. Dey, S. Bhandary, S. P. Thomas, M. A. Spackman and D. Chopra, *Phys. Chem. Chem. Phys.*, 2016, **18**, 31811–31820.
52. Z. Chen, C. S. Wannere, C. Corminboeuf, R. Puchta and P. V. R. Schleyer, *Chem. Rev.*, 2005, **105**, 3842–3888.
53. P. V. R. Schleyer, C. Maerker, A. Dransfeld, H. Jiao and N. J. R. V. E. Hommes, *J. Am. Chem. Soc.*, 1996, **118**, 6317–6318.
54. A. Sirohiwal, V. R. Hathwar, D. Dey and D. Chopra, *ChemPhysChem*, 2017, **18**, 2859–2863.

CHAPTER 4

Experimental Electron Density Studies of Inorganic Solids

SWASTIK MONDAL

CSIR-Central Glass and Ceramic Research Institute,
Jadavpur, Kolkata 700032, India
*E-mail: swastik_mondal@cgcri.res.in

4.1 Introduction

Electron density studies of crystalline materials using X-ray diffraction data have been well established by the development of modern methods of aspherical electron density analysis.[1,2] Using these methods, and with the help of modern diffractometers and high intensity X-ray sources, scientists are now able to reconstruct minute details of the electron density distribution in crystalline materials. Based on the electron density distribution obtained from an aspherical electron density model, like an atom centred multipole (MP) model, a substantial amount of information on interatomic interactions, and chemical and physical properties of crystalline materials can be achieved. Interpretation of static electron densities with the aid of Bader's *quantum theory of atoms in molecules* (QTAIM)[3] has now become the standard method for electron density analysis. At this point, one should notice that the success of modern electron density studies is mainly limited within the realm of organic and metal-organic molecular materials. Electron density analysis of inorganic materials, especially for those that contains heavy elements, is yet far from being routine. Apart from the fact that, historically, methods of

Monographs in Supramolecular Chemistry No. 26
Understanding Intermolecular Interactions in the Solid State: Approaches and Techniques
Edited by Deepak Chopra

Published by the Royal Society of Chemistry, www.rsc.org

electron density studies were mainly developed for molecular materials with light elements, there are several practical reasons why electron density studies of inorganic materials with heavy elements are not trivial. The present chapter briefly reviews the challenges, experimental techniques and procedures for electron density studies of inorganic solids using MP formalism and discusses a few reported case studies.

4.2 Methods for Electron Density Studies

The possibility of observing electron density in crystals using X-ray diffraction techniques has been known since the first half of the 20th century. Direct evaluation of electron densities from an X-ray diffraction experiment was however not possible, due to limitations in experimental measurements. These limitations gave rise to methods for modelling electron densities against X-ray diffraction data. The conventional method according to the (spherical) independent atom model (IAM) tries to model static electron densities and thermal vibrations in terms of atomic coordinates and atomic displacement parameters fitted against experimental data. In the IAM, the spherically averaged electron density of an individual atom in a crystal is considered to be isolated from other atoms present in that crystal. Hence, though the IAM is a tremendously successful method for structural analysis of crystalline materials, it does not provide information on the deformation of the electron density due to chemical bonding and other atomic interactions. Since atomic interactions play a key role in shaping various physical and chemical properties of materials, it is important to determine and characterize the electron density distribution and the deformation of electron density due to various atomic interactions in materials.

Information on the electron density deformation due to atomic interactions can be achieved by using methods of aspherical modelling of the static electron density. In modern aspherical methods, such as the pseudo atom centred MP method, non-spherical density functions are used for describing the atomic electron density.[1] These methods assume that, at sufficiently low temperature, thermal vibrations are minimal and can be safely deconvoluted from the electron densities.[1,4,5] There are various MP formalisms established in the field of electron density studies, namely, Hirshfeld's formalism,[6] Stewart's formalism[7] and the formalism of Hansen and Coppens.[8] Among them, the method of Hansen and Coppens[8] has become the most popular method for static electron density studies nowadays and will be briefly discussed here. More detailed descriptions of these methods can be found elsewhere.[1,6–8]

According to the Hansen and Coppens formalism[8] the density of each atom (ρ_{atomic}) is divided into core, valence and deformation parts (Figure 4.1) and described as follows.

$$\rho_{\text{atomic}}(\mathbf{r}) = P_c\rho_{\text{core}}(r) + P_v\kappa^3\rho_{\text{valence}}(\kappa r) + \sum_{l=0}^{l_{\max}} \kappa'^3 R_l(\kappa' r) \sum_{m=-l}^{+l} P_{lm} y_{lm}\left(\frac{\mathbf{r}}{r}\right).$$

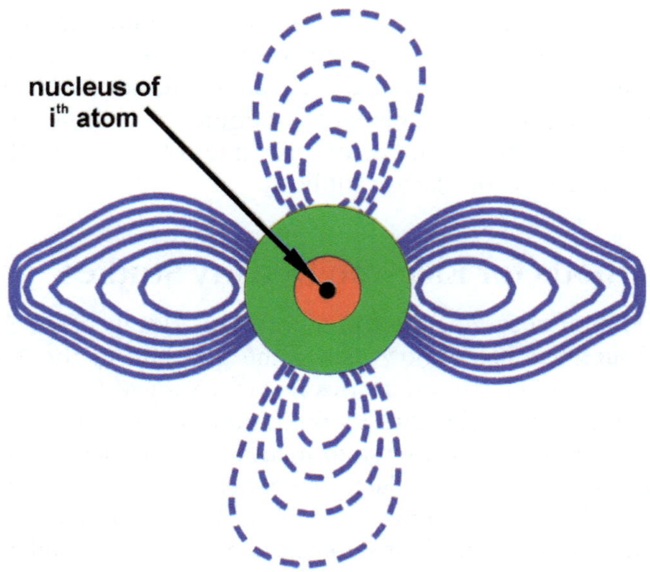

Figure 4.1 Schematic representation of different parts of the pseudo atom electron density: spherical core (red), spherical valence (green) and aspherical deformation density (blue).

The first two terms of the above equation together are known as the κ-formalism, in which P_c and P_v respectively denote the core shell and valence shell population coefficients. The variable κ allows for radial expansion and contraction of the spherical valence shell. The third term describes the aspherical features of the valence electron density. Here, the r-dependent radial function $R_l(r)$ is described as,

$$R_l(r) = \frac{\zeta^{n_l+3}}{(n_l+2)!} r^{n_l} \exp(-\zeta_l r)$$

where n_l can take any positive integer value. κ' is a variable parameter that scales the orbital exponent ζ. The functions y_{lm} is a density normalized real spherical harmonic function, and P_{lm} is the MP population coefficient.

Based on the static electron density distribution obtained from an MP model, a substantial amount of information on interatomic interactions, and the chemical and physical properties of crystalline materials can be obtained with the aid of the QTAIM.[3] According to the QTAIM, the electron density, $\rho(\mathbf{r})$, in a molecule or crystal is divided into non-overlapping atomic basins enclosed by the zero flux surfaces, *i.e.* the gradient of electron density, $\nabla\rho(\mathbf{r})$, has zero flux at any point of those surfaces:

$$\nabla\rho(\mathbf{r})\cdot n(\mathbf{r}) = 0$$

where $n(\mathbf{r})$ is a unit vector normal to the surface at $\mathbf{r}$. From this partitioning scheme, quantitative values for atomic charges, volumes, dipole moments *etc.* can be obtained. In QTAIM, a bonding interaction between two atoms is determined by the presence of a bond critical point (bcp), which is a saddle point in the electron density along the bond path between the bonded atoms. Similarly, the presence of a ring critical point (rcp) and a cage critical point (ccp) indicates the formation of a ring or a three-dimensional cage closed by bond paths. Values of the electron density at the bcp (ρ_{bcp}) and its Laplacian ($\nabla^2\rho_{bcp}$) characterize a chemical bond.[1,3] For example, a chemical bond is considered a closed shell or an ionic bond when ρ_{bcp} is close to zero and $\nabla^2\rho_{bcp}$ has a positive value; on the other hand, electron density should have a large positive value at the bcp and $\nabla^2\rho_{bcp}$ should have a large negative value for covalent or open shell bonding interactions. The characterization of atomic interactions can also be performed by several other indicators, such as the energy density indicators (local kinetic energy density $G(\mathbf{r})$, potential energy density $V(\mathbf{r})$ as well as the total energy density $H(\mathbf{r})$),[9] orientations and positions of valence shell charge concentrations (VSCCs, which are maxima in negative Laplacian), electron delocalization index,[10] source function indicator[11–13] *etc.* Classifications of interatomic interactions based on the topological properties of the electron density, $\nabla^2\rho$ profile and energy densities are given in Table 4.1. More details on the QTAIM and other methods for interpreting electron densities can be found in the literature,[1,3,14–20] as well as in other chapters of this book, and will not be further discussed here in order to avoid redundancy. In this context, it should be noted that there are a few more quite successful indicators for characterizing atomic interactions, such as the electron localization function, the electron localizability indicator *etc.* that have been proposed based on the concept of electron localization.[21–23] However, these descriptors are beyond standardized methods of analysing atomic interactions and will not be covered in this chapter. Readers may consult ref. 21–23 and references therein for further reading on these descriptors.

Apart from the above-mentioned static electron density (deconvoluted from thermal motions) based methods, worth mentioning are methods for studying dynamic electron density, such as the maximum entropy method (MEM).[24,25] To this end, the MP formalisms mentioned above are the established ways for aspherical electron density studies; however the MEM, especially in the light of its model-independent nature, has been found to be promising.[26] Model based dynamic electron densities can also be computed by the inverse Fourier transform of accurately calculated model structure factors.[27] Some recent reports[27,28] have shown that an analysis of dynamic electron densities is capable of producing similar results like the static electron density based methods do. Although the experimental electron density is by definition dynamic, studies of deconvoluted static electron densities remain dominant in the field of electron density studies, perhaps because static electron densities can be interpreted fairly easily within the scope of available theoretical foundations.[2]

Table 4.1 Summary of features characterizing atomic interactions.

Type of elements involved	Type of interaction	ρ_{bcp}	$\nabla^2\rho_{bcp}$	Position of the bcp along bond path with respect to the $\nabla^2\rho(r)$ profile	G_{bcp}/ρ_{bcp}	H_{bcp}/ρ_{bcp}
Light elements	Open-shell covalent bonds	Large	$<<0$	Close to a minimum	<1	$<<0$
	Closed shell ionic bonds, hydrogen bonds, halogen bonds and Van der Waals interactions	Small	>0	Inside a flat region between the two outermost atomic shells	≥ 1	>0
	Intermediate interactions, such as polar bonds	Large	0	Close to a nodal surface	≥ 1	$<<0$
Heavy elements	Open shell	Small	~ 0	Close to a maximum (due to the missing valence shell)	<1	<0
	Donor–acceptor	Small	>0	Close to a nodal surface	~ 1	<0

4.3 Electron Density Studies of Inorganic Crystals

Since most of the modern methods, including experimental strategies for electron density studies, were developed primarily for organic molecular materials with low density, their application to inorganic compounds and extended solids with comparatively higher density is not straightforward. In general, for electron density studies of inorganic solids with heavy elements, care should be taken in each step, starting from data collection and data integration up to aspherical modelling of electron densities. The use of methods for the interpretation of electron density distribution, such as the QTAIM,[3] which has been standardized primarily for organic molecular materials, to inorganic extended solids should also be carefully performed. In this section, challenges regarding experimental techniques and procedures for electron density studies of inorganic solids using MP methods will be discussed. It should be noted that it is beneficial to compare experimentally derived electron densities and associated properties with theoretical calculations. Various aspects of theoretical electron density studies have been covered in other chapters of this book and will not be discussed here.

4.3.1 Experimental Strategies and Challenges

Single crystal X-ray diffraction experiments using monochromatic X-ray radiation are the standard method for experimental electron density studies. There are few examples of electron density studies performed using powder X-ray diffraction data;[29,30] however, use of powder X-ray diffraction data for electron density research is yet non-standard and will not be covered in the present chapter. A single crystal X-ray diffraction data collection strategy for electron density analysis varies based on composition, morphology and degree of crystallinity of the crystal. The composition plays an important role in determining the suitability of a crystal for aspherical electron density analysis. The suitability of a crystal for electron density analysis can be approximately accessed by the suitability factor (S) as given below.[1]

$$S = \frac{V}{\sum_i^{\text{unitcell}} n_{\text{core},i}^2}.$$

Here, V is the volume of the unit cell, and $n_{\text{core},\,i}$ is the number of core electrons of the ith atom. The suitability factor reflects the relative importance of scattering from core electrons. The higher the value of S, the more suitable the crystal is for electron density analysis. Typically the value of S is greater than 3 in the case of organic crystals. On the other hand, in the case of inorganic crystals, the value of S can be quite low (Table 4.2).[31] Accordingly, very accurate data will be required in order to extract meaningful information on aspherical electron density distribution in inorganic crystals with low S values.

Table 4.2 Suitability factors for various crystals.[31]

	V	V/Z	Σn^2_{core}	S
Formamide	223.6	55.9	12	4.7
α-Glycylglycine	579.0	144.8	36	4.0
α-Deuterooxalic acid dihydrate	255.4	127.7	32	4.0
Tetracyanoethylene	897.8	149.6	40	3.7
$Cr(CO)_3C_6H_6$	419.5	209.8	372	0.56
$Cr_2(O_2C_2H_3)_4{\cdot}2H_2O$	1392.2	348.1	720	0.48
$Ni_2(C_2H_2)\,(CsHs)_2$	1058.6	264.7	696	0.38
$C1_2$	225.2	56.3	200	0.28
S_8 (orthorhombic)	3292.9	205.8	800	0.26
Si	157.5	19.7	100	0.19
A1	66.4	16.6	100	0.17
V_3Si	105.7	52.85	1072	0.05

The major factors influencing the quality of a single crystal X-ray diffraction dataset for electron density study (or in general for any kind of structural study) are thermal diffuse scattering, the extinction effect, absorption and the resolution of the data. While using monochromatic X-ray radiation, contamination of harmonics of primary wavelength need to be checked and rectified for better data quality.[32] Among the major factors, the effect of thermal diffuse scattering is somewhat similar for both organic and inorganic materials. It is therefore recommended to collect diffraction data at low temperature ($T \leq 100$ K) in order to avoid thermal diffuse scattering and to reduce the effects of thermal motion on deformation densities due to chemical bonding.[4,33] As far as the resolution is concerned, the thumb rule for all kinds of electron density studies is to get the diffraction data with highest possible resolution. In general, $\sin(\theta)/\lambda = 1$ Å^{-1} is considered to be minimum acceptable value of resolution for diffraction data for electron density studies. Since X-ray scattering power rapidly falls with diffraction angle $\sin(\theta)/\lambda$, accurately collecting high-order data might be technically problematic. In particular, for inorganic materials with heavy elements, high intensity low-order data might overflow (*i.e.* the intensity of the reflection goes beyond the linear range or the maximum detection limit) while trying collecting weak high-order data due to the limited dynamic range of an area detector. One solution to this problem is to collect separate datasets with different exposure times for high- and low-order data, and merge them at a later stage of data processing.

The other two main factors affecting electron density data quality, namely absorption and extinction factors, depend on the morphology, composition and mosaicity of the single crystal sample. Morphology-wise, spherical crystals are best suited for any high quality diffraction data collection and for easy treatment of absorption and extinction effects. Artificially creating spherical crystals by rubbing with damp paper or by grinding for electron density data collection has been reported in the literature.[34]

The composition of a crystal plays an important role in determining the strategy of the data collection. Since number of core electrons is larger in heavy elements compared to light elements, the ratio of the total electron density and the electron density involved in atomic interactions is high for heavy elements compared to light elements. Consequently, the information on electron density deformation due to atomic interactions is contained in intensities of a small number of low angle reflections in the case of inorganic solids containing heavy elements. Accurate measurement of low order data is thus of cardinal importance, particularly in the case of inorganic materials with heavy elements. Again, since the average electron concentration tends to be high in inorganic materials, they are susceptible to show severe extinction and absorption effects.[1] It has been found that the effect of extinction is mainly reducing the intensities of (normally a few) low order reflections.[35] Since the accuracy of the low order data is extremely important for inorganic solids containing heavy elements, corrections of the data for extinction effects (along with absorption effects) is crucial for electron density study. At the experimental level, increasing mosaicity of a crystal by repeated rapid cooling and heating might help in reducing the effects of extinction. Selecting a minute single crystal with dimension less than 50 μm might also help in reducing the extinction effects. Fast data collection using high intensity radiation with short wavelength and a minute crystal is probably the optimum combination possible for reducing extinction and absorption effects. Using high intensity short wavelength synchrotron radiation is thus a good choice for electron density data collection. However, technical problems associated with synchrotron radiations, such as instability of the synchrotron beam, should be accounted for and highly redundant data should be collected for electron density studies.

A final point under this subsection is about the quality of the single crystal required for an electron density study. Particularly for those inorganic solids that often contain defects and disorder, caution should be taken to choose the appropriate single crystals for electron density studies.[36] If necessary, other experimental techniques, such as energy dispersive spectroscopy, electron energy loss spectroscopy, X-ray photoelectron spectroscopy *etc.* should be used in order to verify the actual content and the quality of the crystal. For crystals that often contain defects, disorder or impurities, it may sometimes be necessary to check several electron density models before finalizing a model.[36,37]

4.3.2 Challenges Related to Aspherical Modelling of Electron Densities in Inorganic Solids

One common limiting factor for any kind of MP modelling of electron density is the truncation of MP expansion. Other than this, MP modelling of inorganic solids, particularly those with heavy elements, faces a few extra challenges in comparison to organic compounds. The basic difference between electron densities in light and heavy elements is that the valence

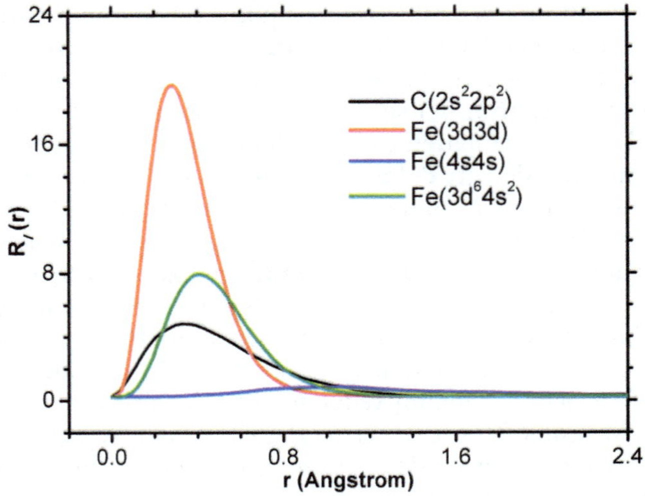

Figure 4.2 Slater type radial density functions (R_l) from the products of elemental valence shells; prepared based on the work by Schmøkel and coworkers.[35]

electron density in light elements stays close to the core, whereas the same in heavy elements extends up to a fairly long distance from the core. In other words, the valence electron densities in light elements are more concentrated around the atomic cores, but in heavy elements valence electron densities are more diffuse and extended radially from the atomic cores (Figure 4.2). This means that refinement of valence functions for heavy elements needs extra caution. For example, as shown in Figure 4.2, in fourth row transition metals, mixing 3d and 4s radial contributions and refining only one radial and one valence population parameter will lead to an averaged description of the electron density (green curve in Figure 4.2) that might not be a good description of the actual electron density.[35] For heavy elements, it is therefore recommended to use a more flexible MP model and to refine different valence shells separately.[35] The diffused and extended nature of the electron density in heavy elements (Figure 4.2), together with the fact that many of these heavy atom containing inorganic solids have small unit cells, cause the low order data to be extremely important for a proper description of the valence densities in these solids. Thus, the accuracy of low order data must be very high and, since extinction primarily affects low order data, special care should be taken to make the experimental data free of extinction effects.

Another important factor for MP modelling is regarding the use of theoretical reference data banks containing wave functions and scattering factors of atoms and ions. Several sets of reference data banks obtained using various levels of theories are available for MP modelling against experimental data.[38–42] Among the available theoretical data banks published in the literature, relativistic effects have only been included in the wave functions calculated by Su *et al.*[41,42] Since relativistic effects for heavy elements are non-negligible,[43] the choice of a theoretical data bank with relativistic effects is better than the others. However, it should be noted that up to now wave functions including relativistic effects have been published for atoms up to $Z = 54$. The only data bank that contains relativistic wave functions for atoms with $Z > 54$ is the VM data bank based on the unpublished work of Volkov and Macchi.[44] The VM data bank as incorporated in the software suite XD2006 contains STO atomic relativistic wave functions calculated at the PBE/QZ4P level for ground state neutral atoms H-Cf ($Z = 1$–98).[44] It is thus recommended to use the VM data bank in order to perform reasonably accurate electron density studies of materials containing heavy atoms with $Z > 54$.

The final point along the same lines is related to the description of core electrons. It should be noted that, in general, for MP modelling, core electrons are kept fixed according to the frozen core approximation. However, this description of core is based on theoretical calculations on gas phase spherical isolated atoms, which might not be true when the atom is participating in chemical bonds.[45] Since the number of core electrons is higher in heavy elements than in light elements, the effects of core polarization due to chemical bonding should be more significant in heavy elements.[35] For an effective description of the core electrons, particularly for heavy elements, it may thus be required to refine radial and MP parameters for some of the core shells.[46,47] However, refining core electron densities needs accurate data up to a very high resolution, which is experimentally difficult to obtain. Even with very high resolution data, meaningful modelling of the core density may sometime be impossible due to high correlations between thermal parameters and MP parameters used for describing the core shells.

As has been mentioned throughout this chapter, as accurate X-ray diffraction data is of cardinal importance for meaningful modelling of electron density, the question arises as to whether there exists a way to do reasonably accurate aspherical modelling of electron density in the absence of high quality data. In the absence of accurate experimental data, one may try to replace MP parameters to reasonable values potentially derived from theory, such as invarioms.[48] Though the application of invarioms has been primarily limited within the realm of organic molecular compounds, there are a few examples where it has been successfully applied to inorganic crystals.[49]

4.3.3 Analysis of Electron Densities in Inorganic Solids

The most popular way of interpreting static electron densities derived from an MP model is using the QTAIM.[3] The QTAIM has been particularly successful for organic compounds and inorganic solids with light elements. However, the application of the QTAIM to inorganic solids with heavy elements should be carefully performed. The simplified classification of bond types by the value of ρ_{bcp} and the sign of Laplacian, as mentioned in Section 4.2, may not be valid for bonds involving heavy elements (fourth row onwards).[14] One reason for this might be that the atomic shell structures normally displayed by the Laplacian are lost or are misleading for many heavy elements, probably due to the large atomic core and diffuse character of the outermost electrons in those heavy elements.[14,50] In cases when characterization of chemical bonds based on ρ_{bcp} and $\nabla^2\rho_{bcp}$ alone is ambiguous, using the electronic energy density at the bcp (H_{bcp}, G_{bcp} or V_{bcp}) as a characterization tool should be helpful.[51]

Furthermore, for rationalizing electron densities and understanding atomic interactions, analysis of $\rho(\mathbf{r})$ and $\nabla^2\rho(\mathbf{r})$ profiles in more than one dimension might be necessary. In particular, the position of $\mathbf{r}_{bcp}$ with respect to the $\nabla^2\rho(\mathbf{r})$ profile along the bond path between two bonded atoms is important for understanding the nature of the interatomic interaction.[51] For example, in K_2SO_4, the bcp is located at a position near to the zero Laplacian line (Figure 4.3),[52] and thus a slight change in the position of the bcp would change the sign of the Laplacian. At times, an investigation of the three-dimensional distribution of the Laplacian might be needed in order to obtain a better picture of the bonding nature with possible polarization.[35] Features of the various above-mentioned bond characterization indicators for chemical bonds involving heavy and light elements are given in Table 4.1. Table 4.1 includes indicators that are generally used for characterizing interatomic interactions in inorganic materials, including metal-organic complexes and organic–inorganic hybrids. A more detailed account of characterizing various kinds of bonds can be found in the review by Macchi and Sironi.[51] Worth mentioning here are the charge shift bonds that have not been considered so far in this chapter. The charge shift bonds have been recently proposed as a distinct class of chemical bonds where bonding energy is governed by the resonance between the ionic and covalent structures of the bond.[53,54] They are distinct from polar bonds and can occur between homo- or hetero-nuclear atoms. In a recent electron density study on fluorite (CaF_2), the electron density distribution in a charge shift bond between fluorine atoms was thoroughly analysed.[55] It was found that the electron density in the bonding region of the homo-nuclear F···F charge shift bond in fluorite is significantly lower than conventional covalent bonds.[55] At this point, it should be noted that since extended and periodic inorganic solids generally do not possess anything like molecular entities, great care should be taken while calculating topological and energy density descriptors in such materials. During electron density analysis, it is necessary to include all atoms up to a sufficient large distance of several angstroms from the point (or the region) where the electron density is being analysed in order to include the contributions of those atoms.

Finally, it should be noticed that many inorganic solids tend to crystallize in high symmetry space groups with several special symmetry positions in the unit cell. Errors in special positions can be several times higher than those at general positions.[31,56,57] Accordingly, in some high symmetry materials residual density can accumulate at special positions. For example, residual density accumulates at a special position (1/2, 1, 0) in an elemental γ-boron structure, which is the midpoint of the boron–boron diatomic dumbbell-type structural unit.[58] Such errors may make it difficult to properly

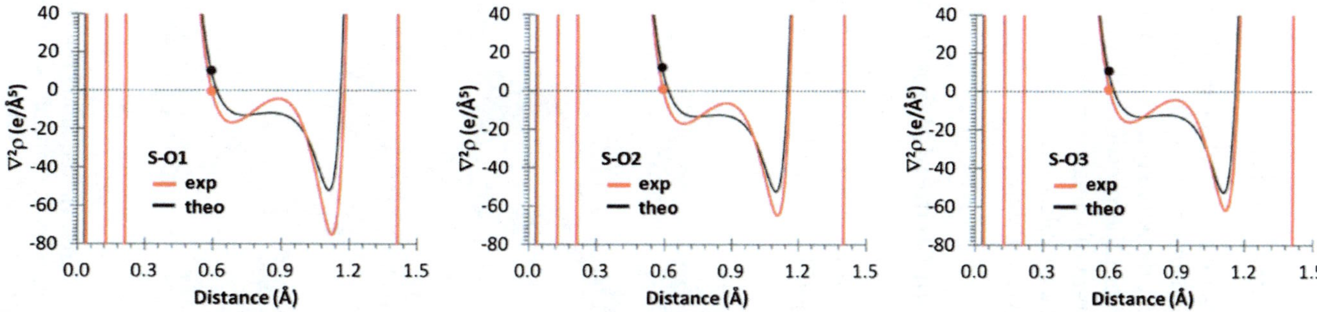

Figure 4.3 Laplacian profiles along the S–O internuclear axes in the K_2SO_4 crystal for the theoretical (black line) and the experimental (red line) data. The black and red dots mark the position of the bcp. On the *x* axes the distance from the S atom is shown. Reprinted with permission from M. S. Schmøkel, S. Cenedese, J. Overgaard, M. R. V. Jørgensen, Y-S. Chen, C. Gatti, D. Stalke, and B. B. Iversen, *Inorg. Chem.* 2012, **51**, 8607, Copyright 2012 American Chemical Society.

model as well as interpret electron densities at or close to those special positions, especially when an atom sits at or close to those positions.[35]

4.4 Few Reported Case Studies

This section discusses a few examples of electron density studies of inorganic solids. Examples are given in order to elucidate the prospects, strategies and challenges of electron density studies of inorganic materials. All examples have been selected from the present decade in order to acquaint the reader with the current state-of-the-art experimental electron density research. For older and further examples, readers are referred to the book by Coppens (ref. 1) and several review articles covering the topic.[15,18,35,51]

4.4.1 Electron Densities in Elemental Boron Allotropes

A study on α-boron was performed by Mondal and co-workers in order to understand the generalized bonding mechanism in icosahedral boron-rich solids.[59] Elemental α-boron (Figure 4.4) crystallizes in a trigonal system with

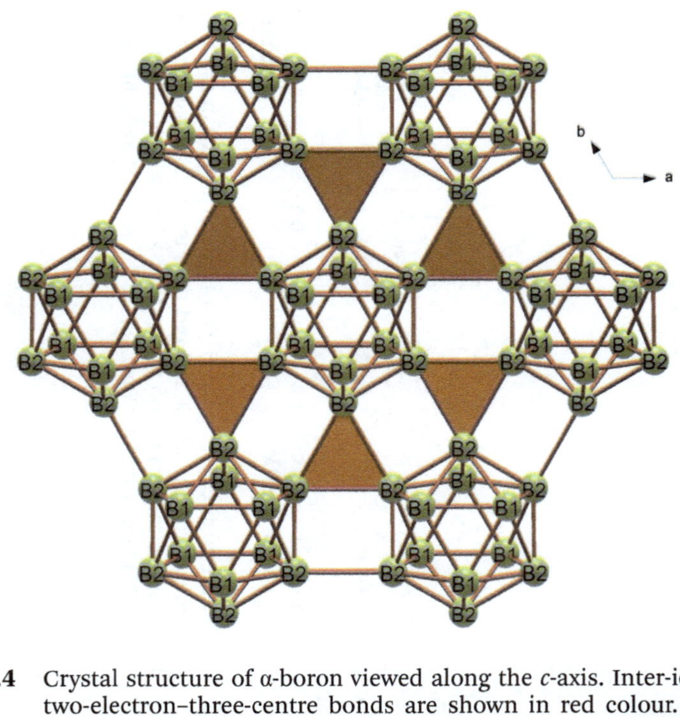

Figure 4.4 Crystal structure of α-boron viewed along the *c*-axis. Inter-icosahedral two-electron–three-centre bonds are shown in red colour. Reprinted with permission from S. Mondal, S. van Smaalen, G. Parakhonskiy, S. J. Prathapa, L. Noohinejad, E. Bykova, N. Dubrovinskaia, D. Chernyshov, L. Dubrovinsky, *Phys. Rev. B*, **88**, 024118, 2013, https://doi.org/10.1103/PhysRevB.88.024118, Copyright 2013 by the American Physical Society.[59]

a = 4.9216(6) Å, c = 12.6019(6) Å and space group *R-3m*.[59] Since boron is a poor scatterer of X-rays, it was difficult to collect accurate high resolution single crystal X-ray diffraction data required for electron density study. Mondal and co-workers collected single crystal X-ray diffraction data of α-boron at 100 K using high intensity synchrotron radiation and long exposure time at beamline F1 of the Hasylab, DESY, Hamburg. High redundancy and high resolution ($\sin(\theta)/\lambda = 1.221$ Å^{-1}) was achieved by collecting the intensity data over several runs using various possible settings of the diffractometer angles. Intensity data were corrected for the oblique incidence and integrated by EVAL15.[60] To improve the data quality, approximately 3% of outlier reflections were rejected using the program EVAL15. The absorption correction, scaling and merging was done by SADABS.[61] The aspherical electron density modelling was performed by using the Hansen and Coppens formalism[8] as implemented in the software XD2006.[44] The final MP model produced excellent reliability factors ($R_F = 0.0116$, $wR_F = 0.0092$, $R_F{}^2 = 0.0146$, $wR_F{}^2 = 0.0185$, GoF = 1.581, $\Delta\rho_{max/min} = 0.139/-0.150$ for reflections with $\sin(\theta)/\lambda \leq 1.06$ Å^{-1} and $R_F = 0.0169$, $wR_F = 0.0110$, $R_F{}^2 = 0.0191$, $wR_F{}^2 = 0.0220$, GoF = 1.6983, $\Delta\rho_{max/min}$ = 0.236/−0.231 for reflections with $\sin(\theta)/\lambda \leq 1.221$ Å^{-1}) indicating correctness of the model. The deviation of the value of GoF from the ideal value of 1 might be due to non-accurate estimations of standard uncertainties of the experimental intensities, especially for very weak high-order reflections of elemental boron. The characterization of atomic interactions was performed using the QTAIM.[3] The topological analysis of the electron density in α-boron revealed the presence of critical points for all expected bonds. The values of ρ_{bcp} and $\nabla^2\rho_{bcp}$ revealed that the chemical bonding in α-boron is predominantly covalent and no evidence has been found for any kind of charge transfer, contrary to what was proposed in an earlier study[62] by Fujimori and co-workers. In order to investigate possibilities of any experimental bias, Mondal and co-workers collected a second dataset using a different single crystal at beamline BM01A of the European Synchrotron Radiation Facility, ESRF, and found similar results from both datasets.

In this work, the importance of high-quality single crystal data for charge density research has been accentuated by revealing the previously proposed[62] peculiar bent nature of the inter-icosahedral two-electron–three-centre bond in α-boron as an artefact caused by the limited information content of the powder X-ray diffraction data used for the previous study. The other major contribution of this electron density study is that it provides experimental proof of the persistent orbital order of the 12 external bonding orbitals of the icosahedral B_{12} cluster (Figure 4.5), the ubiquitous building block of icosahedral boron-rich inorganic materials.

In a different electron density study of another allotrope of boron (Figure 4.6), the high-pressure and high-temperature (metastable) allotrope orthorhombic γ-boron (a = 5.0597(2) Å, b = 5.6304(2) Å, c = 6.9885(2) Å and space group *Pnnm*), Mondal and co-workers have followed similar strategies as mentioned above for the X-ray diffraction experiment and for the MP modelling of the electron density.[58] The final MP model has yielded an excellent

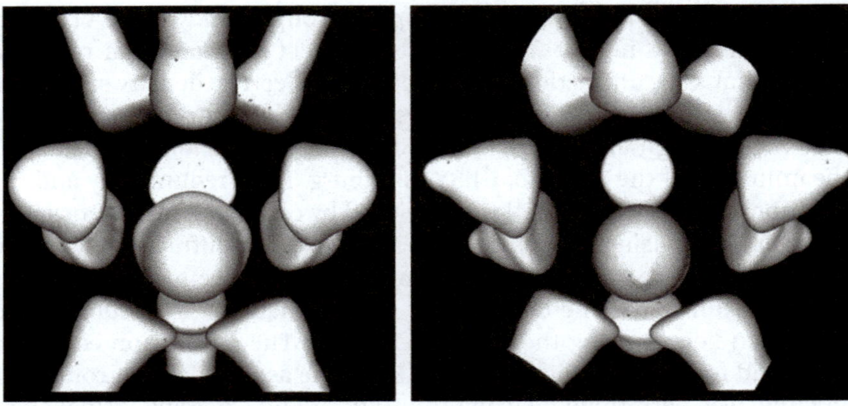

Figure 4.5 Plots of isosurfaces of three-dimensional electron densities (at 0.84 e Å^{-3}) in the B_{12} icosahedral clusters of α-boron (left) and γ-boron (right). They demonstrate the orbital order by lobes of electron densities oriented perpendicular to the surfaces of the B_{12} icosahedral clusters. Reprinted with permission from S. Mondal, S. van Smaalen, G. Parakhonskiy, S. J. Prathapa, L. Noohinejad, E. Bykova, N. Dubrovinskaia, D. Chernyshov, L. Dubrovinsky, *Phys. Rev. B*, **88**, 024118, 2013, https://doi.org/10.1103/PhysRevB.88.024118, Copyright 2013 by the American Physical Society.[59]

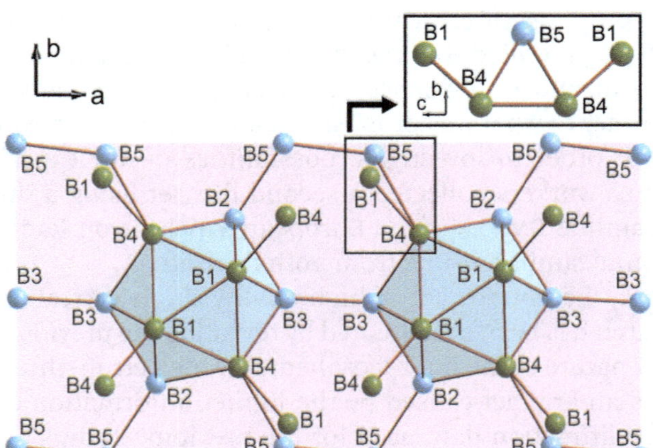

Figure 4.6 Crystal structure of γ-boron along with an atom numbering scheme viewed along the *c*-axis. The polar covalent two-electron–three-centre bond is shown in the inset. Reprinted with permission from S. Mondal, S. van Smaalen, A. Schoenleber, Y. Filinchuk, D. Chernyshov, S. I. Simak, A. S. Mikhaylushkin, I. A. Abrikosov, E. Zarechnaya, L. Dubrovinsky and N. Dubrovinskaia, *Phys. Rev. Lett.*, **106**, 215502, 2011, https://doi.org/10.1103/PhysRevLett.106.215502, Copyright 2011 by the American Physical Society.[58]

fit to the diffraction data with R_F = 0.0120, wR_F = 0.0147, GoF = 0.6272 and $\Delta\rho_{max/min}$ = 0.312/−0.197 e Å^{-3}.[58] The electron density was rationalized by using the QTAIM. The chemical bonding in γ-boron was disputed over its ionic or covalent nature.[63,64] The electron density study[58] by Mondal and co-workers revealed charge transfers between constituent atoms of γ-boron. The study also revealed that there exists a spectrum of chemical bonds in γ-boron, including a 12-atom B_{12} cluster, a one-electron–two-centre bond, a two-electron–two-centre bond and a unique polar-covalent two-electron–three-centre bond involving an edge of the B_{12} icosahedral unit and another boron atom from the two-atom dumbbell unit (Figure 4.7). The two-electron–three-centre bond is unique in the sense that the boron atoms that form this bond contribute unequal amount of electrons. It has been proposed that the charge transfer originates in this polar-covalent two-electron–three-centre bond. No evidence of any ionic bond between constituent boron atoms in γ-boron was found. The analysis of the electron density in γ-boron also reveals the same orbital order in the B_{12} icosahedral group as found in the case of α-boron (Figure 4.5).

4.4.2 Electron Density in Pyrope ($Mg_3Al_2Si_3O_{12}$)

Pyrope ($Mg_3Al_2Si_3O_{12}$) (Figure 4.8) is a garnet that crystallizes in the cubic crystal system with space group *Ia-3d* and lattice parameter a = 11.4405(3) Å.[65] Garnets are a group of silicate minerals and have been used historically as gemstones and abrasives. Various types of garnets have recently gained considerable attention due to their potential ion transport properties.[65,66] Cation disorder and anharmonic thermal motion of cations have been found to influence the ion transport properties in garnets.[67,68] An electron density study of the title garnet, $Mg_3Al_2Si_3O_{12}$ (Figure 4.8), has been performed in order to understand the nature of Mg cation disorder and anharmonic effects.[65]

For electron density data collection, Destro and co-workers used a single crystal ground to a perfect sphere of diameter ~0.45 mm. Intensity data up to $\sin(\theta)/\lambda$ = 1.14 Å^{-1} were collected using a four-circle diffractometer equipped with graphite-monochromated Mo K_α radiation (λ = 0.71073 Å) and a scintillation counter point detector. The data collection temperature was maintained at 30 K. Individual measurements affected by multiple reflections (Renninger reflections)[69] were removed before merging. Since the single crystal was rather big and the absorption coefficient was reasonably high (μ = 1.218 mm^{-1} for Mo K_α radiation), caution was taken to deal with absorption and extinction effects. The data was corrected for absorption using the ($\sin(\theta)/\lambda$)-dependent empirical method.

The MP modelling has been performed according to the pseudo atom formalism of Stewart.[7] The VALRAY2000 suite of programs[70] has been used for the least square refinements. Scattering factors for all atoms have been derived from Stewart's localized atomic orbitals.[71] MP expansions were truncated at

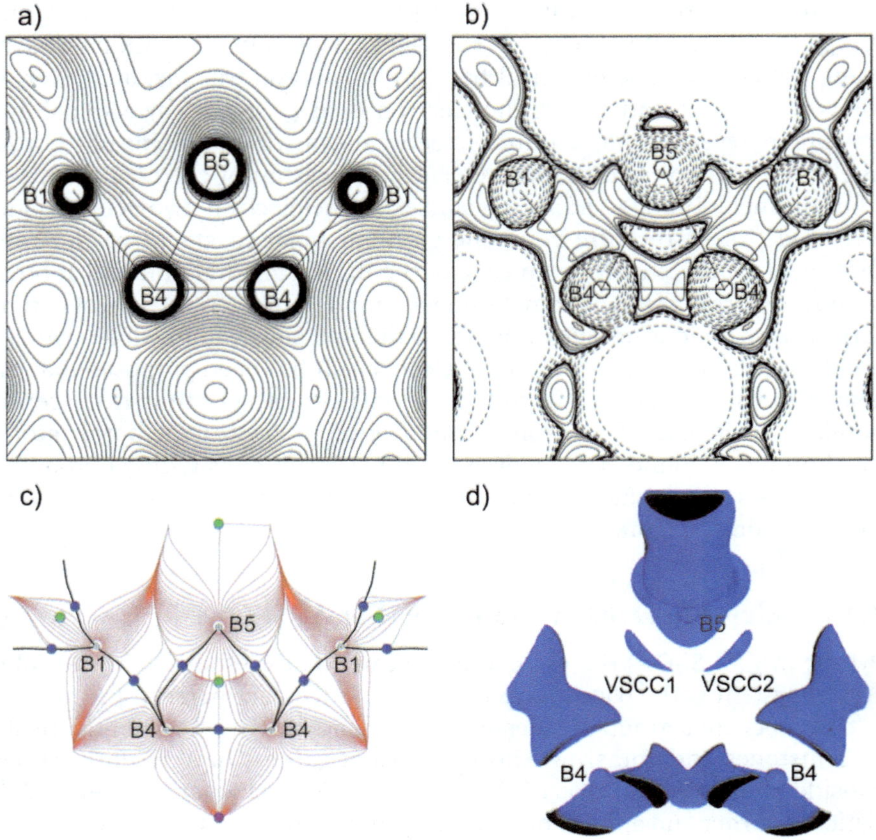

Figure 4.7 Properties on the plane containing the polar covalent two-electron–three-centre bond formed by an edge of the cluster (B4–B4) and a boron atom (B5) from the dumbbell unit. (a) Static electron density; contour lines at 0.05 up to 2.0 e Å^{-3}. (b) Laplacian; contour lines at $\pm(2, 4, 8) \times 10^n$ e Å^{-5} ($-3 \leq n \leq 3$); solid lines indicate positive and dashed lines indicate negative contours. (c) Gradient trajectories of the electron density with bcps (blue), rcps (green), and ccps (purple) indicated. (d) Laplacian at −2.6 e Å^{-5} iso level, showing the VSCCs around the atoms. Reprinted with permission from S. Mondal, S. van Smaalen, A. Schoenleber, Y. Filinchuk, D. Chernyshov, S. I. Simak, A. S. Mikhaylushkin, I. A. Abrikosov, E. Zarechnaya, L. Dubrovinsky and N. Dubrovinskaia, *Phys. Rev. Lett.*, **106**, 215502, 2011, https://doi.org/10.1103/PhysRevLett.106.215502, Copyright 2011 by the American Physical Society.[58]

tricontadipoles (l = 5) level for Mg, Al and Si cations and at hexadecapole (l = 4) level for O anions. In order to understand whether the reported[72] large mean square displacement of Mg in pyrope is due to static disorder or is due to anharmonic thermal motion, two different MP models were constructed. One with only harmonic thermal parameters, and, in the other model, refinements of the Gram–Charlier coefficients (anharmonic thermal parameters) up to the third order (C_{ijk}) for Si and O, and up to the fourth-

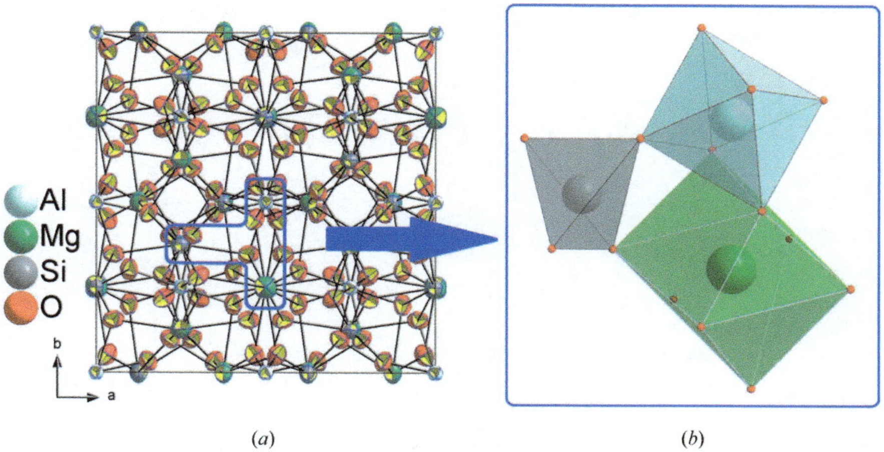

Figure 4.8 (a) View of the crystal structure of the pyrope ($Mg_3Al_2Si_3O_{12}$) along the *c*-axis with atom colour code indicated. (b) Coordination polyhedra of Al (octahedron), Mg (dodecahedron) and Si (tetrahedron) corresponding to the frame highlighted in (a). Reproduced with permission from ref. 65, https://doi.org/10.1107/S2052520617006102, with permission from IUCr Journals.

order (D_{ijkl}) for Mg and Al were performed. The anharmonic model fitted better than the harmonic model to the experimental data. The final MP model with 93 parameters yielded an excellent fit to the experimental data with R_F = 0.0084, $R_F^2 = 0.0081$, $wR_F^2 = 0.0155$, GoF = 0.7598, and $\Delta\rho_{max/min} = 0.12/-0.16$.[65]

In this work, the authors have shown how the inclusion of anharmonic thermal motions improved the MP model of pyrope. As shown in Figure 4.9, the residual densities around the cations disappeared when the refinements of the anharmonic thermal parameters were performed. From the analysis of the electron densities it was concluded that at low temperature there is no static disorder present in the structure. However, the cation Mg has been found to undergo wide displacement from its equilibrium position even at low temperature. The authors showed that the electron density of Mg can be properly modelled by including anharmonicity for Mg. Nevertheless, the author suggested that the diffused nature of the electron density of Mg might also be explained by invoking the presence of some kind of dynamic disorder.[65]

4.4.3 Electron Densities in Pyrite and Marcasite Polymorphs of FeS_2

Experimental determination of electron densities in transition metal based extended solids is not straightforward. Smockel and coworkers took the challenge of determining electron densities in pyrite and marcasite polymorphs of iron sulphide (Figure 4.10) with the aim of answering a number of unsolved

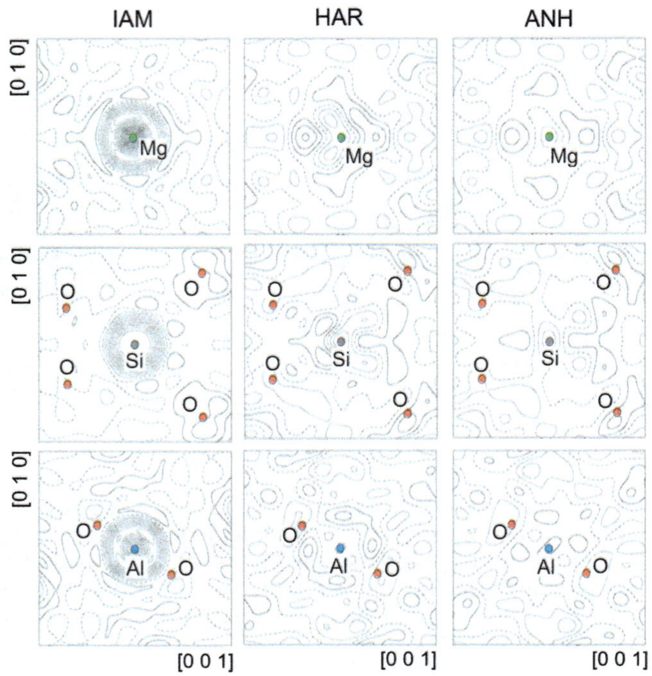

Figure 4.9 3 × 3 Å maps of residual densities on the (100) plane with Mg (first row), Si (second row) or Al (third row) at the centre of each plot. Left, middle and right columns correspond to the IAM, MP model with harmonic thermal parameters (HAR) and the MP model with anharmonic thermal parameters (ANH) respectively. Contour levels are drawn at steps of either 0.2 e $Å^{-3}$ (IAM maps) or 0.05 e $Å^{-3}$ (MP maps), with full (dotted) lines representing positive (negative) values. The dashed line corresponds to the $\Delta\rho = 0$ level. Reproduced with permission from ref. 65, https://doi.org/10.1107/S2052520617006102, with permission from IUCr Journals.

issues regarding the nature of chemical bonding between iron and sulphur.[73] The pyrite polymorph of FeS_2 crystallizes in the cubic space group, *Pa-3* with $a = 5.4155(2)$,[73] whereas marcasite crystallizes in orthorhombic space group *Pnnm* with $a = 4.4330(3)$, $b = 5.4261(3)$, $c = 3.3887(2)$.[73]

Electron density data were collected using synchrotron radiation ($\lambda = 0.4428$ Å) at the beam line ChemMatCARS, 15-ID-B, Advanced Photon Source, ANL, Chicago, USA. A Bruker APEX II CCD detector was used for the data collection. An octahedrally shaped single crystal of approximate dimensions 0.01 mm × 0.01 mm × 0.01 mm was used for the data collection of the pyrite sample. And a plate-like single crystal of approximate dimension 0.005 mm × 0.007 mm × 0.01 mm was used for the data collection of the marcasite sample. The data collection temperature was 10(5) K. The combination of high-intensity synchrotron radiation, very low temperature and very small single crystals was used in order to reduce extinction and absorption effects (in particular because of the presence of iron in the samples). The data integration was

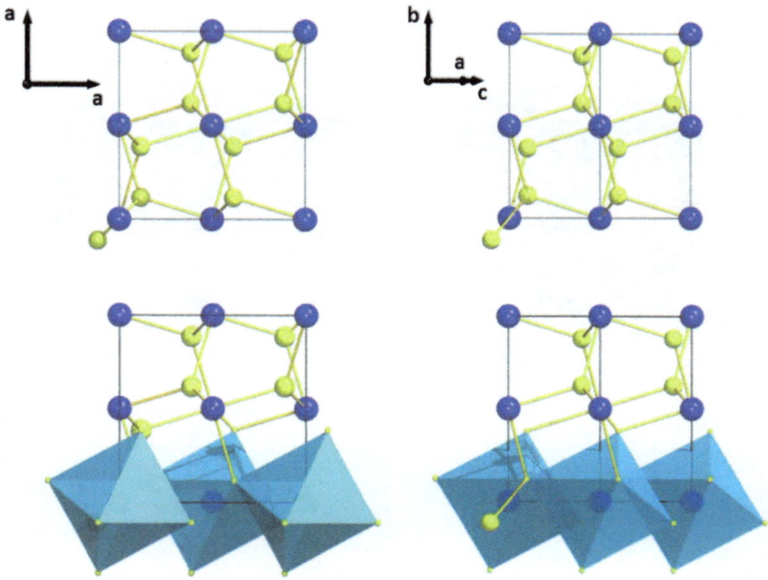

Figure 4.10 Crystal structures of the pyrite (left) and the marcasite (right). Fe and S atoms are represented by blue and yellow spheres respectively. Reproduced from ref. 73 with permission from the Royal Society of Chemistry.

carefully carried out for individual runs using the software SAINT.[74] A few runs were discarded based on high R_{symm} values. The software SADABS[75] was used for scaling, absorption and extinction corrections. The data were averaged using the program SORTAV.[76,77]

MP refinements were performed using the Hansen and Coppens formalism[8] as implemented in XD2006.[44] The VM data bank containing an STO relativistic wave function was used for core and valence orbital description of the atoms.[44] Symmetry allowed MP parameters up to hexadecapole for all atoms to be refined. A type I extinction correction was made for marcasite and a type II extinction correction was made for pyrite. Anomalous dispersion was included in the models and reflections with $I > 2\sigma(I)$ were used for the refinements. Since mixing of 3d and 4s radial functions for Fe may incorrectly describe the electron density (see Figure 4.2 and Section 4.3.2), the valence shell of Fe was split into two components, one with a 3d shell and the other with a 4s shell. A combination of 3s and 3p radial functions ($3s^23p^4$) was used for the description of the sulphur valence shell. The final MP model yielded an excellent fit to the experimental data with $R_F = 0.0073$, $R_{F^2} = 0.0134$, $\Delta\rho_{max/min} = 0.44/-0.40$ for pyrite and $R_F = 0.0082$, $R_{F^2} = 0.0138$, $\Delta\rho_{max/min} = 0.57/-0.55$ for marcasite.[73]

Analysis of the electron density revealed that the character of the Fe–S bonds in both polymorphs is intermediate between shared and closed shell interactions (Figure 4.11) with a non-negligible amount of charge accumulation but significantly negative electronic energy densities at the bond critical

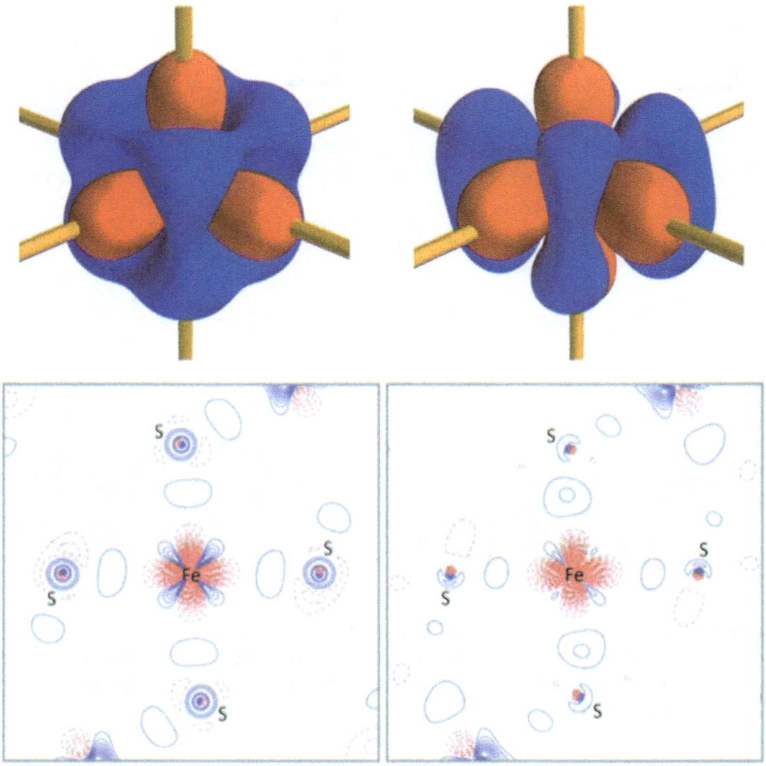

Figure 4.11 Top: isosurface plots of experimental static deformation densities at ± 0.25 e Å^{-3} around Fe atoms in pyrite (left) and marcasite (right). Bottom: static deformation densities for the Fe–S interactions in pyrite (left), and for the axial and equatorial Fe–S interactions in marcasite (right). The increment in the contours is 0.1 e Å^{-3}. Positive contours are plotted with full, blue lines. Negative contours are plotted with dashed, red lines. Reproduced from ref. 73 with permission from the Royal Society of Chemistry.

points. On the other hand, based on the topological and energy density based descriptors, the S–S interactions (Figure 4.12) have been found to be covalent in nature. Between the two polymorphs, the S–S bond in pyrite is 'more' covalent than the same in marcasite.[73] For Fe–S bonds, an opposite trend has been observed.[73]

4.4.4 Electron Density in Caesium Uranyl Chloride ($Cs_2UO_2Cl_4$)

Reports of experimental electron density studies of solids containing very heavy elements are scarce in the literature. The scarcity of experimental electron density studies of inorganic materials with very heavy elements reflects the level of difficulty associated with such studies. Due to the high X-ray

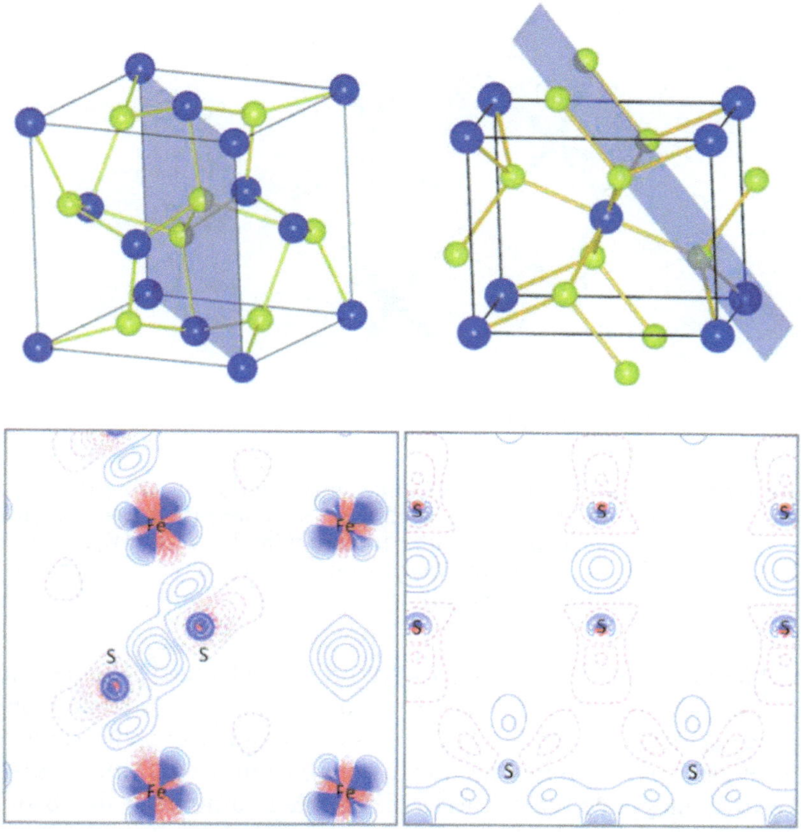

Figure 4.12 Contour plots of the experimental static deformation density (bottom) in a plane (top) containing the S–S dimer in pyrite (left) and marcasite (right). The contour increment is 0.05 e Å^{-3}. Positive contours are plotted with full, blue lines. Negative contours are plotted with dashed, red lines. Reproduced from ref. 73 with permission from the Royal Society of Chemistry.

absorption coefficients of heavy elements, it is difficult to collect high quality X-ray diffraction data for electron density studies. And from the electron density modelling point of view, the main difficulty is that the traditional division of electron density into spherical core, spherical valence and aspherical deformation densities does not work in case of very heavy elements, such as uranium.[46,47] Here the electron density study of caesium uranyl chloride ($Cs_2UO_2Cl_4$) complex (Figure 4.13) as reported by Zhurov and coworkers will be discussed.[46,78]

Caesium uranyl chloride ($Cs_2UO_2Cl_4$) crystallizes in a monoclinic crystal system with space group C2/c and lattice parameters a = 11.7882(3) Å, b = 7.6411(2) Å, c = 5.7686(2) Å, β = 100.438(3)°.[46,78] For electron density data collection, Zhurov and co-workers used an in-house Rigaku diffractometer equipped with an Ultrax-18 Mo rotating anode generator (50 kV, 300 mA),

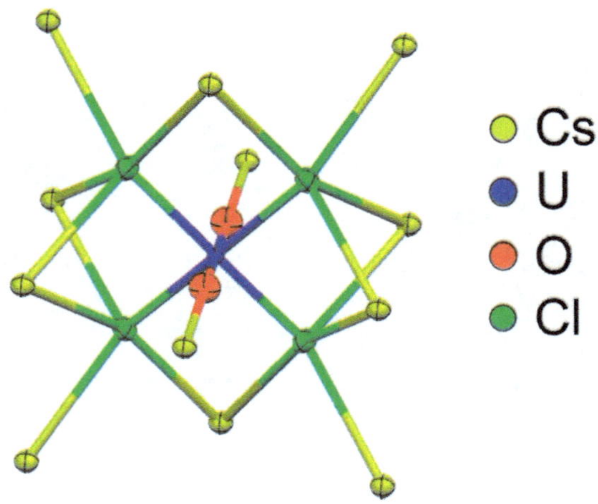

Figure 4.13 Crystal structure of cesium uranyl chloride showing the $[UO_2Cl_4]^{2-}$ anion and the closest neighbour Cs^+ cations with atom colour codes indicated. Reprinted with permission from V. V. Zhurov, E. A. Zhurova, and A. A. Pinkerton, *Inorg. Chem.*, 2011, **50**, 6330, Copyright 2011 American Chemical Society.[78]

a graphite monochromator, and a RAPID cylindrical image plate detector for collecting the single crystal X-ray diffraction data. In order to reduce absorption effects, they collected the data at very low temperature (20 K) and used a small single crystal sample with dimensions 0.04 mm × 0.07 mm × 0.10 mm; afterwards, integrated intensities were corrected numerically for absorption effects.

For MP modelling of the electron density, the Hansen and Coppens formalism[8] as implemented in the software package XD2006 was used. The VM data bank containing a relativistic wave function was used for core and valence orbital description of the atoms.[44] Comparatively light elements present in the structure, namely O and Cl, were treated conventionally using the frozen core approximation. However, the conventional frozen core approximation did not work with U and Cs and produced a significant amount of unfitted residual densities in the vicinities of U and Cs atoms. Even introducing outer core shells to the valence shells and allowing them to deform could not significantly improve the model. Finally, for proper modelling of the electron density, they had to split the U and Cs models into several components, each with its own valence set. The U atom was split into four components at the same position, such that the sum of those four components describes the model for the U atom. The first component was considered as an atom with a core and 6s and 6p orbitals combined in the valence shell. The three other components at the same position were three atoms without any core but only with 7s, 6d, or 5f valence electrons respectively, each in a single shell. Similarly, the Cs atom was split in two components, one with a core and 5s and 5p electrons in the valence shell, and the other component without a core and only a 6s valence shell.

The thermal parameters of O and Cl atoms were refined harmonically. Refinement of anharmonic thermal parameters for the atom Cs yielded a better fit to the experimental data. However, refinement of anharmonic thermal parameters for the U atom resulted in a worse reliability factor. Since anomalous scattering factors may vary depending on the chemical environment,[79–81] they scanned $\Delta f'$and $\Delta f''$ for Cs and U atoms. Based on reliability factors, it was found that the best values of $\Delta f'$ and $\Delta f''$ for the U atom are −9.977 and 9.995 respectively. And for the Cs atom, the best values were found to be −0.288 and 2.199 for $\Delta f'$ and $\Delta f''$ respectively. These values are different from the default values present in the VM data bank as incorporated in the XD2006 software.[44]

The final MP model yielded an excellent fit to the experimental data with $R_F = 0.0072$, $R_F{}^2 = 0.0082$, GoF = 1.1084 and $\Delta\rho_{min/max} = -0.476/0.368$ (for $\sin\theta/\lambda < 1.0$ Å^{-1}).[46,78] The electron density from the final MP model was analysed with the aid of the QTAIM. The features of the static electron density, deformation density and Laplacian in a plane along the two-fold axis of the unit cell that contains the O–U–O line in $Cs_2UO_2Cl_4$ are shown in Figure 4.14. The existence of no significant residual density confirms the reliability of the model. Chemical bonds in $Cs_2UO_2Cl_4$ were characterized based on the electron density distribution, topological descriptors and analysis of electronic energy densities. The U–O chemical bond has been characterized as a triple bond, and a significant covalent nature has been recognized in both the U–O and U–Cl bonds. The major contribution of this study is to establish the strategies for experimental electron density studies in solids containing very heavy elements like actinides.

4.5 Conclusion

This chapter briefly reviews the challenges, strategies and procedures for experimental electron density studies of inorganic solids and discusses a few reported examples. It is pointed out in this chapter that experimental determination of electron densities in inorganic solids is in many ways far from being standardized. In particular, for inorganic solids containing heavy atoms, caution should be taken in each and every step of the electron density study. The chapter has discussed challenges and strategies of all three major steps of electron density studies of inorganic solids, namely, experimental step, modelling step and the analysis step.

It has been suggested that using very small single crystals, low temperature (100 K or below) and high intensity, short wavelength stable radiation will reduce the severity of the few major drawbacks of experimental electron density studies of inorganic solids, such as extinction and absorption effects. It is emphasized to apply extinction and absorption corrections during data processing in order to get accurate experimental data. Considering the limited dynamic range of some detectors, collecting separate datasets with different exposure times when necessary has been suggested, and to merge them later.

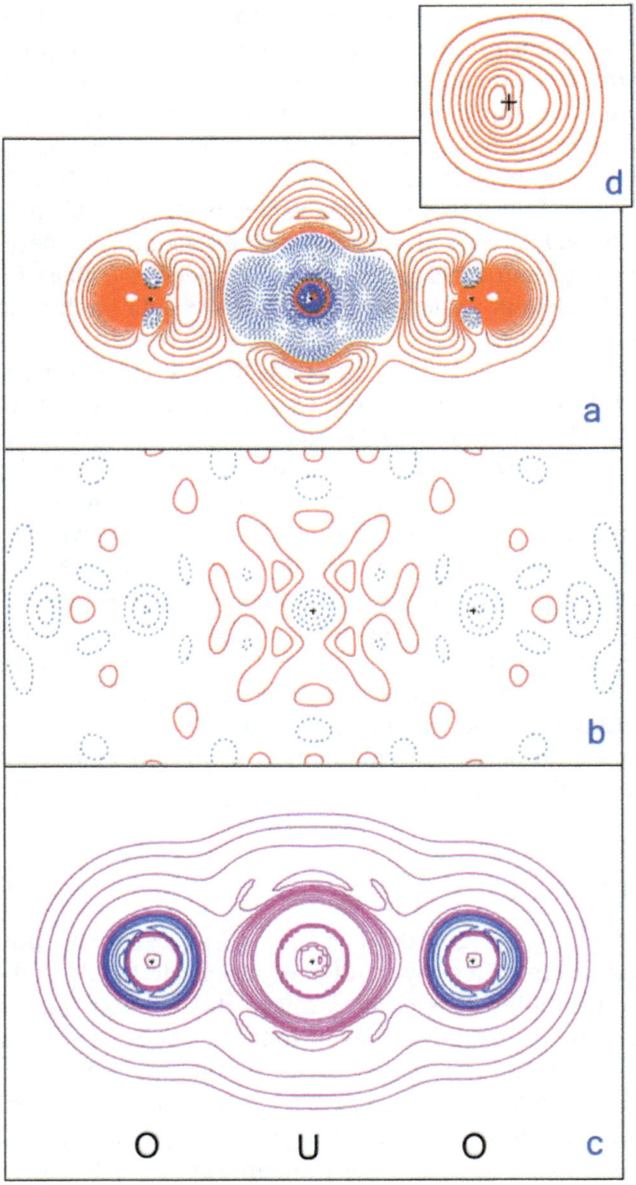

Figure 4.14 (a) The deformation density, (b) residual density, and (c) Laplacian maps for the uranyl ion in a plane through the two-fold axis and the O–U–O line in $Cs_2UO_2Cl_4$, plus (d) the deformation density through the maximum and perpendicular to the U–O bond. Deformation density contour intervals are 0.05 e Å^{-3}. Residual contours are 0.1 e Å^{-3}. Positive values corresponding to an excess of electrons are red; negative ones are blue. The Laplacian contours vary—blue ones corresponding to a negative Laplacian (electron concentration). Reprinted with permission from V. V. Zhurov, E. A. Zhurova, and A. A. Pinkerton, *Inorg. Chem.*, 2011, **50**, 6330, Copyright 2011 American Chemical Society.[78]

As far as MP modelling of the electron density is concerned, for heavy elements, it is better to split atomic density in components with each one describing a part of the valence electron densities. The sum of all those components will describe the total atomic density. And in order to describe core polarizations, some of the spherical core shells might be needed to be included in the valence shells for aspherical modelling. For inorganic solids with light elements the standard frozen core approximation may be used.

The application of the QTAIM to inorganic solids with heavy elements for rationalizing electron densities should also be carefully performed. The simplified classification of bond types based on the values of ρ_{bcp} and $\nabla^2\rho_{bcp}$, and the sign of the Laplacian might not work for many inorganic solids, and thus various other descriptors such as electronic energy densities and analysing three dimensional profiles of the electron density, deformation density and Laplacian have been pointed out as additional tools for characterizing chemical bonds. Additionally, comparing experimentally derived electron densities and associated properties with theoretical calculations would be beneficial.

Acknowledgements

The author gratefully acknowledges the support from the Council of Scientific and Industrial Research (CSIR), India.

References

1. P. Coppens, *X-ray Charge Densities and Chemical Bonding*, Oxford University Press, Oxford, U.K., 1997.
2. P. Coppens and A. Volkov, *Acta Crystallogr., Sect. A*, 2004, **60**, 357.
3. F. W. Bader, *Atoms in Molecules – A Quantum Theory*, Clarendon, Oxford, 1990.
4. F. L. Hirshfeld, *Acta Crystallogr. Sect. A.*, 1976, **32**, 239.
5. T. S. Koritsanszky and P. Coppens, *Chem. Rev.*, 2001, **101**, 1583.
6. F. L. Hirshfeld, *Acta Crystallogr., Sect. B*, 1971, **27**, 769.
7. R. F. Stewart, *Acta Crystallogr., Sect. A*, 1976, **32**, 565.
8. N. K. Hansen and P. Coppens, *Acta Crystallogr., Sect. A*, 1978, **34**, 909.
9. E. Espinosa, E. Molins and C. Lecomte, *Chem. Phys. Lett.*, 1998, **285**, 170.
10. X. Fradera, M. A. Austen and R. F. W. Bader, *J. Phys. Chem. A*, 1999, **103**, 304.
11. R. F. W. Bader and C. Gatti, *Chem. Phys. Lett.*, 1998, **287**, 233.
12. C. Gatti and L. Bertini, *Acta Crystallogr., Sect. A*, 2004, **60**, 438.
13. J. Overgaard, *Acta Crystallogr., Sect. B*, 2016, **72**, 169.
14. P. Macchi, D. M. Proserpio and A. Sironi, *J. Am. Chem. Soc.*, 1998, **120**, 13429.
15. C. Gatti, *Z. Kristallogr.*, 2005, **220**, 399.
16. C. Gatti and P. Macchi, *Modern Charge Density Analysis*, Springer, New York, 2012.

17. P. Macchi, *Crystallogr. Rev.*, 2013, **19**, 58.
18. D. Chopra, *J. Phys. Chem. A*, 2012, **116**, 9791.
19. P. L. A. Popelier, *Atoms in Molecules. An Introduction*, Pearson Education, Harlow, Great Britain, 2000.
20. V. R. Hathwar, *J. Indian Inst. Sci.*, 2017, **97**, 281.
21. A. D. Becke and K. E. Edgecombe, *J. Chem. Phys.*, 1990, **92**(9), 5397.
22. F. R. Wagner, V. Bezugly, M. Kohout and Y. Grin, *Chem.–Eur. J.*, 2007, **13**, 5724.
23. A. I. Baranov, R. Ponec and M. Kohout, *J. Chem. Phys.*, 2012, **137**, 214109.
24. E. T. Jaynes, *Proc. IEEE*, 1982, **70**, 939.
25. S. F. Gull and J. Skilling, *The Maximum Entropy Method in Indirect Imaging*, ed. J. A. Roberts, Cambridge Universtiy Press, U. K., 1984.
26. S. van Smaalen and J. Netzel, *Phys. Scr.*, 2009, **79**, 048304.
27. S. Mondal, S. J. Prathapa and S. van Smaalen, *Acta Crystallogr. Sect. A*, 2012, **68**, 568.
28. C. B. Hubschle and S. van Smaalen, *J. Appl. Crystallogr.*, 2017, **50**, 1627.
29. H. Svendsen, J. Overgaard, R. Busselez, B. Arnaud, P. Rabiller, A. Kurita, E. Nishibori, M. Sakata, M. Takata and B. B. Iversen, *Acta Crystallogr., Sect. A*, 2010, **66**, 458.
30. K. Tolborg, M. R. V. Jørgensen, S. Christensen, H. Kasai, J. Becker, P. Walter, A.-C. Dippel, J. Als-Nielsen and B. B. Iversen, *Acta Crystallogr., Sect. B*, 2017, **73**, 521.
31. E. D. Stevens and P. Coppens, *Acta Crystallogr., Sect. A*, 1976, **32**, 915.
32. C. G. Gianopoulos, Z. Chua, V. V. Zhurov and A. A. Pinkerton, *J. Appl. Crystallogr.*, 2017, **50**, 643.
33. T. Koritsanszky, R. Flaig, D. Zobel, H.-G. Krane, W. Morgenroth and P. Luger, *Science*, 1998, **279**, 356.
34. R. Destro, P. Roversi, M. Barzaghi and R. E. Marsh, *J. Phys. Chem. A*, 2000, **104**, 1047.
35. M. S. Schmøkel, J. Overgaard and B. B. Iversen, *Z. Anorg. Allg. Chem.*, 2003, **639**, 1922.
36. S. Mondal, E. Bykova, S. Dey, I. Sk Ali, N. Dubrovinskaia, L. Dubrovinsky, G. Parakhonskiy and S. van Smaalen, *Sci. Rep.*, 2016, **6**, 19330.
37. S. Mondal, *Chem. Mater.*, 2017, **29**, 6191.
38. E. Clementi and C. Roetti, *At. Data Nucl. Data Tables*, 1974, **14**, 177.
39. E. Clementi and D. L. Raimondi, *J. Chem. Phys.*, 1953, **38**, 2686.
40. C. F. Bunge, J. A. Barrientos and A. V. Bunge, *At. Data Nucl. Data Tables*, 1993, **53**, 113.
41. Z. Su and P. Coppens, *Acta Crystallogr., Sect. A*, 1998, **54**, 646.
42. P. Macchi and P. Coppens, *Acta Crystallogr., Sect. A*, 2001, **57**, 656.
43. K. Batke and G. Eickerling, *Phys. Scr.*, 2016, **91**, 043010.
44. A. Volkov, P. Macchi, L. J. Farrugia, C. Gatti, P. Mallinson, T. Richter and T. Koritsanszky, *XD2016-A Computer Program Package for Multipole Refinement, Topological Analysis of Charge Densities and Evaluation of Intermolecular Energies from Experimental or Theoretical Structure Factors*, 2016.

45. A. Fischer, D. Tiana, W. Scherer, K. Batke, G. Eickerling, H. Svendsen, N. Bindzus and B. B. Iversen, *J. Phys. Chem. A*, 2011, **115**, 13061.
46. V. V. Zhurov, E. A. Zhurova, A. I. Stash and A. A. Pinkerton, *J. Phys. Chem. A*, 2011, **115**, 13016.
47. C. G. Gianopoulos, V. V. Zhurov, S. G. Minasian, E. R. Batista, C. Jelsch and A. A. Pinkerton, *Inorg. Chem.*, 2017, **56**, 1775.
48. B. Dittrich, T. Koritsánszky and P. Luger, *Angew. Chem., Int. Ed.*, 2004, **43**, 2718.
49. Y. V. Nelyubina, A. A. Korlyukov, K. A. Lyssenko and I. V. Fedyanin, *Inorg. Chem.*, 2017, **56**, 4688.
50. R. P. Sagar, A. C. T. Ku, V. H. Smith and A. M. Simas, *J. Chem. Phys.*, 1988, **88**, 4367.
51. P. Macchi and A. Sironi, *Coord. Chem. Rev.*, 2003, **238–239**, 383.
52. M. S. Schmøkel, S. Cenedese, J. Overgaard, M. R. V. Jørgensen, Y.-S. Chen, C. Gatti, D. Stalke and B. B. Iversen, *Inorg. Chem.*, 2012, **51**, 8607.
53. A. Shurki, P. C. Hiberty and S. Shaik, *J. Am. Chem. Soc.*, 1999, **121**, 822.
54. S. Shaik, D. Danovich, W. Wu and P. C. Hiberty, *Nat. Chem.*, 2009, **1**, 443.
55. M. Stachowicz, M. Malinska, J. Parafiniuk and K. Woźniak, *Acta Crystallogr.*, 2017, **B73**, 643.
56. D. W. J. Cruickshank and J. S. Rollett, *Acta Crystallogr.*, 1953, **6**, 705.
57. P. M. De Wolff, *Acta Crystallogr.*, 1966, **20**, 141.
58. S. Mondal, S. van Smaalen, A. Schoenleber, Y. Filinchuk, D. Chernyshov, S. I. Simak, A. S. Mikhaylushkin, I. A. Abrikosov, E. Zarechnaya, L. Dubrovinsky and N. Dubrovinskaia, *Phys. Rev. Lett.*, 2011, **106**, 215502.
59. S. Mondal, S. van Smaalen, G. Parakhonskiy, S. J. Prathapa, L. Noohinejad, E. Bykova, N. Dubrovinskaia, D. Chernyshov and L. Dubrovinsky, *Phys. Rev. B*, 2013, **88**, 024118.
60. A. M. M. Schreurs, X. Xian and L. M. J. Kroon-Batenburg, *J. Appl. Crystallogr.*, 2010, **43**, 70.
61. G. M. Sheldrick, *SADABS 2008/1*, BrukerAXSInst. Inc., Madison, Wisconsin, USA, 2008.
62. M. Fujimori, T. Nakata, T. Nakayama, E. Nishibori, K. Kimura, M. Takata and M. Sakata, *Phys. Rev. Lett.*, 1999, **82**, 4452.
63. A. R. Oganov, J. H. Chen, C. Gatti, Y. Z. Ma, Y. M. Ma, C. W. Glass, Z. X. Liu, T. Yu, O. O. Kurakevych and V. L. Solozhenko, *Nature*, 2009, **457**, 863; Addendum: A. R. Oganov, J. H. Chen, C. Gatti, Y. Z. Ma, Y. M. Ma, C. W. Glass, Z. X. Liu, T. Yu, O. O. Kurakevych and V. L. Solozhenko, *Nature*, 2009, **460**, 292.
64. E. Y. Zarechnaya, L. Dubrovinsky, N. Dubrovinskaia, Y. Filinchuk, D. Chernyshov, V. Dmitriev, N. Miyajima, A. El Goresy, H. F. Braun, S. van Smaalen, I. Kantor, A. Kantor, V. Prakapenka, M. Hanfland, A. S. Mikhaylushkin, I. A. Abrikosov and S. I. Simak, *Phys. Rev. Lett.*, 2009, **102**, 185501.
65. R. Destro, R. Ruffo, P. Roversi, R. Soave, L. a. Loconte and L. L. Presti, *Acta Crystallogr., Sect. B*, 2017, **73**, 722.
66. C. Galven, J. L. Fourquet, M. P. Crosnier-Lopez and F. Le Berre, *Chem. Mater.*, 2011, **23**, 1892.

67. W. Schmidt, P. Bottke, M. Sternad, P. Gollob, V. Hennige and M. Wilkening, *Chem. Mater.*, 2015, **27**, 1740.
68. S. Nishimura, Y. Suzuki, J. Lu, S. Torii, T. Kamiyama and A. Yamada, *Chem. Mater.*, 2016, **28**, 2393.
69. M. Renninger, *Z. Phys.*, 1937, **106**, 141.
70. R. F. Stewart, M. A. Spackman and C. Flensburg, *VALRAY User's Manual, Version 2.1*, Carnegie Mellon University Press, Pittsburgh, USA, and University of Copenhagen Press, Copenhagen, Denmark, 2000.
71. R. F. Stewart, *Electron and Magnetization Densities in Molecules and Solids*, ed. P. Becker, New York, Plenum Press, 1980, pp. 427–431.
72. A. Nakatsuka, M. Shimokawa, N. Nakayama, O. Ohtaka, H. Arima, M. Okube and A. Yoshiasa, *Am. Mineral.*, 2013, **98**, 783.
73. M. S. Schmøkel, L. Bjerg, S. Cenedese, M. R. V. Jørgensen, Y.-S. Chen, J. Overgaard and B. B. Iversen, *Chem. Sci.*, 2014, **5**, 1408.
74. *SAINT*, Bruker, Bruker AXS Inc., Madison, Wisconsin, USA, 2007.
75. *SADABS*, Bruker, Bruker AXS Inc., Madison, Wisconsin, USA, 2001.
76. R. H. Blessing, *J. Appl. Crystallogr.*, 1997, **30**, 421.
77. R. H. Blessing and D. A. Langs, *J. Appl. Crystallogr.*, 1987, **20**, 427.
78. V. V. Zhurov, E. A. Zhurova and A. A. Pinkerton, *Inorg. Chem.*, 2011, **50**, 6330.
79. L. K. Templeton, D. H. Templeton, A. Zalkin and H. W. Ruben, *Acta Crystallogr., Sect. B*, 1982, **38**, 2155.
80. D. H. Templeton, A. Zalkin, H. W. Ruben and L. K. Templeton, *Acta Crystallogr., Sect. C*, 1985, **41**, 1439.
81. L. K. Templeton and D. H. Templeton, *Acta Crystallogr., Sect. C*, 1989, **45**, 675.

CHAPTER 5

Experimental Charge Density Analysis in Organic Solids

VENKATESHA R. HATHWAR

Division of Physics, Faculty of Pure and Applied Sciences,
University of Tsukuba, Tsukuba, 305-8571, Japan
*E-mail: hathwar.rama.fw@u.tsukuba.ac.jp, mudradivenky@gmail.com

5.1 Introduction

Supramolecular chemistry and crystal engineering deal with various kinds of strong and weak intermolecular interactions to design the structure and function of "supermolecules" in a crystal.[1] Intra- and intermolecular interactions in organic compounds play a predominant role in the determination of molecular conformations and close packing of molecules in a crystal. Hydrogen bonds (HBs), halogen bonds (XBs) and van der Waals (vdW) interactions constitute a major part of supramolecular chemistry and modern-day crystal engineering. Understanding the strength and nature of weak interactions is essential to obtain "engineered" interactions between pre-selected building blocks of molecular assemblies. Crystal packing, polymorphism, molecular conformation, self-assembly processes, stabilization of host–guest complexes, ligand binding and molecular folding in biological systems, and the stability and activity of biomolecules are all governed by the strength as well as directionality of molecular interactions in a crystal.[1]

Monographs in Supramolecular Chemistry No. 26
Understanding Intermolecular Interactions in the Solid State: Approaches and Techniques
Edited by Deepak Chopra

Published by the Royal Society of Chemistry, www.rsc.org

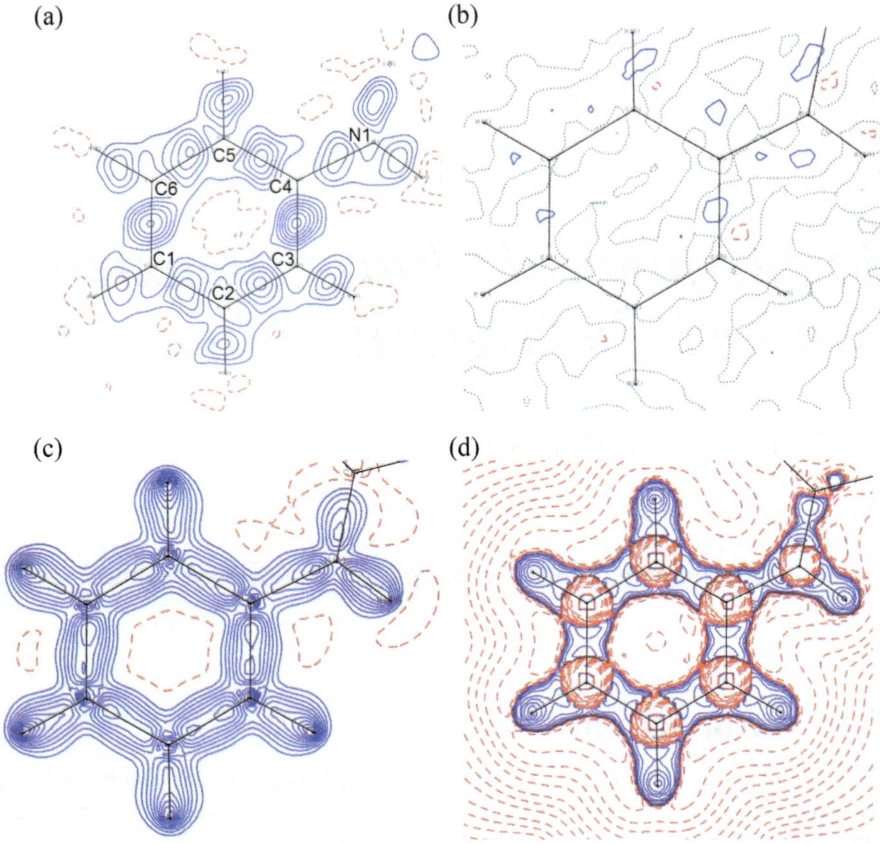

Figure 5.1 Residual density maps (a) from IAM, (b) from MM, (c) static deformation density map and (d) Laplacian map of the phenyl ring in acetanilide from MM using X-ray data collected at 100 K. Positive, negative and zero contours are depicted by red, blue and black colours, respectively. Contours are drawn at the intervals of ±0.1 e Å^{-3} in (a)–(c) and at the logarithmic intervals of $\nabla^2\rho(\mathbf{r})$ e Å^{-5} in (d).

X-ray diffraction (XRD) is found to be the most reliable method for accurate determination of molecular structures in crystalline materials. A precise knowledge of atomic positions in a crystal is obtained by a routine crystal structure determination method based on the independent atom model (IAM). This leads to "ball and stick model" model of atoms in a molecule. Hence, the evaluation of chemical bonding is limited to the analysis of bond lengths, bond angles and molecular conformations to qualitatively describe their importance in the stability of crystal structure. However, the IAM model does not account for the electron density (ED) in the interatomic region due to the chemical bonding. This is revealed by the observation of residual density peaks on the bonding region (Figure 5.1a). After several improvements, a better description of the structural model is provided by the multipole model

(MM).[2] The refinement of MM using high resolution experimental X-ray data results in nearly a featureless residual density map (Figure 5.1b). Aspherical features of ED arising from chemical interactions and lone pair density are visible in the static deformation density maps (Figure 5.1c). Further, quantitative analysis of the chemical bonding is obtained through topological properties of ED extracted from the MM using Bader's quantum theory of atoms in molecules (QTAIM). Hence, quantitative information regarding the nature of chemical bonding and interaction energies can be derived from XRD experiments. The experimental results obtained from the charge density (CD) method are in good agreement with well-established theoretical calculations. Over the last two decades, the field of CD analysis has become an attractive research area due to significant improvements in methodology as well as in instrumentation and detector technology, for example microfocus X-ray tubes for laboratory X-ray diffractometers and the availability of synchrotron facilities.[3]

Recent reviews[3–9] and a textbook[10] highlight the latest developments in the CD method and its reliability for the understanding of bonding interactions in numerous chemical and biological systems. A detailed description of the methodology and theoretical background can be found elsewhere.[11] The present chapter is focused on outlining the application of the CD method for organic compounds using relevant experimental results published in the literature.

5.2 Experimental Requirements

5.2.1 Good Quality Single Crystals and High-resolution X-Ray Data

The primary requirement for the experimental CD analysis is the availability of good quality single crystals that can provide high-resolution X-ray data, normally up to $(\sin\theta/\lambda)_{max} > 1.1$ Å^{-1}. Additionally, crystals should be free from twinning, defects and disorder. X-ray data required for CD analysis is significantly different from routine crystal structure determination. High accuracy and precision in the measurement of X-ray data are desired to obtain a reliable MM by deconvoluting the thermal motion and the ED. Both accuracy and precision in data measurement are essential as the valence electrons contribute to a small portion of the measured intensity and they carry all the chemically most valuable information related to chemical bonding. The main advantage of organic crystals is that valence electrons constitute a relatively large portion of the scattering intensity compared to that of core electrons. Further, recent improvements in detectors for X-rays, such as in charge-coupled devices (CCDs), complementary metal-oxide semiconductors (CMOSs), imaging plates (IPs) and new-generation hybrid pixel detectors (XPAD and Pilatus), and the availability of intense X-ray sources in the form of rotating anodes and microfocus tubes with multilayer mirror optics, provide good-quality X-ray data with high signal-to-noise ratio. In addition, it is important to cool the crystal using either nitrogen or helium cryogenics to increase the data quality and to minimize thermal diffuse scattering. Another advantage of organic crystals is either small or insignificant extinction and absorption effects. Hence, a conventional laboratory diffractometer can provide high-quality X-ray data required for CD modeling. If synchrotron radiation is available for data collection, it is always beneficial to obtain high quality data in a few hours of experimental time. In recent times, synchrotron facilities are optimized for CD studies and even tiny ED features are accurately reproduced in the experimental CD model.

5.2.2 Multipolar Modeling of CD Data

In the XRD experiment, the ED in the unit cell is obtained through an inverse Fourier series summation over all measured structure factors (F_{hkl}) of Bragg reflections such that

$$\rho(x,y,z)=\frac{1}{V}\sum_{h,k,l}F_{hkl}\mathrm{e}^{-2\pi \mathrm{i}(hx+ky+lz)} \tag{5.1}$$

where x, y and z are fractional coordinates of atoms satisfying the relation, $\mathbf{r} = x\mathbf{a} + y\mathbf{b} + z\mathbf{c}$. V and hkl are the unit cell volume and Miller indices, respectively. Aspherical features of the ED in the bonding region are completely neglected in the description of the crystal structure based on IAM. To overcome deficiencies of IAM, the Hansen and Coppens (HC) multipole formalism[2] is used to describe the atomic density as a superposition of pseudoatoms in the CD modelling. The atomic density in the HC model is given by

$$\rho_{\mathrm{atom}}(\mathbf{r})=P_{\mathrm{c}}\rho_{\mathrm{c}}(\mathbf{r})+P_{\mathrm{v}}\kappa^{3}\rho_{\mathrm{v}}(\kappa\mathbf{r})+\sum_{l=0}^{l_{\max}}\kappa'^{3}R_{l}(\kappa'\mathbf{r})\sum_{m=0}^{l}P_{lm\pm}Y_{lm\pm}(\mathbf{r}/r) \tag{5.2}$$

where ρ_c and ρ_v are the spherical core and valence contributions to the atomic density, respectively. P_c, P_v and $P_{lm\pm}$ are the core, valence and multipolar population parameters, respectively. P_v gives a rough estimation of the net atomic charge by $q = N_v - P_v$, where N_v is the number of valence electrons in a free neutral atom. The terms, $Y_{lm\pm}$, and $R_l(\mathbf{r})$, represent real spherical harmonic functions of order l and m corresponding to the orientation of l and the Slater-type radial function, respectively. The coefficients κ and κ' are the contraction–expansion parameters for spherical and multipolar valence densities, respectively.

Multipolar modeling of high-resolution X-ray data at low temperature will efficiently deconvolute the thermal motion and the ED. The quality and correctness of MM are examined by the Hirshfeld rigid-body test,[12] residual density ($\Delta\rho_{\mathrm{residual}}(\mathbf{r}) = \rho_{\mathrm{data}} - \rho_{\mathrm{model}}$ or probability density function (pdf),[13] and a fractal dimension distribution of residual density.[14] Once a reliable MM is obtained, the bonding features are directly analyzed through a static deformation density map, which shows the density deformation with respect to the promolecular density (Figure 5.1c). The static deformation density is described as, $\Delta\rho_{static}(\mathbf{r}) = \rho_{MM} - \rho_{IAM}$, where, ρ_{MM} and ρ_{IAM} are the thermally averaged density obtained from MM and the spherically averaged density from the promolecule, respectively. All these aspects of CD modeling are implemented in *MoPro*[15] and *XD*[16] program suites for the experimental determination of electron density distribution (EDD) using X-ray data.

Another critical issue in the CD study of organic compounds using X-ray data is the treatment of H-atoms in the CD model. Reliable evaluation of ED properties and strength of HBs depends on the accurate determination of the position and anisotropic displacement parameters (ADPs) of the H-atom. In this context, combined X-ray and neutron-diffraction measurements and subsequent derivation of the CD model will be the most accurate. However, there are experimental limitations related to neutron experiments due to inadequate experimental facilities, long data-acquisition times and large crystal size requirement. As an alternative approach, the *SHADE2* web server[17] and Hirshfeld atom refinement (HAR)[18] are found to be powerful in estimating the ADPs of H-atoms in the absence of neutron diffraction data. Another quantum crystallographic method, the X-ray-constrained wavefunction (XCW), has become a promising alternative to MM for CD refinements.[19] In XCW, a theoretically obtained wavefunction can be deformed by the refinement of orbital coefficients against the experimental X-ray data such that the calculated X-ray structure factors best match the experimental ones. XCW provides topological properties of the chemical bonding along with the Roby bond order and electron localizability indicator (ELI).[20,21] In a recent systematic CD study on acetazolamide, the ED results obtained from the XCW were comparable to the MM based analysis.[21]

5.3 Evaluation of ED Features from the Experimental CD Model

5.3.1 Quantum Theory of Atoms in Molecules (QTAIM)

The EDD revealed by static deformation density and Laplacian maps through MM (Figure 5.1c and d) does not explain the nature of chemical interactions, especially the mechanism of electron sharing between atoms. The curvatures of ED in a molecule is analyzed by Bader's QTAIM.[22] According to a unique partitioning scheme of the QTAIM, the total density of the molecule is divided into open subsystems referred to as atomic basins, a physically meaningful partition of the molecule into its atoms. The boundaries among atoms in the molecule are defined by zero-flux surfaces, such that $\nabla\rho(\mathbf{r})\cdot n(\mathbf{r}) = 0$ where $\nabla\rho(\mathbf{r})$ is the gradient of ED and $n(\mathbf{r})$ is the vector normal to the surface at **r**. It has been demonstrated that the atomic regions defined through zero-flux surfaces are non-overlapping and exhaustive. It allows for the determination of properties from individual atomic basins, *e.g.*, net atomic charges, atomic volumes and electrostatic moments. Additionally, the analysis of curvatures of ED include identification of critical points (CPs), where the gradient of the ED vanishes, ($\nabla\rho(\mathbf{r}_{cp}) = 0$). The CP corresponds to the local minima, local maxima and saddle points of the EDD. The rank and signature of the Hessian matrix classify the CPs into nuclear critical points (NCP), bond (saddle) critical points (BCP), ring critical points (RCP) and cage critical points (CCP) with (3, −3), (3, −1), (3, +1) and (3, +3) labels, respectively. The second derivative of the ED given by the Hessian matrix is known as the Laplacian, $\nabla^2\rho(\mathbf{r}_{\mathrm{cp}}) = (\lambda_1 + \lambda_2 + \lambda_3)$, where λ_i defines the eigenvalues of the Hessian matrix. The Laplacian also recovers the shell structure of the atom to reveal a local accumulation/depletion of the ED. If the ED is locally concentrated, then $\nabla^2\rho(\mathbf{r}_{\mathrm{cp}}) < 0$ and $\nabla^2\rho(\mathbf{r}_{\mathrm{cp}}) > 0$ for locally depleted ED at a given point in space. In the case of shared interactions, the value of $\rho(\mathbf{r}_{\mathrm{cp}})$ at the CP is relatively high and $\nabla^2\rho(\mathbf{r}_{\mathrm{cp}}) < 0$, whereas the value of $\rho(\mathbf{r}_{\mathrm{cp}})$ is small and $\nabla^2\rho(\mathbf{r}_{\mathrm{cp}}) > 0$ for closed shell interactions. The presence of a maxima in the negative Laplacian of the ED represents valence shell charge concentrations (VSCCs). Those maxima in the non-bonding regions depict lone pair positions and orientation of the VSCCs indicates the hybridization of the atom. In addition, π-conjugation in the bonding is reflected by the ellipticity value (ε) obtained from eigenvalues of the Hessian matrix such that $\varepsilon = \lambda_1/\lambda_2 - 1$, where λ_1 and λ_2 are the negative curvatures of the ED at the BCP. Ellipticity at the BCP provides a quantitative measure of the π character of a given bond and deviation of the bonding density from cylindrical symmetry. The line of the highest ED between two atoms is defined as the "bond path" and it can deviate from linearity based on the distribution of the ED. Hence, the QTAIM provides a methodology for the quantitative understanding of chemical bonding between two interacting atoms in a molecule, which has gained a lot of importance in the evaluation of the nature of interactions.[4,23]

Bond formation is associated with changes in local potential and kinetic energy densities. The total energy density $H(\boldsymbol{r}_{\mathrm{cp}})$ [the sum of potential energy density, $V(\boldsymbol{r}_{\mathrm{cp}})$ and kinetic energy density, $G(\boldsymbol{r}_{\mathrm{cp}})$] is used to discriminate between covalent and non-covalent interactions.[24] A dominant covalence character in the bonding is observed when $H(\boldsymbol{r}_{\mathrm{cp}}) < 0$ (*i.e.* when the potential energy density is in excess). On the other hand, $H(\boldsymbol{r}_{\mathrm{cp}}) > 0$ indicates a closed shell interaction. Abramov[25] proposed a functional to estimate the kinetic energy density based only on the ED and its derivatives obtained from experimental determinations. Abramov's kinetic energy density functional at the BCP is given by

$$G(\boldsymbol{r}_{\mathrm{cp}}) = \frac{3}{10}(3\pi^2)^{2/3}\rho^{5/3}(\boldsymbol{r}_{\mathrm{cp}}) + \frac{1}{6}\nabla^2\rho(\boldsymbol{r}_{\mathrm{cp}}). \tag{5.3}$$

The local virial theorem provides the corresponding potential energy density at the BCP and it is expressed as

$$V(\boldsymbol{r}_{\mathrm{cp}}) = \frac{1}{4}\nabla^2\rho(\boldsymbol{r}_{\mathrm{cp}}) - 2G(\boldsymbol{r}_{\mathrm{cp}}) \tag{5.4}$$

and the total energy density,

$$H(\boldsymbol{r}_{\mathrm{cp}}) = G(\boldsymbol{r}_{\mathrm{cp}}) + V(\boldsymbol{r}_{\mathrm{cp}}) \tag{5.5}$$

The eqn (5.3)–(5.5) are found to be useful in deriving relationships between topological properties at the BCP and correlated with the nature of chemical bonding. As a word of caution, eqn (5.3) holds well only for low ED regions such as a closed shell interaction. Espinoza *et al.*[26] used the $|V(\mathbf{r}_{cp})|/G(\mathbf{r}_{cp})$ ratio at the BCP to classify interactions such that closed-shell interactions exist when $|V(\mathbf{r}_{cp})|/G(\mathbf{r}_{cp}) < 1$, shared interactions for $|V(\mathbf{r}_{cp})|/G(\mathbf{r}_{cp}) > 2$ and intermediate when the ratio falls between 1 and 2. Similarly, Macchi *et al.*[27] suggested the ratio $G(\mathbf{r}_{cp})/\rho(\mathbf{r}_{cp}) > 1$ for closed-shell interactions and $G(\mathbf{r}_{cp})/\rho(\mathbf{r}_{cp}) < 1$ for shared interactions. The classification of interactions based on energy densities at the BCP is very general and dichotomous in nature.[8] Hence, they should be used to derive interesting and relative trends during the bond formation. The classification of atomic interactions based on topological properties at the BCP is summarized in Table 5.1.

5.3.2 Source Function (SF) Analysis

Bader and Gatti[28] proposed that the ED $\rho(\mathbf{r})$ at any given point $\mathbf{r}$ in a system is determined by contributions from a local source LS($\mathbf{r}$,$\mathbf{r}'$) operating at all other points $\mathbf{r}'$ of the system and is expressed as,

$$\rho(\boldsymbol{r}) = \int \mathrm{LS}(\boldsymbol{r}, \boldsymbol{r}')\mathrm{d}\boldsymbol{r}' \tag{5.6}$$

where,

$$\mathrm{LS}(\boldsymbol{r},\boldsymbol{r}') = \frac{-\nabla^2 \rho(\boldsymbol{r})}{4\pi|\mathbf{r}-\mathbf{r}'|} \tag{5.7}$$

Hence, the ED at $\boldsymbol{r}$ is related to local behavior in terms of the Laplacian weighted by the influence function, $4\pi|\boldsymbol{r}-\boldsymbol{r}'|^{-1}$. It quantifies the influence of all other parts of a system to the local property or chemical behavior of the system. The SF analysis is a powerful ED-based tool to study the localization/delocalization of chemical bonds and can be applied to experimental multipolar density obtained by the CD model. Indeed, SF contributions at the BCP are correlated with the nature and strength of the bond. In general, the contribution from the two bonded atoms to their BCP is higher for a stronger interaction and reflects the localized character of a given interaction and *vice versa*.

5.3.3 Non-covalent Interactions (NCIs) Descriptor

Atomic interactions without any electron sharing to a significant extent are considered as NCIs. They play an important role in many chemical and biological systems. The NCI descriptor was first introduced by Johnson *et al.*[29] for visualizing regions of space involved in either attractive or repulsive interactions. The two quantities that constitute the NCI descriptor are the reduced density gradient (RDG) and the quantity $\rho(\boldsymbol{r})$*signλ_2 where λ_2 is the second greatest eigenvalue of the ED Hessian matrix. The RDG [$s(\boldsymbol{r})$], is a dimensionless quantity and expressed as,

$$s(\boldsymbol{r}) = \frac{|\nabla \rho(\mathbf{r})|}{2(3\pi)^{1/3}\rho(\boldsymbol{r})^{4/3}}. \tag{5.8}$$

Table 5.1 Classification of intra- and intermolecular interactions based on the topological properties of ED at the BCP.

Type of atomic interactions	Topological properties of the ED at the BCP						
				Energy components			
	$\rho(\boldsymbol{r}_{cp})$	$\nabla^2\rho(\boldsymbol{r}_{cp})$	λ_i	$G(\boldsymbol{r}_{cp})/\rho(\boldsymbol{r}_{cp})$	$H(\boldsymbol{r}_{cp})/\rho(\boldsymbol{r}_{cp})$	$\lvert V(\boldsymbol{r}_{cp})\rvert/G(\boldsymbol{r}_{cp})$	$H(\boldsymbol{r}_{cp})$
Shared shell (covalent bonds)	Large	$<<0$	$\lvert\lambda_{1,2}\rvert/\lvert\lambda_3\rvert > 1$	<1	$<<0$	>2	<0
Closed shell (HB, XB, ionic, vdW interactions)	Small	>0	$\lvert\lambda_{1,2}\rvert/\lvert\lambda_3\rvert \leq 1$	≥ 1	>0	<1	>0
Intermediate interactions (polar bonds)	Large	Any value	$\lvert\lambda_{1,2}\rvert/\lvert\lambda_3\rvert > 1$	≥ 1	$<<0$	1 to 2	<0

The ED, $\rho(\boldsymbol{r})$ between interacting atoms can be determined by the experimental CD model. The RDG function is characterized by large values in the region away from the nuclei and vanishes at CP ($\nabla\rho(\boldsymbol{r}) = 0$). Low ED as well as RDG values correspond to non-bonded interactions. The mapping of the quantity, $\rho(\boldsymbol{r})$*signλ_2, on RDG isosurfaces can distinguish between stabilizing [$\rho(\boldsymbol{r})$*sign$\lambda_2 < 0$] and destabilizing [$\rho(\boldsymbol{r})$*sign$\lambda_2 > 0$] interactions in the crystal. Saleh *et al.*[30] developed the NCImilano program to obtain NCI analysis directly from the experimental CD model. It is now further extended to obtain a relationship between stabilization energy and kinetic energy density integrated over volume bounded by RDG isosurfaces.[31]

5.3.4 Lattice and Interaction Energies from the CD Model

One of the key factors that govern molecular assemblies in the crystalline solids is the relative strength of various interactions. The strength of interactions is estimated in terms of total interaction energy (E_{int}) of molecular dimers. Indeed, E_{int} for interacting molecular pairs could be determined from the experimental CD model. It is expressed as the sum of electrostatic (E_{es}), exchange-repulsion ($E_{\text{ex-rep}}$), dispersion (E_{disp}) and induction (E_{ind}) energy contributions and given by

$$E_{\text{int}} = E_{\text{es}} + E_{\text{ex-rep}} + E_{\text{disp}} + E_{\text{ind}}. \tag{5.9}$$

In the CD model, E_{es} is calculated by the exact potential and multipole moment (EPMM) method[32] where it combines the numerical evaluation of the exact Coulomb integral for the short-range interactions between atoms and the Buckingham multipole approximation for the long-range interatomic interactions. The $E_{\text{ex-rep}}$ and E_{disp} terms are approximated with various atom–atom potentials implemented in the XD program suite. For the experimental data, the induction (or polarization) effects of the crystal lattice are already included in the experimentally obtained multipole parameters since these are determined in the crystal environment. Hence, the E_{ind} term is not available independently from the experimental MM. Likewise, the lattice energy is also estimated using the same methodology as described for E_{int} by incorporating all possible intermolecular interactions within a spherical shell centered about the reference molecule in the asymmetric unit. In addition, E_{int} for HBs in organic molecular crystals is calculated using the Espinosa–Molins–Lecomte (EML) relationship ($E_{\text{int}} = 0.5V(\mathbf{r}_{\text{cp}})$).[26] The EML relationship was initially derived for HBs and provides only a rough estimate of interaction energies. Recently, it has been demonstrated that the EML relationship provides an unreliable estimation of interaction energies if it is extended to all types of atom···atom interactions.[33] In many cases, the EML energy provides wrong input on the energetic importance of specific interactions in crystal packing and their subsequent ranking in the crystal structure. Furthermore, it is always beneficial to compare experimentally derived interaction and lattice energies with well-established theoretical calculations and the *PIXEL* method.[34] The program *CrystalExplorer*[35] is also improved further to provide a quantitative description of interactions along with Hirshfeld and electrostatic potential surfaces.

5.3.5 Molecular Electrostatic Potentials

One of the most important properties associated with the charge density distribution in molecular crystals is the electrostatic potential, $V(\boldsymbol{r})$ (ESP). The ESP is related directly to physically observable properties of the ED, and thus provides information about the charge distribution at any point in the crystal. Essentially, it is the energy required to bring a positive unit of charge from infinity to a point in charge distribution space. The ESP provides more insight into the understanding of the chemical reactivity, analysis of molecular binding mechanism and close packing of molecules in a crystal. A negative ESP surface indicates that the point charge is attracted towards it, while a positive ESP surface is an indication of an energetically destabilized system with respect to the point charge at infinity. The ESP at point **r** is defined as

$$V(\boldsymbol{r}) = \sum_j \frac{Z_j}{|\boldsymbol{r} - \boldsymbol{R}_j|} - \int \frac{\rho(\boldsymbol{r})}{|\boldsymbol{r} - \boldsymbol{r}'|} \mathrm{d}\boldsymbol{r}' \tag{5.10}$$

where Z_j is the charge of the jth nucleus at position $\boldsymbol{R}_j$ and $\rho(\boldsymbol{r})$ is the total ED at $\boldsymbol{r}$. It is important to note that the electrostatic forces are long-range forces and therefore it determines the path along which molecules could approach.

Besides these experimentally available CD descriptors, many characterization tools are obtained from theory. The electron localization and delocalization indices,[36] electron localization function (ELF),[37] distributed atomic polarizability[38] and nucleus-independent chemical shifts (NICSs)[39] are helpful to understand the chemical bonding and EDD in a molecule. A detailed description of these methods is available elsewhere.[10]

5.4 Applications

The CD method is extensively applied to address numerous chemical problems, *e.g.* in the evaluation of intra- and intermolecular interactions, the analysis of polymorphs, the study of chemical reactions and even for a detection of ED delocalization effects in aromatic systems. In this section, some representative cases in organic systems are examined based on published results in the literature.

5.4.1 Evaluation of Intra- and Intermolecular Interactions

Quantitative evaluation of interactions in molecular crystals using topological properties of MM has become an area of immediate interest. It provides the basis for understanding supramolecular aggregation and quantitative crystal engineering beyond mere geometrical information. In the study of

HB, Koch and Popelier[23] proposed eight necessary criteria based on QTAIM to distinguish a "true" HB from that of a vdW interaction. The existence of a BCP between hydrogen (H) and acceptor (A) atoms linked through a bond path, the presence of ED at the BCP, a positive value of the Laplacian at the BCP and its correlation with interaction energy constitute the first three criteria. The fourth criterion, the mutual penetration of the H-atom and A-atom basins is considered as "necessary and sufficient" for HB formation. For a true HB, $\Delta r_H = (r_H^0 - r_H) > \Delta r_A = (r_A^0 - r_A)$ and $\Delta r_H + \Delta r_A > 0$ should be obeyed where r_H^0, r_A^0 are non-bonded radii of H- and A-atoms taken as the gas phase vdW radii and r_H, r_A are corresponding bonding radii taken as distances from the BCP to the nuclei. If these conditions are violated then the interaction is essentially vdW in nature. The remaining four criteria are obtained from integration over the atomic basin of the H-atom. They are a loss of charge and energetic destabilization of the H-atom, and a decrease in dipolar polarization and atomic volume of the H-atom. These criteria have been widely applied in the characterization of C–H···O HBs,[40] dihyrogen ($H^{\delta+}\cdots H^{\delta-}$) bonds, strong O–H···O HBs[41] and C–H···F and C–F···F–C interactions.[42] A systematic experimental ED analysis on ionic complexes of 1,8-bis(dimethylamino)-naphthalene has shown an exponential dependence of ED and energy densities at the BCP on the bond path length and a Morse-type dependence for Laplacian values for several intermolecular interactions. In another study, a clear distinction between C–H···O HB and C–H···π vdW interaction is demonstrated from the experimental CD studies on coumarin, 1-thiocoumarin, and 3-acetylcoumarin using X-ray data collected at 90 K.[40] Indeed, a well-defined "region of overlap" between C–H···O HBs and C–H···π vdW interactions has been determined and the corresponding R_{ij} value falls in the range 2.75–2.85 Å. In the extended study,[43] numerous intra- and intermolecular interactions in molecular crystals are quantitatively classified as strong HBs, weak HBs and vdW interactions based on the KP criteria (Figure 5.2). Nowadays, the experimental CD results on HBs are often very consistent with high level theoretical calculations. Even small values of ED (~0.01 e $Å^{-3}$) at the BCP of intermolecular interactions from experiments can be reproduced with high accuracy when compared with theoretical values.[44]

In the study of low-barrier hydrogen bonds (LBHB), short intramolecular O–H···O HBs are subjected to CD studies using combined X-ray and neutron diffraction data collected at very low temperatures. The neutron data are necessary to obtain the accurate determination of the best structural model, especially in relation to the position and ADPs of H-atoms. High-resolution X-ray data are required to obtain complementing information on ED features. The CD study on benzoylacetone was performed with 8.4 K X-ray data and 20 K neutron data to investigate an unsymmetrical *cis*-enol system with a short intramolecular O–H···O HB of distance 1.329(11) Å.[45] This study provided a direct quantitative evidence for extensive π-density delocalization in the keto-enol group. The short O–H···O HB is characterized by both electrostatic and covalent bonding contributions. Indeed, this type of short

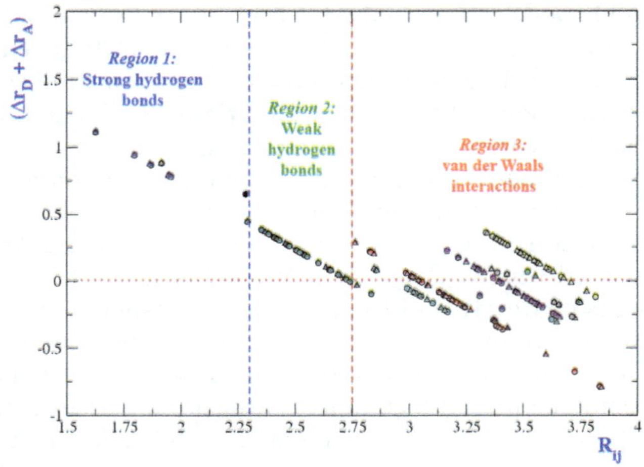

Figure 5.2 Classification of intermolecular interactions based on topological properties at the BCP using the KP criteria. Adapted with permission from ref. 43 with permission from the Royal Society of Chemistry.

O–H···O HBs are described as resonance-assisted hydrogen bonds (RAHB)[46] where substantial and equal negative charges are found on both O atoms accompanied by the positive charge on the H-atom. In another interesting ED study on short intermolecular N–H···O HBs of the catalytic triad in serine proteases, the absence of an LBHB for short N–H···O HBs has been confirmed from the topological properties of ED.[47] The study indicated that H-atoms in all short intermolecular N–H···O HBs of the compound are localized near the N-atoms. Additionally, it has shown that short and strong HBs in molecular crystals need not be LBHBs.

Another important concept, the source function (SF) analysis is used to characterize the chemical bonding.[28] According to the SF analysis, the ED at a given point in space is determined from the contribution of every atomic basin within the molecule. It can be applied to the experimental CD model and provides an additional quantitative tool to evaluate the nature of HBs[47] as well as conjugation/delocalization effects in aromatic systems.[48] These effects are determined from the sum of atomic SF contributions to a reference point (normally at the BCP of a bond). In the study of HBs,[47] the SF contribution from the H-atom has been a strong indicator of the nature of a given HB, namely a weak HB, a strong HB, a single well HB or a RAHB. It has been shown that the SF contribution from the H-atom to the BCP of H···O bonding changes from negative to positive with increasing strength of the HB.[47] In a recent CD study on urea based anion-receptor complexes,[49] the SF analysis has been applied to classify HB contributions as strong and weak HBs in the crystal structure. It was revealed that the acidic nature of H-atoms participating in different HBs is exhibited by SF contributions of H-atoms. Thus, the SF results could provide subtle changes in electronic distributions, which go beyond the routine topological analysis of HBs. Recently, SF analysis was used to reveal electron delocalization effects in aromatic systems based on experimental multipolar density of benzene and naphthalene.[48] They were selected as examples of unperturbed aromatic systems. Experimental SF% values from X-ray data were found to be consistent with those estimated from *ab initio* periodic wavefunctions. This study displayed the ability of experimental MM to accurately reproduce the subtle electronic effects induced by electron delocalization on the atomic SF% contributions.

An important purpose of the CD analysis in molecular crystals is to obtain quantitative information on energies of various interactions in the crystal packing. One of the early attempts to determine interaction and lattice energies from the experimental X-ray data was performed by Abramov *et al.*[50] on glycylglycine, DL-histidine, and DL-proline. The procedure of calculations was described in Section 5.3.4. It was established that the lattice energy obtained from the experimental MM shows a significant discrepancy when compared with theoretical values, even at the HF level. A similar discrepancy for lattice energies from the CD model is highlighted in recent studies during the comparative study with well-established theoretical methods.[51,52] Hence, special attention is required while interpreting the lattice energy results from the CD method. However, those studies suggest that dimer interaction energies are in good agreement with other theoretical methods, for example *PIXEL*.[34]

It is always desirable to establish a correlation between topological properties of ED and physical/chemical properties of a material. In this context, organic semiconductor rubrene[52] was studied by combined X-ray and neutron diffraction data. The electronic delocalization of π-density on the conjugated teracene backbone of the rubrene molecule and π···π stacking interactions of distance 3.706(1) Å in a crystal have been experimentally confirmed in rubrene. Indeed, these two factors are responsible for the observed high mobility of charges and good semiconducting properties of rubrene. Likewise, the ED results on rubrene endoperoxide[53] show a loss of delocalization of π-density after the formation of an O–O bridge in rubrene molecules. Recently, optical properties such as polarizability and the refractive index of organic molecular crystals have been determined from the refinement of experimental X-ray diffraction data using the XCW method.[54,55] The optical properties estimated from the XCW method in several non-linear optical materials are consistent with *ab initio* MP2 estimates that incorporate crystal field effects and experimental measurements at optical frequencies.[56]

5.4.2 Chemical Reactivity in Organic Solids

Another successful application of the experimental CD method is its ability to describe chemical reactivity *via* EDDs in solids. The difference in catalytic activity of bis(pentafluorophenyl)(*N*-pyrrolyl) borane [$(C_6F_5)_2BNC_4H_4$] and bis(pentafluorophenyl)(*N*-pyrrolidinyl)borane [$(C_6F_5)_2BNC_4H_8$] compounds has been investigated using the CD method.[57] Among the three VSCCs around the N-atom in both compounds, the most pronounced VSCC is facing towards the B-atom in $(C_6F_5)_2BNC_4H_8$, whereas it is found in a heterocyclic

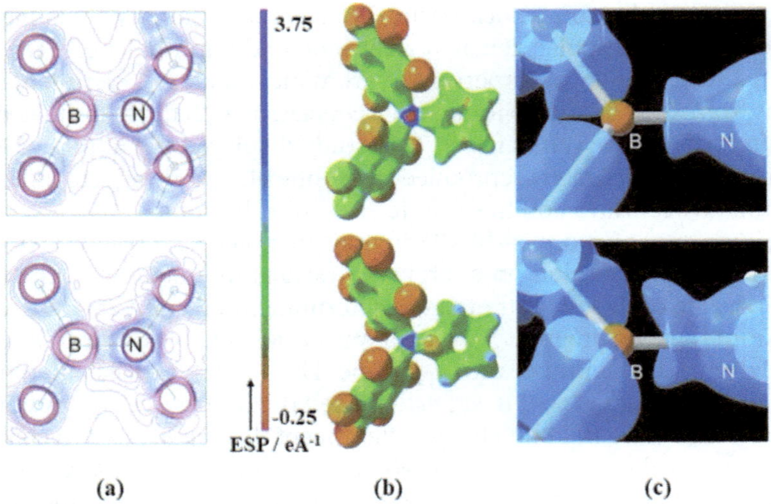

Figure 5.3 (a) The Laplacian map, (b) ESP isosurface ranging from 0.25 to +3.75 e Å^{-1} mapped on $\rho(\mathbf{r}) = 0.65$ e Å^{-3} and (c) reactive surface ($\nabla^2(\mathbf{r}) = 0$ e Å^{-5}) in the B–N–C$_{ortho}$ plane of [$(C_6F_5)_2BNC_4H_4$] (top row) and [$(C_6F_5)_2BNC_4H_8$] (bottom row). Reproduced from ref. 9 with permission from John Wiley and Sons, Copyright © 2011 WILEY-VCH Verlag GmbH & Co. KGaA, Weinheim.

ring for $(C_6F_5)_2BNC_4H_4$ (Figure 5.3a). Experimental results have validated an earlier assumption that the ED is donated from the N-atom of the heterocycle into the vacant "*p*" orbital of the B-atom, which is more pronounced in the case of [$(C_6F_5)_2BNC_4H_8$]. In addition, distinct negative charges are found at the N-atom due to the electron-withdrawing effect of the C_6F_5 group, which is counterbalanced by both the B-atom and the heterocyclic ring in [$(C_6F_5)_2BNC_4H_8$] while the B-atom alone in [$(C_6F_5)_2BNC_4H_4$]. This is clearly visualized in the ESP map as there are more positive ESP regions on the B-atom in [$(C_6F_5)_2BNC_4H_4$] than on [$(C_6F_5)_2BNC_4H_8$] (Figure 5.3b). The reactive surface plots ((Figure 5.3c) indicate that the B-atom in [$(C_6F_5)_2BNC_4H_4$] is less shielded for a potential nucleophilic attack, which further confirms a better catalytic property of [$(C_6F_5)_2BNC_4H_4$] when compared to its saturated analog, [$(C_6F_5)_2BNC_4H_8$].

Recently, the nature of dative N–B bonding and dihydrogen ($H^{\delta+}\cdots H^{\delta-}$) bonding have been examined in several Lewis acid-base adducts using experimental EDD.[58] The dative N–B bond has displayed both covalent and ionic character coupled with a charge transfer of 0.03–0.14e. It is also shown that the crystal structure of Lewis acid-base adducts is stabilized by dihydrogen bonding. Dihydrogen interactions are weak but attractive electrostatic interactions. In recent studies, the ESP is found to be more useful in the study of the chemical reactivity of molecules, where topological properties fail to provide any distinct differences. The ESP obtained from the experimental ED has been applied to study the activity of two penicillin derivatives, namely the active penamecillin and the inactive penamecillin-1β-sulfoxide.[59] The ESP maps show that the entire phenylalanyl residue is negatively charged and negative ESP extends over one side of the lactam ring in penamecillin-1β-sulfoxide. Hence, the negative ESP region blocks the site of nucleophilic attack on the carbonyl bond and results in inactivity of penamecillin-1β-sulfoxide. In another interesting study, differences in the chemical reactivity of α, β-unsaturated carbonyl and hydrazone systems have been evaluated using synchrotron X-ray data collected at 12 K.[60] The topological and ELI results have demonstrated that the reactivity differences in the carbon atom at the β-position in α, β-unsaturated carbonyl and hydrazone compounds are related to opposed resonance effects of the electron-withdrawing carbonyl group and electron-donating hydrazone group. Hence, the carbon atom at the β-position in the α, β-unsaturated carbonyl compound is available for a nucleophile attack, whereas the same position is more reactive towards an electrophilic attack in the α, β-unsaturated hydrazone compound. In a recent CD study,[61] the Se–N bond cleavage mechanism and drug action were described in the antioxidant compound ebselen. A significantly high $\nabla^2\rho(\mathbf{r})$ value for the Se–N bond when compared to the Se–C bond suggests that the Se–N bond is more ionic in nature, whereas the Se–C bond has a relatively stable covalent feature. This observation supports an earlier assumption that the Se–N bond cleaves during the interaction of nucleophilic molecules with ebselen. It is also responsible for the low-toxicity of the drug during the biotransformations. In another systematic study, ESP and molecular dipole moments obtained from the experimental MM have provided differences in relative binding affinities (RBA) between estrogen compounds (α- and β-estradiol, estrone), the phytoestrogen genistein and the synthetic estrogens, diethylstilbestrol (DES) and dienestrol (DNS).[62] It has been illustrated that the synthetic estrogens (DES and DNS) have the largest negative areas in the ESP map, which envelop the whole aromatic rings. Therefore, both molecules show the highest RBA values to estrogen receptors. The estrone and α-estradiol molecules have large negative ESP areas only at one side of the molecule, which result in intermediate values of RBA. However, estriol has one of the lowest RBA values among all molecules due to its less pronounced negative ESP areas.

5.4.3 Polymorphs and Cocrystals

The CD method has received considerable importance in the analysis of polymorphism and cocrystals, due in particular to their impact on recent pharmaceutical research. The CD descriptors are useful to reveal subtle electronic differences of polymorphs as well as to understand their stability and reactivity in the solid state. A key aspect of "drug design" is an unequivocal characterization of all possible polymorphs of a drug compound and assessing their stability. It is equally challenging to predict a stable polymorphic form of a drug compound from theory since the energy differences between different polymorphic forms may only be on the order of a few kJ mol^{-1}. Some interesting experimental CD investigations include concomitant polymorphism in 3-acetylcoumarin,[63] conformational polymorphism in famotidine,[64] coumarin 314 dye[65] and piracetam,[66] packing polymorphism in 5-nitrouracil,[67] *p*-nitrophenol,[68] paracetamol[69] and sulfathiazole.[51] The CD investigations on polymorphs of ionic compounds are rare and only few studies have been performed, for example on hydrated 1,8-bis(dimethylamino) naphthalene hydrochloride polymorphs (in *P*-1 and $P2_1/n$)[70] and benzidine dihydrochloride polymorphs (in *P*-1 and *Pbcn*).[71] In an interesting study, the stability of paracetamol polymorphs [monoclinic ($P2_1/n$) and orthorhombic (*Pbca*)] has been explained based on CD results.[69] The monoclinic form of paracetamol is more stable than its orthorhombic form even though it has a lower density. It has been demonstrated that strong H-bonds are responsible for the higher stability of the monoclinic form, whereas weak interactions are responsible for the higher density of the orthorhombic form. The diuretic drug acetazolamide (AZM), has been explored to understand the unusual phenomenon where kinetically stable polymorph (form II in $P2_1/n$) is obtained over a thermodynamically stable polymorph (form I in *P*-1) upon slow cooling of boiling aqueous solution.[72] The VSCCs of the N-atom in the S–N bond show a distinct lone pair (LP) for I but the corresponding LP is drawn into the density of the S–N bond in II (Figure 5.4a). Hence, a double bond character is enhanced for the S–N bond in II. The topological results are further supported by the 3D ELF isosurfaces (Figure 5.4b) and the natural bond orbital (NBO) analysis. NBO results demonstrate a pure "*p*" character of the LP orbital confirming the prominent sp^2 character for the N-atom in II. A significant overlap between LP and anti-bonding σ* S–C orbitals is seen in II whereas an ineffective overlap is displayed for I (Figure 5.4c). This indicates that the LP of the N-atom in I has a mixture of "*s*" and "*p*" characters. Thus, the CD study has established that the change in hybridization on the N-atom of the sulphonamide group is responsible for polymorphic modifications and resulted in "hybridization induced polymorphism".

The role of C=O···C=O, C=O···N–O and N–O···N–O dipole–dipole interactions in the crystal packing of 3,3-dinitrobenzophenone polymorphs (in $P2_12_12$ and *P*-1) is evaluated using topological properties of ED and NCI descriptors.[73] The EDD has revealed that the molecular pairs held by antiparallel dipole–dipole interactions contribute significantly toward the stability of the *P*-1 polymorph. Among all dipole–dipole interactions in the polymorph *P*-1, the antiparallel C=O···C=O interaction is energetically most stable followed by antiparallel C=O···N–O, antiparallel N–O···N–O and orthogonal N–O···N–O type dipole–dipole interactions. However, the crystal structure of the $P2_12_12$ polymorph is mainly stabilized by C–H···O interactions and characterized by the absence of stabilizing dipole–dipole interactions in the crystal.

One of the recent developments in the pharmaceutical industry is "cocrystals" where the physical and chemical properties of active pharmaceutical ingredients (APIs) are enhanced by forming multicomponent molecular crystals of API with a suitable coformer.[74] Most APIs are administered in the

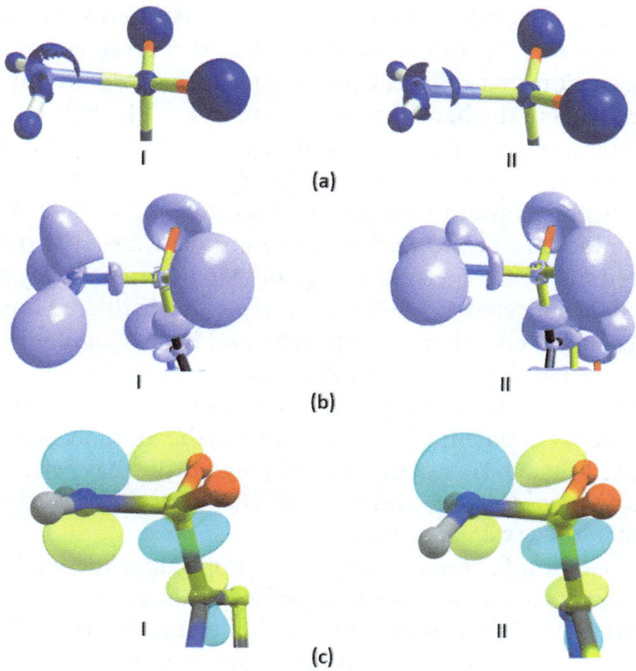

Figure 5.4 (a) Experimental 3D Laplacian isosurfaces of AZM polymorphs I and II around the nitrogen atom plotted at −40 e Å^{-5}. (b) 3D ELF surfaces at an isovalue of 0.85 e Å^{-3}. (c) NBO analysis portraying charge-transfer interaction between the lone-pair (LP) orbital of N with the σ* S–C antibonding orbital in I and II respectively. The positive and negative lobes are depicted by the blue and yellow colors, respectively. Reproduced from ref. 72 with permission from the Royal Society of Chemistry.

solid form and it is desirable to extract the maximum benefit from a cocrystal formation. It has been demonstrated that cocrystals of APIs show improved chemical, physical and pharmacological properties such as solubility, stability and bioavailability over their individual components.[74] If multicomponent molecular crystals contain ionic components due to a proton transfer between two components, then it is called as a "salt". Even some APIs exhibit a tendency to proton transfer in multicomponent crystals due to external perturbations, such as temperature. Hence, a complete understanding of intermolecular interactions is necessary while discussing the stability of APIs as well as in rationalizing the design of cocrystals. A CD analysis was performed on two molecular complexes, one was a cocrystal of nicotinamide-salicylic acid and the other was a salt of nicotinamide-oxalic acid, to distinguish a cocrystal from salt based on the topological properties of ED and molecular ESP.[75] The difference in topological properties between N···H–O and N–H···O HBs in two complexes demonstrated a partial transfer of proton, even in the case of a nicotinamide-salicylic acid cocrystal. In addition, the CD results confirmed an earlier hypothesis that the formation of cocrystals and salts depends on the ΔpK_a value of the individual constituents. Similarly, the CD analysis on polymorphs of urea–barbituric acid cocrystals has validated an earlier observation that different polymorphs are obtained due to different mesomeric forms of the barbituric acid molecule.[76] The net charges and ESP from the experimental MM clearly indicate the shift of ED in the *Cc* polymorph towards the mesomeric form B, whereas it is towards form E in the $P2_1/c$ polymorph. In another study, two polymorphs of the 2 : 1 cocrystal of isonicotinamide and oxalic acid in space groups *C*2/*c* (I) and *P*-1 (II) were characterized by the CD method using both single crystal X-ray and neutron diffraction data collected at 100 K.[77] A short, strong hydrogen bond (SSHB) of O–H···N was observed in both polymorphs and from a topological analysis it was found to be covalent in nature. The topological analysis of ED confirmed the formation of an unusual quasi-centered heteronuclear intermolecular SSHB of O···H···N in polymorph II. In addition, the SF analysis from ED models established that O–H···N HBs in both polymorphs have characteristics of strong HBs whereas C–H···O and N–H···O HBs have either moderate or weak HB features in the crystal structure.

A word of caution for readers may be appropriate while analyzing the lattice energy results from the MM on polymorphs. Recent CD studies highlight a large discrepancy in the estimation of lattice energies from the CD model when compared to well-established theoretical methods.[51,52,78] Additionally, the lattice energy estimation from the MM is sensitive to the CD model. It has been illustrated in the study of polymorphs of sulfathiazole[51] and carbamazepine[78] that different CD models provide completely different lattice energy values, which might affect the stability order of polymorphs.

5.4.4 Halogen Bonding (XB) and Other σ-Hole Bonding

XBs are analogous interactions to the HB. They play a key role in many chemical and biochemical processes and are extensively utilized in supramolecular chemistry and crystal engineering.[79] The XB is characterized by an anisotropic charge distribution of ED over covalently bound halogen (X) atoms such that the ED is depleted along the C–X bond and resulting in accumulation of ED on its sides (Figure 5.5a). Hence, it produces a positive electrostatic potential along the C–X bond and acts as Lewi's acid center for bonding. This effect is called the "polar flattening effect". Therefore, a negatively charged X-atom can behave simultaneously as a Lewis acid and a Lewis base. The formation of positive electrostatic potential along the C–X bond was referred as a σ-hole by Politzer *et al.*,[80] and the subsequent interaction is known as σ-hole bonding. The σ-hole site interacts with the negative sites (nucleophiles) in an attractive manner. Among the halogens, the polar flattening effect is enhanced along the F < Cl < Br < I series such that the F-atom has minute σ-hole features.

Early experimental CD studies on the C–I···N,[81] C–I···O[82] and C–Br···N XBs in organic molecular complexes were performed using X-ray data collected at 90 K. However, the static deformation density and Laplacian maps were

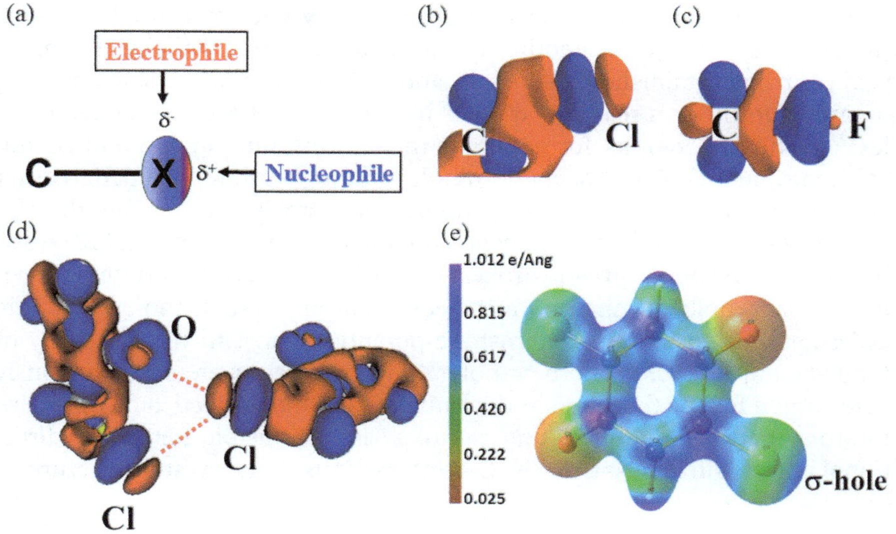

Figure 5.5 (a) A schematic representation of the "σ-hole or polar flattening effect" on halogens. Experimental observation of the "σ-hole" in 3D static deformation density maps of (b) Cl-atom in 2,5-dichloro-1, 4-benzoquinone,[86] (c) F-atom in 2-chloro-4-fluorobenzoic acid,[88] (d) 3D static deformation density map showing Cl···O and Cl···Cl XBs in 2,5-dichloro-1, 4-benzoquinone and (e) the "σ-hole" on the Cl-atom in the experimental ESP isosurface mapped on $\rho(\mathbf{r}) = 0.5$ e Å^{-3}.

not clearly demonstrating polar flattening effects on these heavier halogens. This might be due to a considerable amount of absorption and anomalous scattering effects on the measured X-ray data. However, recent CD studies in organic compounds containing the Cl-atom have shown clear experimental evidence for the anisotropic charge distribution (σ-hole formation) on the Cl-atom (Figure 5.5b).[83–85] The attractive nature of Cl···Cl interactions in C_6Cl_6 is explained as an electrostatic interaction between the charge depletion region (δ^+) and the charge concentration region (δ^-) on the Cl-atoms.[83] Indeed, similar CD features have been observed for the C–Cl···O=C XB in 2,5-dichloro-1,4-benzoquinone[86] and supporting the earlier "bumps-in hollow" hypothesis for XBs (Figure 5.5d). The σ-hole formation on the Cl-atom is experimentally confirmed by the ESP isosurface map and validates the C–Cl···O=C XB as an attractive electrostatic interaction (Figure 5.5e). The topological properties of ED and individual components of total interaction energy have quantitatively confirmed the dominant electrostatic contribution to XB. However, the participation of "organic fluorine" in interactions and their structure stabilizing role are controversial in the literature due to their small size, high electronegativity and smaller polarizability of the ED.[87] Even ESP estimated from high level theoretical calculations could not reveal any σ-hole formation on the organic F-atom.[80] For the first time, a tiny σ-hole region on the organic fluorine was observed in the experimental CD analysis of C–Cl···F–C and C–F···F–C interactions in 2-chloro-4-fluorobenzoic acid and 4-fluorobenzamide, respectively, using X-ray data measured at 100 K (Figure 5.5c).[88] Likewise, another CD study in pentafluorophenyl 2,2′-bithiazole further established C–F···F–C and C–F···S–C interactions involving organic fluorine as a realistic "σ-hole" interactions.[89] Recent extended studies[90,91] validated "σ-hole" formation on the fluorine atom in different chemical environments. The concept of "σ-hole bonding" is now extended to other similar interactions in organic compounds, namely "carbon bonding",[92] "chalcogen bonding"[93] and "pnicogen bonding".[94] An experimental ED study in dimethylammonium 4-hydroxybenzoate[95] provided experimental evidence for "σ-hole" formation on the carbon atom of $-CH_3$ moiety in static deformation density and electrostatic potential maps. In the CD study of Fmoc-Leu-ψ[CH_2NCS],[96] both "σ-holes" and "π-holes" were simultaneously determined for the first time in the same compound. Indeed, a co-operative interplay between "σ-holes" and "π-holes" is responsible for highly directional and stabilizing N=C=S···N=C=S interactions in the crystal structure.

5.4.5 Validating the Concept of Charge Shift Bonding (CSB)

CSB has recently been proposed as a third chemical bonding scenario in addition to known, classical covalent and ionic bonding where electrons are shared or completely transferred, respectively.[97] The CSB is identified by a fluctuating electron-pair density between two bonded atoms. It is further characterized by large covalent–ionic resonance interaction energy and depleted ED in the bonding region. A benchmark molecule for exhibiting CSB is fluorine F_2 where the F–F bonding is described as CSB. This bonding mechanism is found in numerous molecules by theoretical calculations and experimental CD analysis. An experimental ED investigation in [1.1.1]propellane has established that the central C···C bonding ($d = 1.579$ Å) between two "inverted" bridgehead carbon atoms is CSB.[98] In the experiment, the $\rho(\mathbf{r})$ and $\nabla^2\rho(\mathbf{r})$ values were 1.31(3) e Å^{-3} and 10.3 (1) e Å^{-5}, respectively, for the central C···C bonding, whereas those values for wing bonds are in the range, 1.55(2) to 1.79(2) e Å^{-3} and −13.0(1) to −11.6(1) e Å^{-5}, respectively. This indicates that the C···C bonding has high $\rho(\mathbf{r})$ and positive $\nabla^2\rho(\mathbf{r})$ values at the BCP, which is unusual for a short C–C bond. Indeed, the topological properties of ED clearly bring out the differences between the classical covalent bond and CSB in [1.1.1]propellane. A recent elaborate CD study in rubrene endoperoxide illustrated CSB characteristics for the O–O bond in the compound.[53] The $\rho(\mathbf{r})$ and $\nabla^2\rho(\mathbf{r})$ values from the experimental model are 1.716(1) e Å^{-3} and 15.106(0) e Å^{-5}, respectively. Generally, positive $\nabla^2\rho(\mathbf{r})$ values are often found in significantly polarized bonds, for example in C=O and C–F bonds. However, the O–O bond is a completely nonpolar homonuclear bond. In addition to topological features, the O–O bond is further characterized by the electron localizability indicator, (ELI-D) analysis.[36] The ELI-D iso-surfaces show the loss of electronic localization in the O–O bonding region by the lack of a significant bonding basin (Figure 5.6). The ELI-D isosurface depicts two tiny O–O bonding basins with the hourglass shaped feature along the O–O bond.

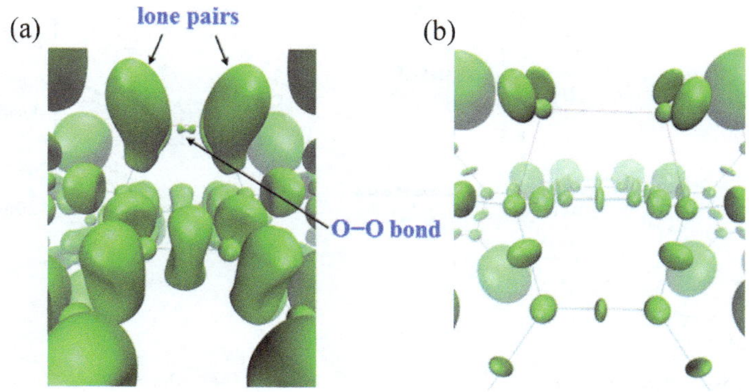

Figure 5.6 ELI-D isosurfaces depicting the O–O CSB bond with isovalues of (a) 1.35 and (b) 1.8. The top horizontal line represents the O–O bond. Reprinted with permission from V. R. Hathwar, M. K. Thomsen, M. A. H. Mamakhel, M. O. Filso, J. Overgaard and B. B. Iversen, *J. Phys. Chem. A.*, 2016, **120**, 7510–7518. Copyright 2016. American Chemical Society.[53]

These two oxygen basins contain in total only 0.55 e, which is inconsistent with a mechanism of normal two-electron bonding. The lack of charge along the O–O bond is found to reside in the four lone pair basins, two on each oxygen atom, which in total contain 10.4 e, or 2.4 e more than expected.

5.4.6 Phase Transitions in Organic Solids

Organic functional materials are found to be an attractive alternative to inorganic counterparts. They are explored due to their low-cost synthesis and flexibility to tune their physical properties. Among them, charge-transfer (CT) organic complexes show a ferroelectric structural transformation due to the broken symmetry under the influence of external stimuli, such as temperature, pressure and photo-irradiation.[99] This structural transformation is known as neutral to ionic (NI) phase transition and resulting in a partial degree of charge transfer between electron donor (D) and acceptor (A) molecules. For example, the tetrathiafulvalene (TTF)-*p*-chloranil (CA) molecular complex has the mixed stack of TTF and CA molecules ($D^oA^oD^oA^o$...) at an equal distance of 3.611(1) Å along the stacking axis in the neutral phase at 105 K.[100] The stacking axis in the TTF–CA complex coincides with the crystallographic "*a*" axis in the crystal structure. It undergoes a first order NI transition upon cooling at 84 K, where a partial electron transfer occurs between the TTF and CA molecules ($D^+A^-D^+A^-$...).[101] The NI transition also results in the structural transformation from the $P2_1/n$ to Pn in the crystal. After the NI phase transition, a stacking of the TTF–CA complex has two distinct inter-stacking distances of 3.504(4) Å and 3.685(4) Å at 15 K (Figure 5.7). Hence, weakly bound TTF–CA dimers are formed at the shorter stacking distance [3.504(4) Å] due to the NI phase transition. Figure 5.7 summarizes the structural changes across the NI transition in the compound. Indeed, the

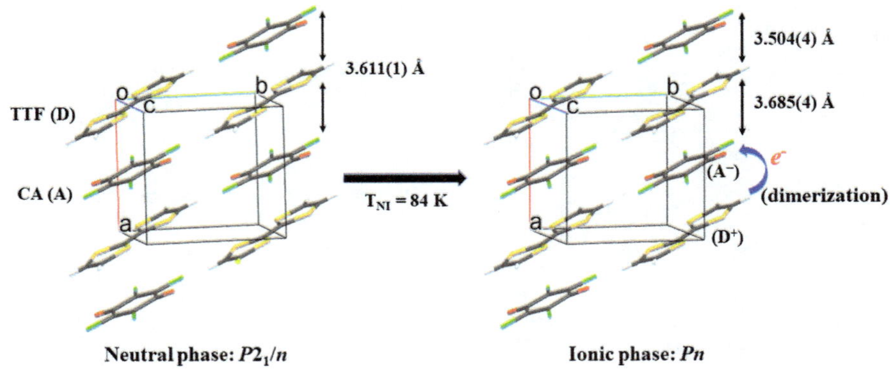

Figure 5.7 The structural transformations in the TTF–CA molecular complex during the NI phase transition.

TTF-CA molecular compound acquires ferroelectric properties due to the charge transfer process and broken symmetry in the crystal.

High-resolution CD experiments were performed on the TTF-CA crystal at 105 K and 15 K (both above and below the NI transition) to quantify the amount of charge transfer between TTF and CA molecules and to evaluate the role of intra- and inter-stack intermolecular interactions in the NI transition.[100] Both X-ray datasets were collected on a four-circle diffractometer equipped with a 2D detector. The structural analysis from two data demonstrated that the C=C bond lengths were increased in TTF whereas the C–S bond lengths were decreased upon cooling. Similarly, the C–C bonds were elongated in CA while the C=C, C=O and C–Cl bonds were found to be shortened. This observation was correlated with the CT between the LUMO of CA and the HOMO of TTF. Furthermore, the amount of charge transfer (Δq_{CT}), was determined from the integration of experimental ED over all atomic basins. The net charges (q_{CT}) of the complex were 0.21(2)e and 0.74 (2)e at 105 K and 15 K, respectively. Hence, the charge transfer (Δq_{CT}) was found to be 0.53e. This was further supported by the values of Δq_{CT} determined from the experimental kappa and multipole methods. An evaluation of topological properties of intermolecular interactions had suggested that C–H⋯O inter-stack contacts were the strongest interactions in both phases and further reinforced during the NI transition, whereas the Cl⋯Cl interactions remain unchanged across the phase transition.

5.4.7 CD Studies Under High Pressure

In recent times, the number of high pressure structural studies has significantly increased due to advances made in diamond anvil cells (DACs), the availability of brighter radiation sources and improved experimental procedures. In fact, pressure is an efficient thermodynamic variable for generating polymorphism, phase transitions and polymerization by conformational and structural transformations.[102] This might result in new crystal phases that are not stable at ambient pressure conditions. If the material properties are evaluated with the structure at high pressure, key information about the structure–property relationship can be derived. Indeed, some chemical transformations under pressure are irreversible and they produce interesting material properties at ambient conditions. However, a main concern for the CD analysis using high-pressure X-ray diffraction (HPXRD) is the lower quality of measured data. The DAC used to generate pressure drastically reduces the resolution, completeness and quality of X-ray data, which are critical requirements for the reliability of the experimental CD model.[102] In addition, unavoidable hindrances include high background intensity from either a gasket or the pressure-transmitting medium, limited opening angle of the DAC, the DAC absorption and overlapped high-intensity diamond scattering reflections. Often, high-resolution and good-quality data from HPXRD are obtained using synchrotron radiations only and demands extensive experimental preparation and data analysis.

Recently, a pioneering high-pressure charge density (HPCD) experiment was carried out on single crystals of an aromatic compound, syn-1,6 : 8,13-bis-carbonyl[14] annulene (BCA) at the I15 beamline of a diamond light source using a pinhole defined monochromatic beam with wavelength of 0.31 Å and an Atlas CCD detector.[103] The X-ray data was collected at 7.7 GPa on two crystals such that the two crystals were separately exposed to collimated X-rays by a pinhole of 30 mm in diameter. A methanol–ethanol 4 : 1 mixture was used as pressure transmitting medium. The resolution and redundancy of the data were $d = 0.5$ Å and 4.7, respectively. The compound BCA has two resonant electronic configurations and the aim of the experiment was to evaluate the changes in aromaticity of a molecule with applied pressure. In this interesting HPCD study, the localization of the ED in one of the resonant configurations is observed under high pressure and it has been experimentally characterized by the topological analysis of ED. Reliability of experimental results is further validated by the XCW method[55] and periodic density functional theory (PDFT) calculations (Figure 5.8).

5.4.8 CD Databases

The experimental CD analysis is not feasible when adequate quality crystals are not available to obtain high resolution X-ray data. In the case of large molecules and lower crystal symmetry, the time required to obtain high resolution data is also detrimental. As described in Section 5.3.2, P_v, P_{lm}, κ and κ' are refinable parameters, along with atomic coordinates and ADPs for each atom, by the least-squares procedure in the CD model. Hence, CD modeling demands superior quality redundant X-ray data for the refinement of an enormous number of parameters per atom. Additionally, the resolution of diffraction data should be at least 0.5 Å for a full multipolar refinement. The resolution of measured experimental data limits the refinement of number of multipolar parameters in the CD model. In the case of macromolecules, theoretical calculations have also become necessarily approximate due to

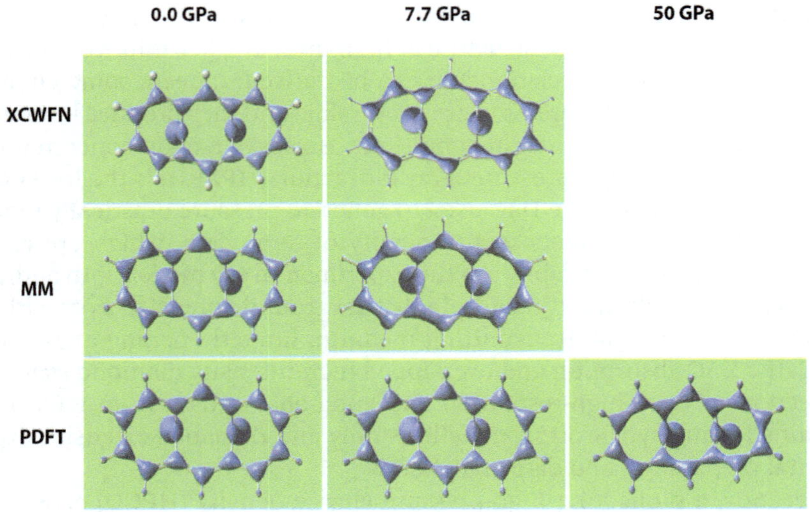

Figure 5.8 Demonstration of localization of the ED in one of the resonant configurations for BCA obtained from XCW, MM and PDFT methods at different pressure points. Adapted from N. Casati, A. Kleppe, A. P. Jephcoat and P. Macchi, *Nat. Commun.*, 2016, 7, 10901, http://dx.doi.org/10.1038/ncomms10901, Copyright © 2016.[103] Published under the terms of the CC BY 4.0 license, https://creativecommons.org/licenses/by/4.0/.

the complexity involved in the crystal structure. Hence, the idea of building CD databanks based on pseudoatoms has been contemplated. The main reason to build a CD database is for obtaining the EDD from routine X-ray data (resolution, d = 0.84 Å) by transferability of multipole parameters on a target molecule. On the other hand, it has been demonstrated that multipolar parameters (P_v, P_{lm}, κ and κ') for atoms are also indistinguishable in chemically identical environments. Based on this assumption, there are four CD databases, namely UBDB (University at Buffalo Databank),[104] Invariom,[105] ELMAM (experimental library multipolar-atom model)[106] and SBFA (supramolecular synthon-based fragments approach)[107] are available for use. UBDB and Invariom have multipolar parameters obtained from theoretical calculations whereas ELMAM and SBFA have the same from high resolution X-ray experiments. There are pros and cons for both experimental and theoretical databases. Theoretical databases are more flexible and can be readily extendable to include all functional groups in any chemical environment. On the other hand, experimental databases include crystal field effects but are subject to experimental errors and incompleteness in the measurement. Therefore, both theoretical and experimental databases are useful for an improvement of the CD model by cross-validating the two approaches.

All pseudoatom databases help in obtaining the improved structural model using the aspherical structure factors calculated from the transferable aspherical atom model. The structure refinement statistics and ADPs are superior when compared to the IAM refinement. Additionally, the quality of the estimated structural, electrostatic and topological properties from each database is comparable with high resolution CD results. The transferability of the pseudoatom CD features has been exploited in the recent literature to derive ED properties such as interaction energies and ESPs on several target molecules. The success of databases is realized in the accurate structural refinements of both small and protein molecules,[108] evaluation of topological analysis and ESP in a protein–cofactor complex,[109] and calculation of electrostatic interaction energies in an inhibitor-enzyme complex.[110]

5.5 Conclusions

As described in the above sections, the experimental CD method is now well established for organic systems. Its ability to unravel subtle electronic differences in the EDD of a crystalline material has played a significant role in the classification of HBs, quantification of XB, detecting unique bonding mechanisms like CSB and investigation of stable polymorphs. Indeed, qualitative descriptions of chemical bonding used in the field of "crystal engineering" and "drug design" could be supplemented by quantitative inputs related to the strength and nature of chemical bonding. The quality and reliability of the experimental CD method have been drastically improved in recent times by advanced instruments and synchrotron radiations, which have facilitated the use of smaller single crystals for the X-ray experiment. Nowadays, X-ray data up to $(\sin\theta/\lambda)_{max} = 1.51$ Å^{-1} with high redundancy are easily feasible in hours of experimental time for even organic compounds.[52] A subsequent use of additional CD descriptors from theory such as ELI-D, ELF and NICS provide render additional support to the experimental results. The CD modeling based on a database approach has been found to be effective in cases where it is impossible to obtain good quality crystals and/or high-resolution X-ray data. The CD results obtained from the database approaches are promising toward quantitative crystal engineering.

Acknowledgements

I thank Professor Eiji Nishibori, Professor Bo B. Iversen and University of Tsukuba for all research facilities and funding.

References

1. G. R. Desiraju, *The Crystal as a Supramolecular Entity*, John Wiley & Sons, Chichester, 1996.
2. N. K. Hansen and P. Coppens, *Acta Crystallogr., Sect. A*, 1978, **34**, 909–921.
3. M. R. V. Jorgensen, V. R. Hathwar, N. Bindzus, N. Wahlberg, Y. S. Chen, J. Overgaard and B. B. Iversen, *IUCrJ*, 2014, **1**, 267–280.
4. A. Krawczuk and P. Macchi, *Chem. Cent. J.*, 2014, **8**, 68.

5. D. Chopra, *J. Phys. Chem. A*, 2012, **116**, 9791–9801.
6. P. Macchi, *Crystallogr. Rev.*, 2013, **19**, 58–101.
7. V. R. Hathwar, *J. Indian Inst. Sci.*, 2017, **97**, 281–298.
8. C. Gatti, *Z. Kristallogr.*, 2005, **220**, 399–457.
9. D. Stalke, *Chem.–Eur. J.*, 2011, **17**, 9264–9278.
10. C. Gatti and P. Macchi, *Modern Charge Density Analysis*, Springer, Newyork, 2012.
11. P. Coppens, *X-ray Charge Densities and Chemical Bonding*, Oxford University Press, Oxford, U.K., 1997.
12. F. L. Hirshfeld, *Acta Crystallogr., Sect. A*, 1976, **32**, 239–244.
13. R. Herbst-Irmer, J. Henn, J. J. Holstein, C. B. Hubschle, B. Dittrich, D. Stern, D. Kratzert and D. Stalke, *J. Phys. Chem. A*, 2013, **117**, 633–641.
14. K. Meindl and J. Henn, *Acta Crystallogr., Sect. A*, 2008, **64**, 404–418.
15. B. Guillot, L. Viry, R. Guillot, C. Lecomte and C. Jelsch, *J. Appl. Crystallogr.*, 2001, **34**, 214–223.
16. A. Volkov, P. Macchi, L. J. Farrugia, C. Gatti, P. Mallinson, T. Richter and T. Koritsanszky, *XD2016-A Computer Program Package for Multipole Refinement, Topological Analysis of Charge Densities and Evaluation of Intermolecular Energies from Experimental or Theoretical Structure Factors*, 2016.
17. A. Ø. Madsen, *J. Appl. Crystallogr.*, 2006, **39**, 757–758.
18. S. C. Capelli, H. B. Burgi, B. Dittrich, S. Grabowsky and D. Jayatilaka, *IUCrJ*, 2014, **1**, 361–379.
19. D. Jayatilaka and D. J. Grimwood, *Acta Crystallogr., Sect. A*, 2001, **57**, 76–86.
20. S. Grabowsky, P. Luger, J. Buschmann, T. Schneider, T. Schirmeister, A. N. Sobolev and D. Jayatilaka, *Angew. Chem., Int. Ed.*, 2012, **51**, 6776–6779.
21. S. P. Thomas, D. Jayatilaka and T. N. G. Row, *Phys. Chem. Chem. Phys.*, 2015, **17**, 25411–25420.
22. R. F. W. Bader, *Atoms in Molecules - a Quantum Theory*, Clarendon, Oxford, 1990.
23. U. Koch and P. L. A. Popelier, *J. Phys. Chem.*, 1995, **99**, 9747–9754.
24. D. Cremer and E. Kraka, *Angew. Chem., Int. Ed.*, 1984, **23**, 627–628.
25. Y. A. Abramov, *Acta Crystallogr., Sect. A*, 1997, **53**, 264–272.
26. E. Espinosa, E. Molins and C. Lecomte, *Chem. Phys. Lett.*, 1998, **285**, 170–173.
27. P. Macchi, L. Garlaschelli and A. Sironi, *J. Am. Chem. Soc.*, 2002, **124**, 14173–14184.
28. R. F. W. Bader and C. Gatti, *Chem. Phys. Lett.*, 1998, **287**, 233–238.
29. E. R. Johnson, S. Keinan, P. Mori-Sanchez, J. Contreras-Garcia, A. J. Cohen and W. T. Yang, *J. Am. Chem. Soc.*, 2010, **132**, 6498–6506.
30. G. Saleh, L. Lo Presti, C. Gatti and D. Ceresoli, *J. Appl. Crystallogr.*, 2013, **46**, 1513–1517.
31. G. Saleh, C. Gatti and L. Lo Presti, *Comput. Theor. Chem.*, 2015, **1053**, 53–59.

32. A. Volkov, T. Koritsanszky and P. Coppens, *Chem. Phys. Lett.*, 2004, **391**, 170–175.
33. M. A. Spackman, *Cryst. Growth Des.*, 2015, **15**, 5624–5628.
34. A. Gavezzotti, *New J. Chem.*, 2011, **35**, 1360–1368.
35. S. K. Wolff, D. J. Grimwood, J. J. McKinnon, M. J. Turner, D. Jayatilaka and M. A. Spackman, *CrystalExplorer 3.2*, University of Western Australia, Perth, 2015.
36. M. Kohout, *Int. J. Quantum Chem.*, 2004, **97**, 651–658.
37. A. D. Becke and K. E. Edgecombe, *J. Chem. Phys.*, 1990, **92**, 5397–5403.
38. A. Krawczuk, D. Pérez, K. Stadnicka and P. Macchi, *Trans. Am. Crystallogr. Assoc.*, 2011, **42**, 1.
39. P. V. Schleyer, C. Maerker, A. Dransfeld, H. J. Jiao and N. J. R. V. Hommes, *J. Am. Chem. Soc.*, 1996, **118**, 6317–6318.
40. P. Munshi and T. N. G. Row, *J. Phys. Chem. A*, 2005, **109**, 659–672.
41. P. R. Mallinson, G. T. Smith, C. C. Wilson, E. Grech and K. Wozniak, *J. Am. Chem. Soc.*, 2003, **125**, 4259–4270.
42. D. Chopra, T. S. Cameron, J. D. Ferrara and T. N. G. Row, *J. Phys. Chem. A*, 2006, **110**, 10465–10477.
43. P. Munshi and T. N. G. Row, *CrystEngComm*, 2005, **7**, 608–611.
44. V. V. Zhurov and A. A. Pinkerton, *J. Phys. Chem. A*, 2015, **119**, 13092–13100.
45. G. K. H. Madsen, B. B. Iversen, F. K. Larsen, M. Kapon, G. M. Reisner and F. H. Herbstein, *J. Am. Chem. Soc.*, 1998, **120**, 10040–10045.
46. G. Gilli, F. Bellucci, V. Ferretti and V. Bertolasi, *J. Am. Chem. Soc.*, 1989, **111**, 1023–1028.
47. J. Overgaard, B. Schiøtt, F. K. Larsen and B. B. Iversen, *Chem.–Eur. J.*, 2001, **7**, 3756–3767.
48. C. Gatti, G. Saleh and L. Lo Presti, *Acta Crystallogr., Sect. B*, 2016, **72**, 180–193.
49. I. L. Kirby, M. Brightwell, M. B. Pitak, C. Wilson, S. J. Coles and P. A. Gale, *Phys. Chem. Chem. Phys.*, 2014, **16**, 10943–10958.
50. Y. A. Abramov, A. Volkov, G. Wu and P. Coppens, *J. Phys. Chem. B*, 2000, **104**, 2183–2188.
51. I. Sovago, M. J. Gutmann, J. G. Hill, H. M. Senn, L. H. Thomas, C. C. Wilson and L. J. Farrugia, *Cryst. Growth Des.*, 2014, **14**, 1227–1239.
52. V. R. Hathwar, M. Sist, M. R. V. Jorgensen, A. H. Mamakhel, X. P. Wang, C. M. Hoffmann, K. Sugimoto, J. Overgaard and B. B. Iversen, *IUCrJ*, 2015, **2**, 563–574.
53. V. R. Hathwar, M. K. Thomsen, M. A. H. Mamakhel, M. O. Filso, J. Overgaard and B. B. Iversen, *J. Phys. Chem. A*, 2016, **120**, 7510–7518.
54. A. E. Whitten, D. Jayatilaka and M. A. Spackman, *J. Chem. Phys.*, 2006, **125**, 174505.
55. D. Jayatilaka, *Phys. Rev. Lett.*, 1998, **80**, 798–801.
56. D. Jayatilaka, P. Munshi, M. J. Turner, J. A. K. Howard and M. A. Spackman, *Phys. Chem. Chem. Phys.*, 2009, **11**, 7209–7218.
57. U. Flierler, D. Leusser, H. Ott, G. Kehr, G. Erker, S. Grimme and D. Stalke, *Chem.–Eur. J.*, 2009, **15**, 4595–4601.

58. S. Mebs, S. Grabowsky, D. Forster, R. Kickbusch, M. Hartl, L. L. Daemen, W. Morgenroth, P. Luger, B. Paulus and D. Lentz, *J. Phys. Chem. A*, 2010, **114**, 10185–10196.
59. A. Wagner, R. Flaig, B. Dittrich, H. Schmidt, T. Koritsanszky and P. Luger, *Chem.-Eur. J.*, 2004, **10**, 2977–2982.
60. S. Grabowsky, M. Weber, D. Jayatilaka, Y. S. Chen, M. T. Grabowski, R. Brehme, M. Hesse, T. Schirmeister and P. Luger, *J. Phys. Chem. A*, 2011, **115**, 12715–12732.
61. S. P. Thomas, K. Satheeshkumar, G. Mugesh and T. N. G. Row, *Chem.-Eur. J.*, 2015, **21**, 6793–6800.
62. E. A. Zhurova, V. V. Zhurov, P. Kumaradhas, S. Cenedese and A. A. Pinkerton, *J. Phys. Chem. B*, 2016, **120**, 8882–8891.
63. P. Munshi and T. N. G. Row, *Cryst. Growth Des.*, 2006, **6**, 708–718.
64. A. Overgaard and D. E. Hibbs, *Acta Crystallogr., Sect. A*, 2004, **60**, 480–487.
65. P. Munshi, C. Jelsch, V. R. Hathwar and T. N. G. Row, *Cryst. Growth Des.*, 2010, **10**, 1516–1526.
66. M. H. Chambrier, N. Bouhmaida, F. Bonhomme, S. Lebegue, J. M. Gillet, C. Jelsch and N. E. Ghermani, *Cryst. Growth Des.*, 2011, **11**, 2528–2539.
67. R. S. Gopalan, G. U. Kulkarni and C. N. R. Rao, *ChemPhysChem*, 2000, **1**, 127–135.
68. G. U. Kulkarni, P. Kumaradhas and C. N. R. Rao, *Chem. Mater.*, 1998, **10**, 3498–3505.
69. Y. V. Nelyubina, I. V. Glukhov, M. Y. Antipin and K. A. Lyssenko, *Chem. Commun.*, 2010, **46**, 3469–3471.
70. A. A. Hoser, L. Dobrzycki, M. J. Gutmann and K. Wozniak, *Cryst. Growth Des.*, 2010, **10**, 5092–5104.
71. A. A. Hoser, K. N. Jarzembska, L. Dobrzycki, M. J. Gutmann and K. Wozniak, *Cryst. Growth Des.*, 2012, **12**, 3526–3539.
72. S. Sarkar, M. S. Pavan, S. Cherukuvada and T. N. G. Row, *Chem. Commun.*, 2016, **52**, 5820–5823.
73. T. S. Thakur and S. S. Singh, *Cryst. Growth Des.*, 2015, **15**, 3280–3292.
74. N. Schultheiss and A. Newman, *Cryst. Growth Des.*, 2009, **9**, 2950–2967.
75. V. R. Hathwar, R. Pal and T. N. G. Row, *Cryst. Growth Des.*, 2010, **10**, 3306–3310.
76. M. Gryl, A. Krawczuk and K. Stadnicka, *Acta Crystallogr., Sect. B*, 2011, **67**, 144–154.
77. M. Schmidtmann, L. J. Farrugia, D. S. Middlemiss, M. J. Gutmann, G. J. McIntyre and C. C. Wilson, *J. Phys. Chem. A*, 2009, **113**, 13985–13997.
78. I. Sovago, M. J. Gutmann, H. M. Senn, L. H. Thomas, C. C. Wilson and L. J. Farrugia, *Acta Crystallogr., Sect. B*, 2016, **72**, 39–50.
79. G. Cavallo, P. Metrangolo, R. Milani, T. Pilati, A. Priimagi, G. Resnati and G. Terraneo, *Chem. Rev.*, 2016, **116**, 2478–2601.
80. T. Clark, M. Hennemann, J. S. Murray and P. Politzer, *J. Mol. Model.*, 2007, **13**, 291–296.
81. R. Bianchi, A. Forni and T. Pilati, *Chem.-Eur. J.*, 2003, **9**, 1631–1638.

82. R. Bianchi, A. Forni and T. Pilati, *Acta Crystallogr., Sect. B*, 2004, **60**, 559–568.
83. T. T. T. Bui, S. Dahaoui, C. Lecomte, G. R. Desiraju and E. Espinosa, *Angew. Chem., Int. Ed.*, 2009, **48**, 3838–3841.
84. V. R. Hathwar and T. N. G. Row, *J. Phys. Chem. A*, 2010, **114**, 13434–13441.
85. M. V. Vener, A. V. Shishkina, A. A. Rykounov and V. G. Tsirelson, *J. Phys. Chem. A*, 2013, **117**, 8459–8467.
86. V. R. Hathwar, R. G. Gonnade, P. Munshi, M. M. Bhadbhade and T. N. G. Row, *Cryst. Growth Des.*, 2011, **11**, 1855–1862.
87. D. Chopra and T. N. G. Row, *CrystEngComm*, 2011, **13**, 2175–2186.
88. V. R. Hathwar and T. N. G. Row, *Cryst. Growth Des.*, 2011, **11**, 1338–1346.
89. M. S. Pavan, K. D. Prasad and T. N. G. Row, *Chem. Commun.*, 2013, **49**, 7558–7560.
90. V. R. Hathwar, D. Chopra, P. Panini and T. N. G. Row, *Cryst. Growth Des.*, 2014, **14**, 5366–5369.
91. D. Dey, S. Bhandary, A. Sirohiwal, V. R. Hathwar and D. Chopra, *Chem. Commun.*, 2016, **52**, 7225–7228.
92. D. Mani and E. Arunan, *Phys. Chem. Chem. Phys.*, 2013, **15**, 14377–14383.
93. S. P. Thomas, S. P. K. P. Veccham, L. J. Farrugia and T. N. G. Row, *Cryst. Growth Des.*, 2015, **15**, 2110–2118.
94. S. Sarkar, M. S. Pavan and T. N. G. Row, *Phys. Chem. Chem. Phys.*, 2015, **17**, 2330–2334.
95. S. P. Thomas, M. S. Pavan and T. N. G. Row, *Chem. Commun.*, 2014, **50**, 49–51.
96. R. Pal, G. Nagendra, M. Samarasimhareddy, V. V. Sureshbabu and T. N. G. Row, *Chem. Commun.*, 2015, **51**, 933–936.
97. S. Shaik, D. Danovich, B. Silvi, D. L. Lauvergnat and P. C. Hiberty, *Chem.–Eur. J.*, 2005, **11**, 6358–6371.
98. M. Messerschmidt, S. Scheins, L. Grubert, M. Patzel, G. Szeimies, C. Paulmann and P. Luger, *Angew. Chem., Int. Ed.*, 2005, **44**, 3925–3928.
99. M. Filatov, *Phys. Chem. Chem. Phys.*, 2011, **13**, 144–148.
100. P. Garcia, S. Dahaoui, C. Katan, M. Souhassou and C. Lecomte, *Faraday Discuss.*, 2007, **135**, 217–235.
101. J. B. Torrance, A. Girlando, J. J. Mayerle, J. I. Crowley, V. Y. Lee, P. Batail and S. J. Laplaca, *Phys. Rev. Lett.*, 1981, **47**, 1747–1750.
102. A. Katrusiak, *Acta Crystallogr., Sect. A*, 2008, **64**, 135–148.
103. N. Casati, A. Kleppe, A. P. Jephcoat and P. Macchi, *Nat. Commun.*, 2016, **7**, 10901.
104. P. M. Dominiak, A. Volkov, X. Li, M. Messerschmidt and P. Coppens, *J. Chem. Theory Comput.*, 2007, **2**, 232–247.
105. B. Dittrich, C. B. Hubschle, P. Luger and M. A. Spackman, *Acta Crystallogr., Sect. D*, 2006, **62**, 1325–1335.
106. B. Zarychta, V. Pichon-Pesme, B. Guillot, C. Lecomte and C. Jelsch, *Acta Crystallogr., Sect. A*, 2007, **63**, 108–125.

107. V. R. Hathwar, T. S. Thakur, T. N. G. Row and G. R. Desiraju, *Cryst. Growth Des.*, 2011, **11**, 616–623.
108. J. M. Bak, S. Domagala, C. Hubschle, C. Jelsch, B. Dittrich and P. M. Dominiak, *Acta Crystallogr., Sect. A*, 2011, **67**, 141–153.
109. B. Zarychta, A. Lyubimov, M. Ahmed, P. Munshi, B. Guillot, A. Vrielink and C. Jelsch, *Acta Crystallogr., Sect. D*, 2015, **71**, 954–968.
110. B. Fournier, E. E. Bendeif, B. Guillot, A. Podjarny, C. Lecomte and C. Jelsch, *J. Am. Chem. Soc.*, 2009, **131**, 10929–10941.

CHAPTER 6

Charge Density Studies and Topological Analysis of Hydrogen Bonds in Proteins

SUMAN KUMAR MANDAL AND PARTHAPRATIM MUNSHI*

Department of Chemistry, School of Natural Sciences,
Shiv Nadar University, Tehsil Dadri, Uttar Pradesh-201314, India
*E-mail: parthapratim.munshi@snu.edu.in

6.1 Introduction

Proteins, made of polypeptide chains, are one of the building blocks of life and play an important role in living organisms. Protein transforms to its native state from its unfolded state *via* an intermediate state. The structure of the protein molecule arises due to the folding of polypeptide chains consisting of particular sequences of amino acids to generate compact domains with specific three-dimensional orientation. The function of a protein is determined by its structure. During the folding process, the non-polar side chains get buried due to hydrophobic effects and the main chains form N–H···O=C hydrogen bonds (HBs). The two main stabilizing forces contributing to the conformational stability of proteins are the hydrophobic effects and the HBs.[1,2] However, there is a continuing debate to answer convincingly which of these two energetic factors predominates during folding to stabilize

Monographs in Supramolecular Chemistry No. 26
Understanding Intermolecular Interactions in the Solid State: Approaches and Techniques
Edited by Deepak Chopra

Published by the Royal Society of Chemistry, www.rsc.org

the protein. While some earlier studies[3–6] have been directed towards the understanding of hydrophobic effects, recent experimental[7–10] and theoretical[11] evidence points to H-bonding.

The importance of HBs in the protein structure was realized long ago by Mirsky and Pauling.[12] The crucial H-bond in a protein structure occurs between the N–H and C=O groups of the main-chain peptide. Moreover, these are the major HBs that are found in globular protein[13] and are the key building blocks of both α-helix and β-sheet structures. As estimated by Pauling's group, these HBs contribute about 5–8 kcal mol^{-1} to the stability of the protein.[14,15] There is compelling evidence that the major contribution to the H-bond energy comes from electrostatic interaction energy,[16] which provides by far the best structure–function correlation for proteins.[17] Knowledge of such H-bonding networks and the local electrostatics in proteins is essential for proper thermodynamic modelling, prediction of protein folding pathways, binding interactions and molecular recognition.

Accurate study of H-bonds in proteins demands experimental localization of H-atoms, which can be achieved up to a certain extent using high-resolution X-ray,[18] but precisely using neutron diffraction experiments.[19] Since the determination of the first protein structure in 1958,[20] the protein crystallography field has evidenced tremendous developments. In the past two decades, techniques have massively evolved in every stage of protein crystallography starting from crystallization to data collection to data quality and structure refinement. These have been possible due to the advent of third and fourth generation synchrotron sources, automated data collection techniques using large and highly efficient fast detectors and methodological developments. Further, pharmaceutical industries showed great interest in protein crystallography soon after the introduction of high-throughput crystallography. Moreover, with the advancement in data-refinement programs, structural modelling and refinements are also now becoming almost automated. Recently, the availability of modern synchrotron radiation sources have facilitated the data collection and structure solution of proteins and nucleic acids at high (sub-atomic) resolution. High-resolution X-ray crystallography is now a mature branch of modern crystallography.

The Hansen and Coppens multipole[21] formalism is the widely accepted approach for performing experimental change density analysis (ECDA) using high-resolution X-ray diffraction data collected at low temperature. Within this formalism, the total atomic density $\rho(\mathbf{r})$ is divided broadly into two parts; spherical and aspherical and expressed as follows:

$$\rho(\boldsymbol{r}) = P_c\rho_{\text{core}}(r) + P_c\kappa^3\rho_{\text{val}}(\kappa r) + \sum_{l=0}^{l\max}\kappa'^3 R_l(\kappa' r)\sum_{m=0}^{l} P_{lm\pm}d_{lm\pm}(\Theta,\phi).$$

The core population (P_c) and density (ρ_{core}) along with valence population (P_v), density (ρ_{valence}) and expansion/contraction coefficient κ (kappa) comprise the spherical part. The aspherical part, described using polar coordinates (r, θ, ϕ), which takes into account the aspherical deformation of valence

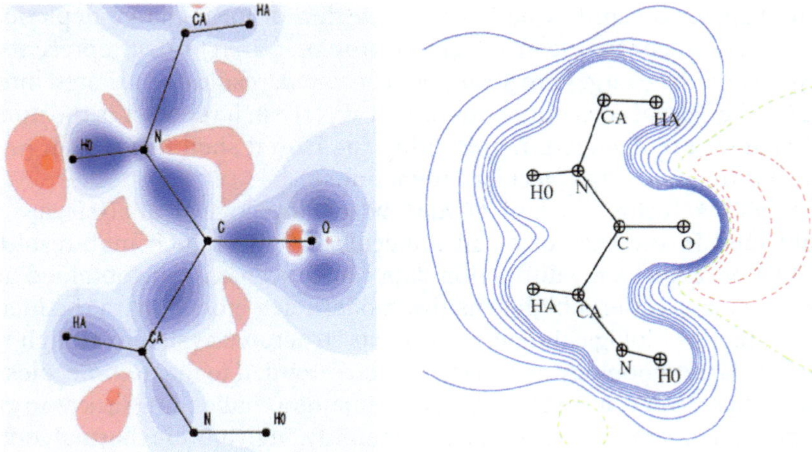

Figure 6.1 Showing deformation of electronic density of atoms (left) and the electrostatic potentials (right) in a peptide molecule. Positive and negative contours at ±0.05 e Å^{-3} are shown in blue and red, respectively.

electron density, is given in terms of normalized spherical harmonics ($d_{lm\pm}$), multipole population ($P_{lm\pm}$), radial function (R_l) and a scaling parameter for radial expansion and contraction κ'.

Although a large number of studies on molecular modelling reported in the literature are based on point-charge potentials, it is recognised that they have inherent limitations, such as modelling of lone pairs and aromatic rings.[22] Multipole modelling based ECDA is a powerful technique to provide an accurate analytical model for the electron density distribution in a crystal. Simple crystal structure analysis can only provide the geometrical features, whereas ECDA, using high-resolution X-ray diffraction data and if possible in combination with neutron diffraction data, allows extracting precise structural information in terms of its detailed electron density distributions. At sub-atomic resolution the deformation of the electronic cloud of atoms due to chemical bonding and interactions is highlighted *via* multipole modelling of electron densities, $\rho(\mathbf{r})$, which can be utilized to calculate electrostatic potentials, $V(\mathbf{r})$ (Figure 6.1). The molecular electronic density and the derived electrostatic potential play a paramount role in the modelling of physical and chemical properties of materials and biomolecules.

Further, the derived $\rho(\mathbf{r})$ and $V(\mathbf{r})$ can be utilized together to estimate electrostatic interaction energies.[23] HBs can be characterized by performing topological analysis of $\rho(\mathbf{r})$ *via* Bader's quantum theory of atoms in molecules (QTAIM).[24] Using QTAIM one can identify the bond critical point (BCP), where the gradient of electron density vanishes, *i.e.* $\nabla\rho(\mathbf{r}) = 0$. The bond path (BP) can also be traced along the electron density gradient, which is a tensor and is represented as R_{ij} and is not necessarily equal to the inter-nuclear distance. The second derivative of $\rho(\mathbf{r})$, *i.e.* $\nabla^2\rho(\mathbf{r})$, is called a Laplacian, which is defined as the sum of principal curvatures ($\lambda_1 + \lambda_2 + \lambda_3$). The values and signs of the Laplacian signify whether it is a region of local charge depletion or accumulation. For the electron density profile, if $\nabla^2\rho(\mathbf{r}) < 0$, it represents the maxima and the charges are locally concentrated, resulting in shared interaction (in the case of covalent bonds) and if $\nabla^2\rho(\mathbf{r}) > 0$, it represents the minima and charges are depleted, representing the closed shell interaction (in the case of intra and intermolecular interactions).

The ECDA technique and QTAIM based topological analysis are well established in the case of small molecules,[25,26] for which high-resolution (~0.40 Å or better) X-ray diffraction data can conveniently be obtained using laboratory sources and high-sensitive modern detectors.[27] With the increasing number of biological macromolecular structures available at high resolution, this methodology has recently been extended to macromolecules.[28–30] Nevertheless, ECDA on protein systems remains a challenging task as in growing good quality protein crystals, their stability, high atomic thermal motion, high solvent content, large number of H-atoms *etc.* are some common detrimental factors, due to which crystals often diffract to a limited resolution. Up to now, although there is a total of about 120 000 X-ray protein structures deposited in the Protein Data Bank (PDB: http://www.rcsb.org/pdb/home/home.do), only ~36 unique X-ray protein structures are reported to have resolutions higher than 0.80 Å. However, so far, only six of those high-resolution X-ray protein structures ranging from 4.7 kDa to 55 kDa have been subjected to ECDA.[28,29,31–34] Details of those protein systems along with the resolutions of the diffraction data are listed in Table 6.1. In all these cases, the initial

Table 6.1 Details of the protein systems (as deposited in PDB) subjected to ECDA.

Protein system	PDB ID	Number of atoms	Resolution (Å)	Source	Detector	Deposition year
High-potential iron–sulfur (HiPIP)[31]	5D8V	920	0.48	Spring-8 (BL41XU)	MAR CCD	2016
Crambin[28]	1EJG	424	0.54	Desy, Hamburg (BW7A)	MAR IMAGE PLATE	2000
Hen egg white lysozyme (HEWL)[32]	2VB1	1375	0.65	APS (19-ID)	ADSC CCD	2007
Human aldose reductase (hAR)[29]	1US0	3066	0.66	APS (19-ID)	ANL CCD 1.02M	2004
Cholesterol oxidase (ChOx)[33]	4REK	4290	0.74	SSRL (BL9-2)	MARMOSAIC 325 mm CCD	2015
NADH-cytochrome b5 reductase (b5R)[34]	5GV8	2490	0.78	Spring-8 (BL41XU)	RAYONIX MX225HE CCD	2017

charge density models were constructed based on the experimental electron densities transferred from the database. The database transfer approach is discussed in Section 6.2.3. In this context, it is noteworthy that for small molecules the multipole model is usually constructed based on step-wise multipole refinements of high-resolution X-ray diffraction data.

In this chapter, we will provide an overview of ECDA in protein systems and its basic requirements, the available tools and a brief protocol for successful multipole modelling. We will also provide a brief review of the reported ECDA on six selected proteins and discuss the key outcomes of those analyses. To end, we will discuss the topological properties of N–H···O=C hydrogen bonds in both α-helices and β-sheets of protein human aldose reductase (hAR), based on experimental charge density database transfer on its high-resolution (0.66 Å) structure. Thereby, we estimate the H-bond energies (electrostatic and dissociation) and discuss the relationship between these two energies and their relationship with various topological properties. The relationship among various topological properties will also be discussed.

6.2 Protein Charge Density Analysis

6.2.1 Approach

As discussed above, unlike small molecules, ECDA in protein systems faces certain experimental challenges. Moreover, given the large dimension of proteins, performing multipole refinements based on the least-squares method and within the Hansen–Coppens formalism is almost impractical. Because multipole modelling involves refinements of a large number of parameters (as many as 34 for non-H-atoms) using a large number of structure factors. However, protein is made of repeated units of small peptides consisting of common functional groups. Therefore, the usual practice is to determine charge densities based on multiple modelling of these peptides or other small molecules using either very high-resolution diffraction data or high-resolution theoretical structure factors and build a database. Thereby, the multipole-based electron densities of various functional groups are transferred from the database to construct the initial multipole model of a protein structure. The transferability approach is applied with the assumption that the same chemical groups possess the same charge density information in different chemical environments.

6.2.2 Basic Requirements

In case of ECDA, for constructing an accurate multipole model, the subjected protein needs to meet the following conditions, which are highly demanding.

- Low thermal displacement parameters, $B < 8.0$ Å^2.
- Least number of disordered atoms or residues.

- Low solvent proportion.
- Availability of stable, excellent quality and highly diffracting single crystals.
- Diffract to high-resolution (~0.80 Å or better) at low temperature (100 K or less).
- Availability of multipole information of certain functional groups in the database for the construction of an initial model.

6.2.3 Methodologies and Tools

Currently, there are three charge density databases available for transferring and building an initial model of electron densities for protein systems. Two of these are constructed following multipole modelling based on theoretical structure factors obtained *via* quantum calculations on gas phase small molecules. These databases are known as Invariom (invariant atoms)[35] and UBDB (University at Buffalo Pseudoatom Databank).[36] The Invariom approach was successfully applied on an 18-residue peptaibol antibiotic trichotoxin A50E macromolecule for the calculation of its improved electrostatic potentials.[37] Use of the UBDB database on Sunitinib and its protein complexes helped to estimate accurate electrostatic interaction energies.[38] The advantage of these theoretical databases is that the multipole information of a new chemical group in case it is required for the construction of a protein charge density model can be derived quickly by performing an *ab initio* quantum calculation and need not depend on its high-resolution X-ray diffraction experiment. However, currently, given the advancements in technologies, achieving excellent quality high-resolution diffraction data for small molecules is becoming routine and the number of high-resolution protein structures is increasing steadily. The third charge density database is the Experimental Library of Multipolar Atom Model (*ELMAM2*),[39] in which the aspherical pseudoatom parameters are stored from highly accurate ECDA on all 20 amino acids, some small peptides, nucleic acids and small molecules with common chemical groups. *ELMAM2* has been successfully employed to a number of proteins: crambin,[28] hAR,[29] DING protein[30] and cholesterol oxidase.[33] This database was also employed for the multipole modelling and subsequent probing of active site interactions of the triclinic form of HEWL.[32] In the case of DFPase,[40] some of the H-atoms were visible at high-resolution; this provided new insights on the catalytic mechanism. Structural and electrostatic properties derived using all these three pseudoatom databases have been compared and verified.[41,42] The outcomes from these databanks appeared to be comparable with each other. However, as expected, the atomic polarization effect due to the crystal field was prominent in the case of experimental charge density database multipole modelling using *ELMAM2*. It is to be noted that, for ECDA, the transferability would work best and the accuracy of the multipole modelling could be achieved provided the subjected protein structure is also determined at a high resolution, preferably at 0.80 Å or better.

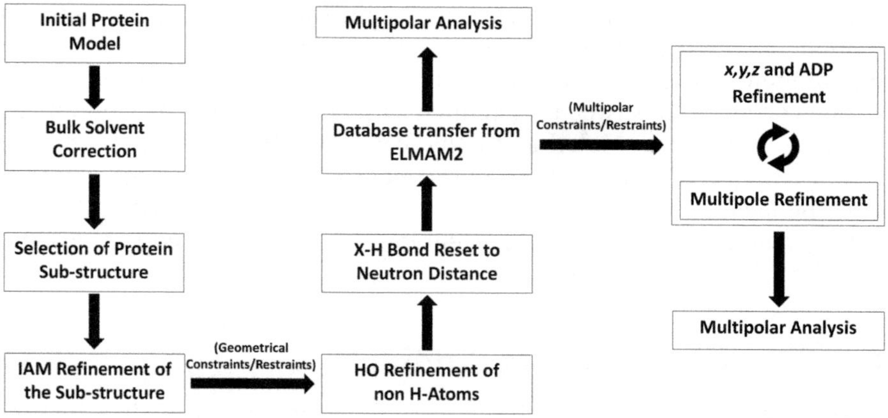

Figure 6.2 Flowchart for following the steps of multipolar refinement of proteins using *MoPro*.

6.2.4 Multipolar Refinement

With the availability of a large number of biological macromolecule structures at high resolution, and given the technical advances, the experimental charge density method is extended from well-established small molecules to macromolecules. Consequently, the program *MoPro* is dedicated to perform the charge density study *via* multipole refinement at sub-atomic resolution of structures ranging from small molecules to biological macromolecules.[43,44] As per the need of users, some complex refinement strategies can also be written within the program *MoPro*. Applicability of numerous analysis tools, restraints and constraints makes this program highly efficient for structure and electron-density refinement.[44] The multipole refinement using *MoPro* is further simplified *via* its user-friendly Java language based GUI. Although the details of the multipole refinement protocols for protein structures using *MoPro* have been discussed earlier,[29] here we provide a brief discussion based on a flowchart, as shown in Figure 6.2.

Often due to the higher thermal motion of the solvent molecules, the option solvent flattening[45] for masking solvent reflections is applied in the initial high-resolution structure and thereby the sub-structure of protein is selected for independent atom model (IAM) refinement. Subsequently, a higher order (HO) refinement with diffraction data of $d \sim 0.9$ Å or less is performed to refine only the positional (x, y, z) and atomic displacement parameters (ADPs) of non-H-atoms. At this stage the hydrogen ADPs are refined isotropically using the lower order diffraction data, while the H-bond length (X–H) is set to a standard neutron bond distance.[46] Thus accuracy in atomic positions is achieved and the multipolar parameters are then transferred from *ELMAM2* with appropriate constraints and restraints applied for multipoles, κ and κ' parameters, atomic positions and ADPs. For proper redistribution of valence densities and for the deconvolution of thermal motion, positional and ADP refinements are performed together and then multipolar refinements are performed in cycles until convergence is reached. Imposing different constraints (hard limit) or restraints (soft limit) helps to avoid strong correlations between the refined variables and thus the smooth convergence in refinement is achieved. In the case of diffraction data with insufficient resolution (not sub-atomic) for modelling protein charge densities, only the position and ADPs are refined using all reflections within the IAM approximation and the multipolar parameters are then fixed to the values transferred from the database.

6.3 Six Selected ECDA Studies

As pointed out above, although there are so far about 36 protein structures have been reported at, or better than, a resolution of 0.8 Å, but to date the ECDA has been performed on only six of them. The protein systems ranging from small to moderate to large in size and with different numbers of α-helices and β-strands are displayed in Figure 6.3. Here, we briefly discuss their structural details (Table 6.2) and the key features on which these systems were chosen for ECDA.

The high-potential iron–sulfur protein (HiPIP) from *Thermochromatium tepidum* comprising 83 residues crystallizes in space group $P2_12_12_1$ has B_{avg} of only ~6.5 Å^2 and is diffracted up to a very high-resolution of 0.48 Å using X-ray in its reduced form.[31] This is the highest resolution diffraction data used to date for ECDA in a protein system. This is also the first case of ECDA on

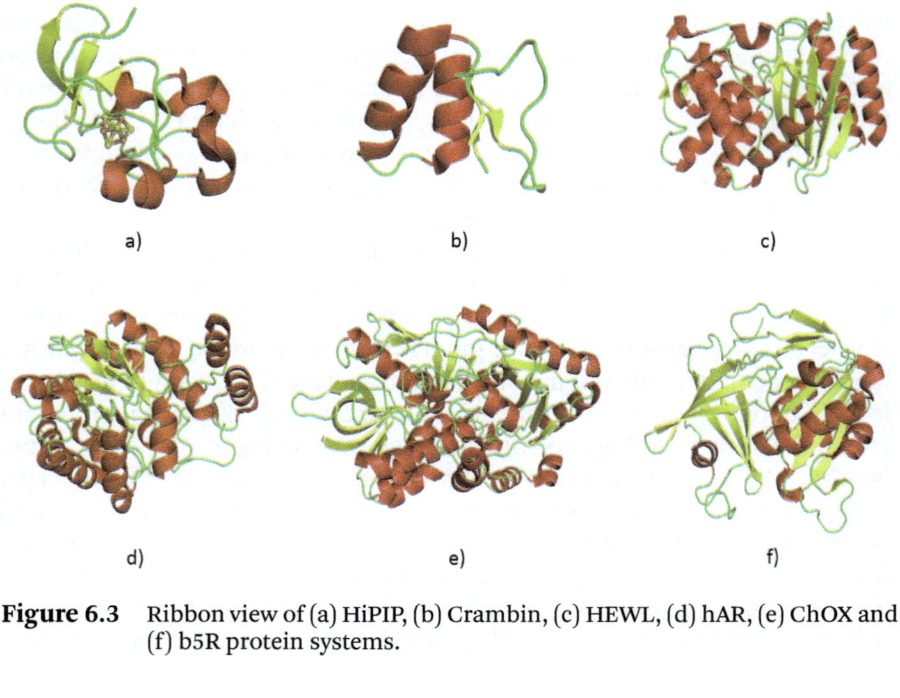

Figure 6.3 Ribbon view of (a) HiPIP, (b) Crambin, (c) HEWL, (d) hAR, (e) ChOX and (f) b5R protein systems.

Table 6.2 Structural details of the protein systems subjected to ECDA.

PDB ID	5D8V	1EJG	2VB1	1USO	4REK	5GV8
Molecular weight (kDa)	8.8	4.7	14	36	55	34
Number of residues	83	46	129	316	499	272
Number of α-helix residues	19	19	52	142	144	63
Number of β-sheet residues	13	4	8	37	106	92
B_{eq} (Å^2)	~6.5	2.8	8.3	~6	7.3	-
Cell parameters a, b, c (Å)	46.48, 58.9, 23.44	40.82, 18.5, 22.37	27.07, 31.25, 33.76	49.82, 66.59, 47.26	51.24, 72.92, 63.01	48.48, 72.11, 84.91
α, β, γ (°)	90, 90, 90	90, 90.47, 90	87.98, 108, 112.1	90, 92.4, 90	90, 105.13, 90	90, 90, 90
Space group	$P2_12_12_1$	$P1\ 2_1\ 1$	$P1$	$P1\ 2_1\ 1$	$P1\ 2_1\ 1$	$P2_12_12_1$

metalloproteins. The centre of the protein holds one Fe_4S_4 cluster of which iron 3*d* and sulfur 3*p* electron densities were visualized based on QTAIM analysis of electron densities. However, for the Fe_4S_4 cluster the multipolar values were not available in the database. Hence, for multipolar refinement initial values of P_{val} and κ were obtained using the grid search method[31] for Fe, S and Cys-S. After that, all the multipole values $P_{lm\pm}$ were refined except for P_{00}. The ECDA further revealed certain fine features at the metal binding site that were inaccessible from earlier structural studies. Electron density lobes were found to be symmetrically distributed around Fe atoms corresponding to 3*d* orbital. The S and Cys-S were found to exist with diffused electron densities corresponding to the 3*p* orbital. In some cases, the 3*d* electron density of Fe pointed towards the 3*p* electron density of S. The 3*d*–3*p* overlap was noticed to be smaller for short Fe–S bonds and *vice versa* (please refer to Figure 6.3 of ref. 31). However, it was observed that the 3*d*–3*p* overlap was proportional to the Fe–S bond length. Thereby the authors observed asymmetry in the iron–sulfur bond of the Fe_4S_4 cluster and demonstrated the structural basis of charge storage on electron transfer in this redox protein, which plays a critical role for bacterial photosynthesis.

The small protein crambin (4.7 kDa), consisting of only 46 residues, was considered as a first case study for ECDA in protein systems.[28] This protein, with a very low B_{avg} of 2.5 Å^2 and only about 30% of its unit cell volume occupied by solvent molecules, diffracted to a very high resolution of 0.54 Å. Through this study, the authors demonstrated that the average multipole values transferred from the database is a viable option for protein ECDA as the refined multipoles values correlated up to 25% with the transferred multipole values.

HEWL[32] having 129 residues (14 kDa) is known to exist in various crystal forms—triclinic, monoclinic, orthorhombic and tetragonal. However, only high-quality triclinic crystals are known to diffract to high-resolution (0.65 Å) using X-rays. Thus, this form is also chosen for ECDA. This study has shown that the covalent bond features of HEWL resemble those of small molecules. Further, the existence of some weak C–H···O interactions, which could be responsible for the protein stability, are also analysed from the ECDA point of view.

Human aldose reductase (hAR) complexed with $NADP^+$ (cofactor) and IDD594 (inhibitor) is reported to have diffracted to a high resolution of 0.66 Å, which is the highest resolution achieved to date for a protein system consisting of 316 residues or more.[18] This large enzyme has only B_{avg} of ~6 Å^2, which is well within the preferred limit for protein CD analysis. This was then the largest protein system considered for ECDA using *MoPro*.[29] The very high-resolution diffraction data and low atomic thermal motion allowed refining its multipole parameters after constructing its initial multipole model based on the *ELMAM2* database. This study focused on the estimation of electrostatic potentials (ESPs) and protein–ligand interactions at the active site. The comparative study based on database transfer, DFT calculation and from the atomic point charge taken in the *AMBER*[47] software dictionary indicated that the results are in excellent agreement. Moreover, the results based on ECDA revealed that there are significant electron lone pair contributions towards the protein–ligand interactions, which helped in understanding the atomic polarization effects due to interactions.

The largest protein system (499 residues) so far employed for ECDA is cholesterol oxidase (ChOx); the 55 kDa bacterial enzyme binds to a FAD (flavin adenine dinucleotide) cofactor diffracted up to a high resolution of 0.74 Å.[33] In this case the multipole model was constructed based on the electron densities transferred from the *ELMAM2* database and no further multipole modelling was performed. This study discussed the effects of thermal motion on the atomic densities. Atoms with higher thermal motion depicted the least bonding densities and *vice versa*. The multipole refinement statistics were also correlated with the atomic thermal motions. Bonds consisting of atoms with lower thermal motion exhibited better statistics. Further, ESPs at the FAD surface derived based on the multipole atom model were compared with those obtained from AMBER charges. ESPs from the multipole model showed nice complementarity between the FAD surface and the active site cavity. Further, based on QTAIM analysis of electron densities a total of 66 interactions were computed at the active site of the ChOx–FAD complex. All the H-bond paths were found to be highly directional and pointed towards the lone pair of electrons of the respective acceptor atoms. The conformation of FAD was found to be governed by these key interactions at the active site. The most important part of FAD is the isoalloxazine three-ring moiety, which plays an important role in redox reaction/electron transfer. This moiety is noticed to be involved in four strong H-bond interactions of which three were believed to be responsible for the deformation of the FAD ring. Detailed topological analysis of H-bond interactions at the active site revealed some key features of electron densities, which were believed to be responsible for holding the FAD molecule at the ChOx active site.

The medium sized (272 residues) flavoprotein NADH-cytochrome b5 reductase (b5R) also contains an FAD molecule as a co-factor. In order to understand the role of the flavin molecule in the function of the flavoproteins, and to derive its fine structural features, the protein complex was subjected to ECDA. For this, multipole parameters were transferred from the *ELMAM2* database to its high-resolution structure (0.78 Å). The study was mainly focused on the topological analysis of covalent bonds of the FAD co-factor and its classical (N–H···O) and non-classical (C–H···O/N *etc.*) HBs with the protein environment. The high-quality multipole model based on systematic refinement allowed the authors to compare their results with the ECDA of small molecules. The topological properties of the H-bonds displayed a similar trend to that observed in the case of small molecules.[48] Thereby, the importance of non-classical HBs for the stability of the protein structure is deduced. A comparison of classical *vs.* non-classical HBs revealed that the energies of the latter ones are less. Overall, the study highlighted the electronic structure of FAD, its fine features and its interactions with the b5R protein environment. Lastly, the topological analysis revealed that the FAD molecule in this complex structure possess significantly different interactions than its structure with ChOx.

6.4 Use of Neutron Diffraction Data

More than half of the atoms in proteins are H-atoms,[49] which often form bonds to direct protein stability. Unfortunately, due to the weak interaction of X-rays with light elements, H-atoms position are almost inaccessible from routine X-ray diffraction experiments. However, accurate study of H-bonding in proteins demands experimental localization of H-atoms, which is mostly accessible from neutron diffraction experiments and to some extent from ultra-high-resolution X-ray diffraction experiments. The importance of accuracy in H-positions and their ADPs for proper multipole modelling has been discussed and practiced in the case of small molecules.[50,51] On the other hand, refining protein structure *via* joint refinement of X-ray and neutron diffraction data is not uncommon.[52–56] However, use of neutron data for ECDA of proteins is rare, and to the best of our knowledge only one such study on heart fatty acid binding protein (H-FABP) complexed with oleic acid has been reported to date.[57] As discussed above, for ECDA of protein, the usual practice is to mask the bulk solvent before proceeding for modelling, as the information of H-atoms of water molecules are poorly determined using X-ray data. However, for H-FABP, the diffraction data were collected at room temperature using both X-ray (0.98 Å) and neutron (1.90 Å). Joint refinement of X-ray/neutron data was performed before the multipolar electron-density parameters were transferred from the *ELMAM2* database. The study was focused on obtaining precise information on the water cluster and the bound oleic acid at the active site. The analysis provided accurate information of orientation of water molecules, the electrostatic potentials and the intermolecular interactions involving water clusters, oleic acid and the protein at the active site. Further, this ECDA study combining X-ray and neutron discussed the role of water in the hydration sphere and how the binding of FA to the protein cavity affects the hydration sphere. Out of a total of 14 water molecules in the cluster two strong interactions were formed with the FA at the carboxylate group, which was identified and followed by calculations on the electrostatic interaction energy with and without the FA. These two water molecules were believed to provide more stability to the FA in the cavity. Moreover, one water molecule is seen ideally positioned to stabilize the U-shape of the FA as it interacted with both the ends of FA. This study is in agreement with the results of Matsuoka *et al.*,[58] which suggested that the water cluster in the cavity of H-FABP is present to distinguish between the various size (C10–C18 FA Chain) and rigidity of the FA and allowed for the correctly fit ones to bind. Lastly, this study suggested that the determination of the joint refinement structure with high-resolution X-ray diffraction (for accurate positions of non-H-atoms) and high-resolution neutron (for precise location of H-atoms) is necessary for accurate determination of electron densities in protein.

6.5 Topological Analysis of Hydrogen Bonding

Research in the area of protein charge density analysis has recently been stimulated, in particular with the interest in the topological analysis of HBs, protein–ligand interactions and their electrostatics.[28–30,59,60] A topological analysis of HBs, based on multipolar atoms to model the total electron density, yields information on their electrostatic interaction energies and dissociation energies.[61] Several empirical laws, either knowledge based or realized through quantum calculations, have also been proposed to reproduce the energetics of HBs depending on the geometry of the donor-hydrogen···acceptor moieties.[62,63] *VMoPro* is the tool for the visualization of the derived molecular properties, such as electron density, electrostatic potential, critical points and for the interaction energy calculations. The computation of the different components of the electrostatic energy between interacting molecules has been implemented in the *MoPro* package using efficient numerical integration algorithms.

6.5.1 Computation of Electrostatic Interaction and Dissociation Energies

Based on charge densities and electrostatic potentials, the electrostatic interaction energy between two entities A and B is computed as follows:

$$E_{\text{elec}} = \int \phi_B(\boldsymbol{r})\rho_A(\boldsymbol{r})\mathbf{dr} = \int \phi_A(\boldsymbol{r})\rho_B(\boldsymbol{r})\mathbf{dr}$$

where $\phi_B(\boldsymbol{r})$ is the electrostatic potential due to the charge distribution of entity B and ρ_A is the charge distribution of entity A and *vice versa*. For the calculation of the electrostatic interaction energy of N–H···O=C H-bonds the C=O group was considered as entity A and N–H groups as entity B, and *vice versa*. The integration over the whole space with ρ as non-zero was perfomed using program *VMoPro*.

For the estimation of the overall energy of the H-bonds (E_{HB}) the following equation[64] was used,

$$E_{HB} = -D_E = \frac{1}{2}V(\mathbf{r}_{cp})$$

where $V(\mathbf{r}_{cp})$ is the potential energy density at the BCP and D_E is the dissociation energy. $V(\mathbf{r}_{cp})$ can be estimated from the values of $\rho(\mathbf{r})$ and $\nabla^2\rho(\mathbf{r})$ using the formula given by Abramov.[65]

$$V(\mathbf{r}_{cp}) = \frac{1}{4}\nabla^2\rho(\mathbf{r}_{cp}) + 2G(\mathbf{r}_{cp})$$

where, $G(\mathbf{r}_{cp})$ is the kinetic energy density and is expressed as follows:

$$G(\mathbf{r}_{cp}) = \frac{3}{10}(3\pi^2)^{2/3}\rho^{5/3}(\mathbf{r}_{cp}) + \frac{1}{6}\nabla^2\rho(\mathbf{r}_{cp})$$

6.5.2 The Case of Human Aldose Reductase (hAR)

In order to estimate the E_{elec} and E_{HB} or D_e of main-chain N–H···O=C hydrogen bonds, here we considered the high-resolution (0.66 Å) structure of hAR as deposited in the PDB (ID 1US0). An earlier study of multipole based topological analysis of HBs in hAR was restricted to only the main chain of α-helices and involved both $i \rightarrow i + 4$ and $i \rightarrow i + 3$ types of N–H···O=C bonds and $i \rightarrow i + 3$ type of N···O contacts.[66] hAR consisting of 316 residues has 14 α-helices (146 residues) and 10 β-sheets (37 residues) and we considered both the secondary structures for the topological analysis of main-chain N–H···O=C hydrogen bonds using multipole based electron densities. For this purpose, the multipole parameters were directly transferred from the *ELMAM2* database after adjusting the X–H bond lengths to standard values of neutron distances.[46] Subsequently, the search for BCPs performed for $i \rightarrow i + 4$ type of H-bonds within the 10 α-helices resulted in a total of 62 N–H···O=C hydrogen bonds (Table 6.3), whereas a total of 20 N–H···O=C hydrogen bonds were found within the two of parallel and eight of antiparallel β-sheets. The $i \rightarrow i + 3$ type of N–H···O=C hydrogen bonds present within the four α-helices are not considered in our analysis. For the BCP search, we allowed up to 2.6 Å (sum of van der Waals radii of H and O) distance between the H and O atoms and a bond angle (∠N–H···O) of more or equal to 120°. The topological parameters $\rho(\mathbf{r})$, $\nabla^2\rho(\mathbf{r})$ and thus the $V(\mathbf{r})$ at the BCPs and R_{ij} were then calculated for the estimation of E_{elec} and E_{HB} or D_e of main-chain N–H···O=C hydrogen bonds.

Representative strong and weak N–H···O=C hydrogen bonds both in α-helices and β-sheets were mapped in terms of the static deformation densities and are displayed in Figure 6.4. The corresponding Laplacian maps are displayed in Figure 6.5. These maps clearly show the different directionality of lone pairs of electrons and also the varying strength of HBs in the secondary structures.

Table 6.3 Summary of the electrostatic and dissociation energies of H-bonds in the α-helices and β-sheets of hAR.

hAR (1us0)	α-helix	β-sheet
Number of H-bonds	62	20
Electrostatic interaction energy (kJ mol^{-1})	−37.16	−39.24
Dissociation energy (kJ mol^{-1})	−21.79	−23.80

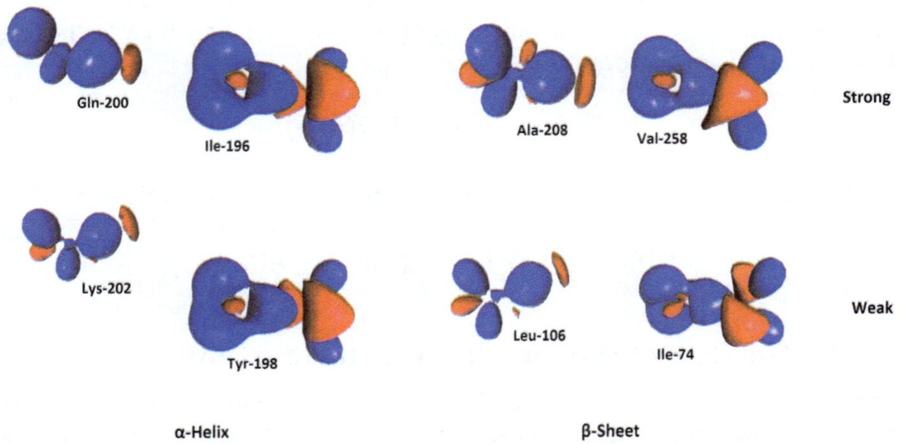

Figure 6.4 Representative 3D deformation density maps, drawn for the main chain N–H···O=C hydrogen bonds, for strong (top) and weak (bottom) H-bonds in the α-helices (left) and β-sheets (right) of hAR. Positive and negative contours at ±0.1 e Å^{-3} are shown in blue and red, respectively.

The exponential relationship between ρ_{BCP} and R_{ij} for N–H···O=C hydrogen bonds in both α-helices and β-sheets as given in Figure 6.6 appear to follow a similar trend to that observed in the case of small molecules.[67,68] The R_{ij} values range from 1.75 Å to 2.6 Å and ρ_{BCP}values range from 0.045 e Å^{-3} to 0.28 e Å^{-3}. However, the majority of the data points are found to lie at R_{ij} of between 1.75 Å and 2.3 Å, and the corresponding ρ_{BCP} values are within the range of 0.08 e Å^{-3} to 0.25 e Å^{-3}, which is in accordance with the characteristics of strong HBs in small molecules.[68] As expected, some of the N–H···O hydrogen bonds are found to exist at longer O···H distances (2.45 Å to 2.6 Å) with ρ_{BCP} values of <0.05 e Å^{-3} in α-helices than those in β-sheets. The remarkable fitting parameters for the N–H···O=C hydrogen bonds in both α-helices and β-sheets (Figure 6.6) and the excellent agreement of these results with those in small molecule claims the reliability of the present database transferred multipole based topological analysis of hAR. Subsequently, we went ahead and looked at the other relationships between the various topological parameters.

The values of E_{elec} and R_{ij} also followed an exponential relationship and displayed excellent fitting for both α-helices and β-sheets (Figure 6.7). Although the values of E_{elec} range from −14 kJ mol^{-1} to −65 kJ mol^{-1} but the majority

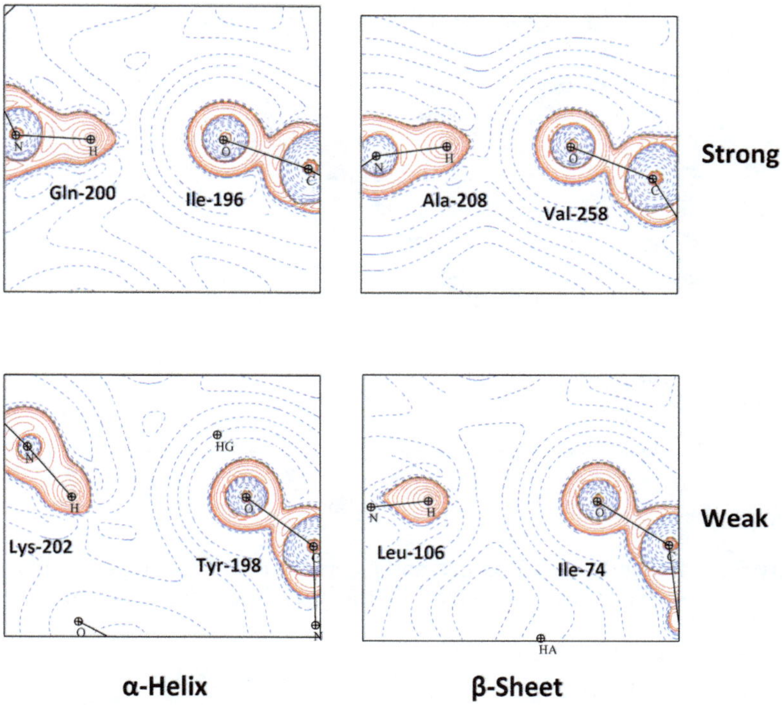

Figure 6.5 Representative Laplacian map, drawn using the C, O and H atoms of the main chain N–H···O=C hydrogen bonds, for strong (top) and weak (bottom) H-bonds in the α-helices (left) and β-sheets (right). Positive and negative contours drawn at logarithmic intervals (in e Å^{-5}) are shown by red solid and blue dashed lines, respectively.

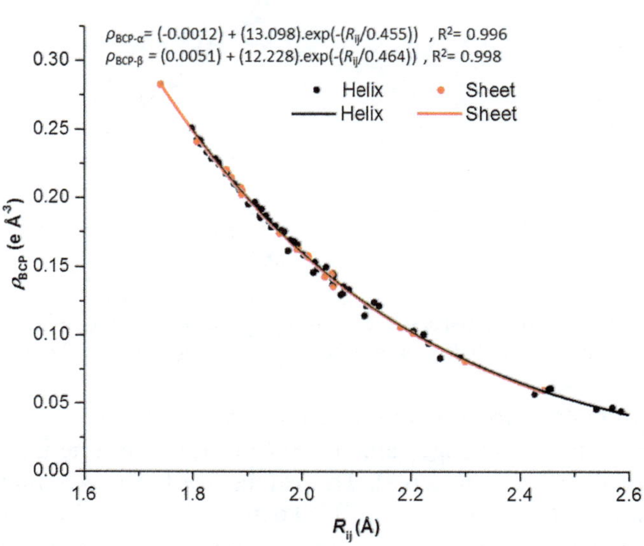

Figure 6.6 Dependence between R_{ij} and ρ_{BCP} of N–H···O=C hydrogen bonds in hAR. Curve fitting parameters and the legends are given in the inset.

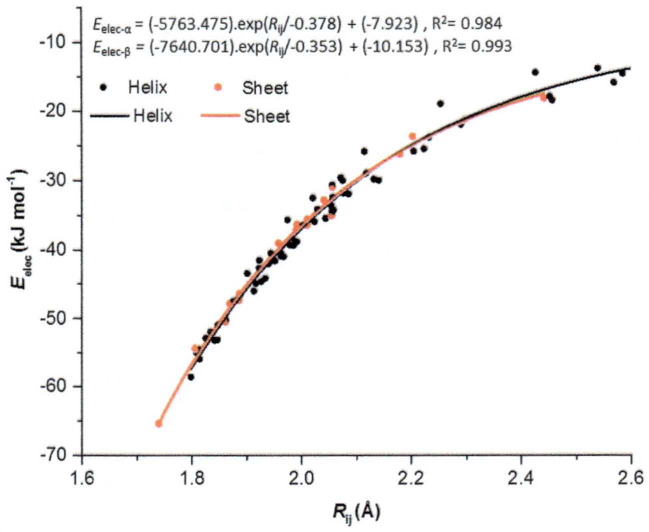

Figure 6.7 Dependence between R_{ij} and E_{elec} of N–H···O=C hydrogen bonds in hAR. Curve fitting parameters and the legends are given in the inset.

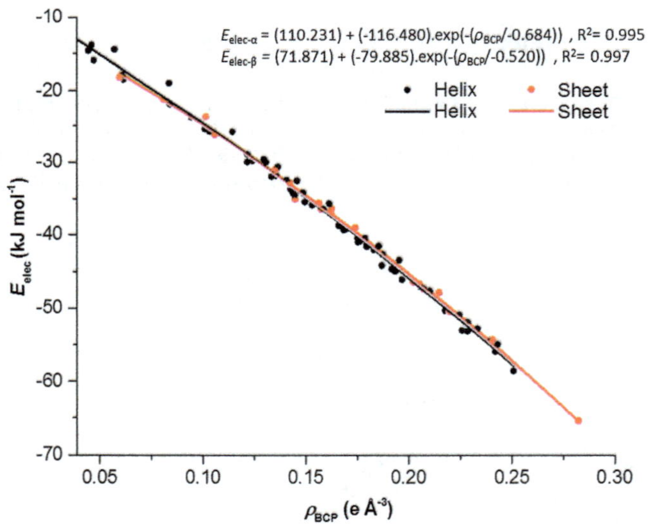

Figure 6.8 Dependence between ρ_{BCP} and E_{elec} of N–H···O=C hydrogen bonds in hAR. Curve fitting parameters and the legends are given in the inset.

of the N–H···O=C hydrogen bonds have E_{elec} values between −23 kJ mol^{-1} to −60 kJ mol^{-1} with an average value of −37 kJ mol^{-1} for α-helices and −39.24 kJ mol^{-1} for β-sheets (Table 6.3). The values are highly comparable to the experimental values (34–35 kJ mol^{-1}).[14] Further, the E_{elec} values were found to have an exponential relationship with ρ_{BCP} and the values correlated exceptionally well (Figure 6.8). Furthermore, the E_{elec} values varied linearly with

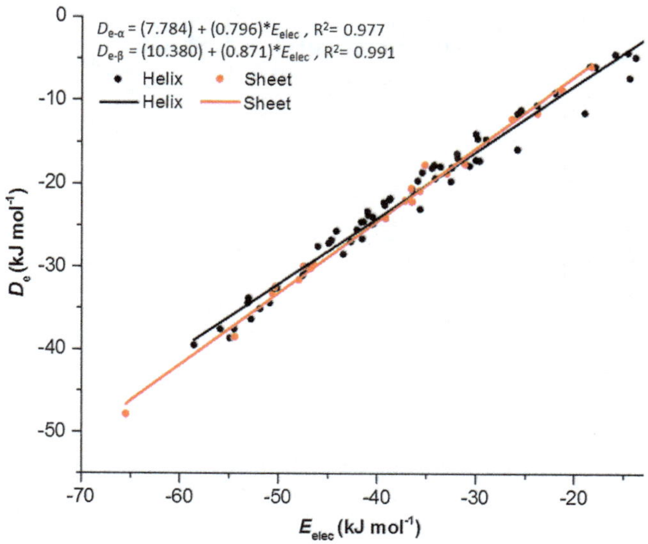

Figure 6.9 Dependence between E_{elec} and D_e of N–H···O=C hydrogen bonds in hAR. Curve fitting parameters and the legends are given in the inset.

the values of D_e and were found to correlate exceptionally well (Figure 6.9). However, the E_{elec} values for both α-helices and β-sheets were found to be systematically higher than the corresponding D_e values. The average values of E_{elec} for both α-helices and β-sheets were about 16 kJ mol^{-1} higher than those of D_e (Table 6.3). The D_e values are found to be within the range of −4 kJ mol^{-1} to −40 kJ mol^{-1} for N–H···O=C bonds of α-helices while for β-sheets the values are between −6 kJ mol^{-1} to −48 kJ mol^{-1}, and these were found to vary exponentially with the values of R_{ij} (Figure 6.10). However, the D_e values for the strong H-bonds (1.75 Å < R_{ij} < 2.3 Å) were found to be in the range of −8 kJ mol^{-1} to −40 kJ mol^{-1}. Like E_{elec} the values of D_e and ρ_{BCP} are also found to vary exponentially and the values correlated very well for both α-helices and β-sheets (Figure 6.11).

6.6 Final Remarks

The importance of HBs in proteins was realized long ago, since the determination of its first structure. However, the proper analysis of HBs in proteins became possible only recently, when the charge density method of analysis was extended from small molecules to macromolecules, thanks to the advancement in technology and also the advent of multipolar based databases. Here we provided an overview of the current status, challenges and scope for charge density analysis in protein systems. Our discussion covered the approach, basic requirements, limitations, available methodologies and tools, and a brief multipole refinement protocol for performing protein charge density analysis. Although there are three charge density databases

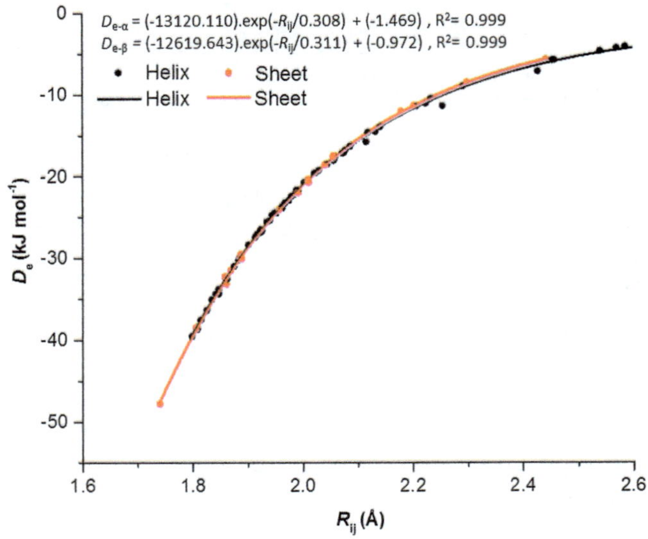

Figure 6.10 Dependence between R_{ij} and D_e of N–H···O=C hydrogen bonds in hAR. Curve fitting parameters and the legends are given in the inset.

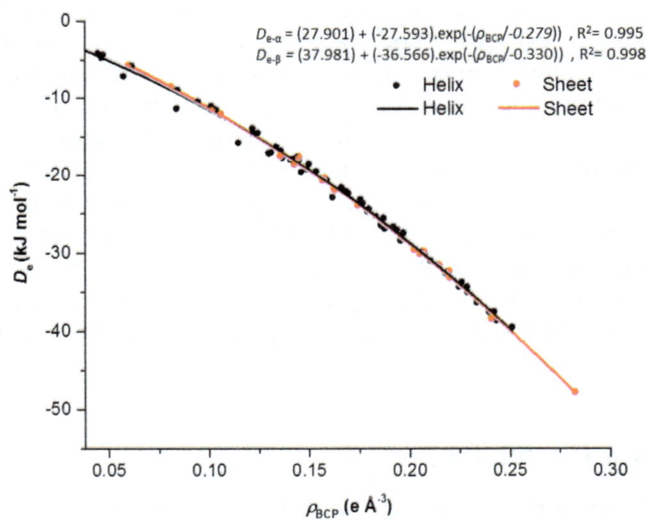

Figure 6.11 Dependence between ρ_{BCP} and D_e of N–H···O=C hydrogen bonds in hAR. Curve fitting parameters and the legends are given in the inset.

available for multipole modelling of macromolecules, to the best of our knowledge no protein charge density has been obtained using the two theoretical databases Invariom and UBDB. Thus our discussion focuses on ECDA of proteins using only the experimental database *ELMAM2*. Here, we briefly reviewed the reported ECDA on six protein systems ranging from 4.7 kDa to 55 kDa and with resolutions of 0.8 Å or better, and the key outcomes, which were inaccessible from the routine structural studies. The use of neutron diffraction data and its joint refinement with X-ray diffraction data applied in the lone case of H-FABP complexed with oleic acid[57] has been recognized as an advanced approach for multipole modelling of electron densities in protein. The study helped to quantify the interactions among the water molecules in the hydration sphere and with the protein residues at the active sites. Topological analysis of protein main-chain N–H···O=C H-bonds, both in α-helices and β-sheets of protein hAR based on electron densities from the *ELMAM2* database transferred on its high-resolution (0.66 Å) structure, revealed some fine features of weak and strong HBs. The quantitative characterization of HBs in terms of their topological properties revealed that the N–H···O=C hydrogen bonds in proteins follow the same trend as those in small molecules.[67,68] The values of topological properties, $\rho(\mathbf{r})$ and R_{ij} correlated exceptionally well with both the electrostatic, E_{elec} and dissociation D_{e} energies of HBs. The average values of E_{elec} were found to be comparable with those of reported values from experiment.[14] Both the E_{elec} and D_{e} of N–H···O=C hydrogen bonds as estimated here are sufficiently accurate, and extending such studies to both canonical and non-canonical HBs in proteins might be beneficial for understanding key interactions in the protein microenvironment. Many more such quick studies using a multipole-based database may pave the way to understanding the energies involved in protein folding and thereby its mechanism. With the increasing number of high-resolution X-ray protein structures and also, to some extent, the protein structures determined from neutron diffraction, it is expected that ECDA in proteins will become more feasible and will reach the current status of small molecule ECDA.

Acknowledgements

We are grateful to Shiv Nadar University for infrastructure and high-performance computer cluster facility. PM thanks SERB, India Govt. for research grant (EMR/2014/000491). SKM thanks UGC for a Junior Research Fellowship.

References

1. C. N. Pace, *Nat. Struct. Mol. Biol.*, 2009, **16**, 681–682.
2. C. N. Pace, B. A. Shirley, M. McNutt and K. Gajiwala, *FASEB J.*, 1996, **10**, 75–83.
3. J. D. Bernal, *Nature*, 1939, **143**, 663–667.
4. J. Chen and W. E. Stites, *Biochemistry*, 2001, **40**, 15280–15289.
5. W. Kauzmann, *Adv. Protein Chem.*, 1959, **14**, 1–63.
6. C. Tanford and J. Reynolds, *Nature's Robots: A History of Proteins*, OUP Oxford, 2003.
7. A. R. Fersht, *Trends Biochem. Sci.*, 1987, **12**, 301–304.
8. C. N. Pace, S. Trevino, E. Prabhakaran and J. M. Scholtz, *Philos. Trans. R. Soc. London, Ser. B*, 2004, **359**, 1225–1235.

9. P. J. Fleming and G. D. Rose, *Protein Sci.*, 2005, **14**, 1911–1917.
10. G. D. Rose, P. J. Fleming, J. R. Banavar and A. Maritan, *Proc. Natl. Acad. Sci. U. S. A.*, 2006, **103**, 16623–16633.
11. J. Kyte, *Structure in Protein Chemistry*, Garland Science, 2006.
12. A. E. Mirsky and L. Pauling, *Proc. Natl. Acad. Sci. U. S. A.*, 1936, **22**, 439–447.
13. D. F. Sticke, L. G. Presta, K. A. Dill and G. D. Rose, *J. Mol. Biol.*, 1992, **226**, 1143–1159.
14. L. Pauling, R. B. Corey and H. R. Branson, *Proc. Natl. Acad. Sci. U. S. A.*, 1951, **37**, 205–211.
15. L. Pauling and R. B. Corey, *Proc. Natl. Acad. Sci. U. S. A.*, 1951, **37**, 729–740.
16. S. Scheiner, in *Reviews in Computational Chemistry*, ed. D. Boyd and K. Lipkowitz, VCH Publishers, New York, 1991, pp. 165–218.
17. A. Warshel, P. K. Sharma, M. Kato and W. W. Parson, *Biochim. Biophys. Acta, Proteins Proteomics*, 2006, **1764**, 1647–1676.
18. E. I. Howard, R. Sanishvili, R. E. Cachau, A. Mitschler, B. Chevrier, P. Barth, V. Lamour, M. Van Zandt, E. Sibley, C. Bon, D. Moras, T. R. Schneider, A. Joachimiak and A. Podjarny, *Proteins: Struct., Funct., Genet.*, 2004, **55**, 792–804.
19. P. Munshi, S. L. Chung, M. P. Blakeley, K. L. Weiss, D. A. A. Myles and F. Meilleur, *Acta Crystallogr., Sect. D: Biol. Crystallogr.*, 2012, **68**, 35–41.
20. J. C. Kendrew, G. Bodo, H. M. Dintzis, R. G. Parrish, H. Wyckoff and D. C. Phillips, *Nature*, 1958, **181**, 662–666.
21. N. K. Hansen and P. Coppens, *Acta Crystallogr. Sect. A*, 1978, **34**, 909–921.
22. D. J. Willock, S. L. Price, M. Leslie and C. R. A. Catlow, *J. Comput. Chem.*, 1995, **16**, 628–647.
23. B. Fournier, E.-E. Bendeif, B. Guillot, A. Podjarny, C. Lecomte and C. Jelsch, *J. Am. Chem. Soc.*, 2009, **131**, 10929–10941.
24. R. F. W. Bader, *Atoms in Molecules*, Wiley Online Library, 1990.
25. P. Coppens, *X-ray Charge Densities and Chemical Bonding*, International Union of Crystallography, 1997, vol. 4.
26. P. Munshi, PhD Thesis, Indian Institute of Science, 2005.
27. R. Herbst-Irmer and D. Stalke, *Acta Crystallogr., Sect. B: Struct. Sci., Cryst. Eng. Mater.*, 2017, **73**, 531–543.
28. C. Jelsch, M. M. Teeter, V. Lamzin, V. Pichon-Pesme, R. H. Blessing and C. Lecomte, *Proc. Natl. Acad. Sci. U. S. A.*, 2000, **97**, 3171–3176.
29. B. Guillot, C. Jelsch, A. Podjarny and C. Lecomte, *Acta Crystallogr., Sect. D: Biol. Crystallogr.*, 2008, **64**, 567–588.
30. D. Liebschner, M. Elias, S. Moniot, B. Fournier, K. Scott, C. Jelsch, B. Guillot, C. Lecomte and E. Chabrière, *J. Am. Chem. Soc.*, 2009, **131**, 7879–7886.
31. Y. Hirano, K. Takeda and K. Miki, *Nature*, 2016, **534**, 281–284.
32. J. Held and S. Van Smaalen, *Acta Crystallogr., Sect. D: Biol. Crystallogr.*, 2014, **70**, 1136–1146.
33. B. Zarychta, A. Lyubimov, M. Ahmed, P. Munshi, B. Guillot, A. Vrielink and C. Jelsch, *Acta Crystallogr., Sect. D: Biol. Crystallogr.*, 2015, **71**, 954–968.

34. K. Takaba, K. Takeda, M. Kosugi, T. Tamada and K. Miki, *Nature*, 2017, **7**, 1–10.
35. B. Dittrich, C. B. Hübschle, K. Pröpper, F. Dietrich, T. Stolper and J. J. Holstein, *Acta Crystallogr., Sect. B: Struct. Sci., Cryst. Eng. Mater.*, 2013, **69**, 91–104.
36. P. M. Dominiak, A. Volkov, X. Li, M. Messerschmidt and P. Coppens, *J. Chem. Theory Comput.*, 2007, **3**, 232–247.
37. B. Dittrich, C. S. Bond, R. Kalinowski, M. A. Spackman and D. Jayatilaka, *CrystEngComm*, 2010, **12**, 2419.
38. M. Malińska, K. N. Jarzembska, A. M. Goral, A. Kutner, K. Woźniak and P. M. Dominiak, *Acta Crystallogr., Sect. D: Biol. Crystallogr.*, 2014, **70**, 1257–1270.
39. S. Domagała, B. Fournier, D. Liebschner, B. Guillot and C. Jelsch, *Acta Crystallogr., Sect. A: Found. Crystallogr.*, 2012, **68**, 337–351.
40. M. Elias, D. Liebschner, J. Koepke, C. Lecomte, B. Guillot, C. Jelsch and E. Chabriere, *BMC Res. Notes*, 2013, **6**, 308.
41. V. Pichon-Pesme, C. Jelsch, B. Guillot and C. Lecomte, *Acta Crystallogr., Sect. A: Found. Crystallogr.*, 2004, **60**, 204–208.
42. A. Volkov, T. Koritsanszky, X. Li and P. Coppens, *Acta Crystallogr., Sect. A: Found. Crystallogr.*, 2004, **60**, 638–639.
43. B. Guillot, L. Viry, R. Guillot, C. Lecomte and C. Jelsch, *J. Appl. Crystallogr.*, 2001, **34**, 214–223.
44. C. Jelsch, B. Guillot, A. Lagoutte and C. Lecomte, *J. Appl. Crystallogr.*, 2005, **38**, 38–54.
45. A. Fokine and A. Urzhumtsev, *Acta Crystallogr., Sect. D: Biol. Crystallogr.*, 2002, **58**, 1387–1392.
46. F. H. Allen, *Acta Crystallogr., Sect. B: Struct. Sci.*, 1986, **42**, 515–522.
47. D. A. Case, T. A. Darden, T. E. Cheatham, C. L. Simmerling, J. Wang, R. E. Duke, R. Luo, M. Crowley, R. C. Walker, W. Zhang and others, *Amber 10*, 2008.
48. E. Espinosa, M. Souhassou, H. Lachekar and C. Lecomte, *Acta Crystallogr., Sect. B: Struct. Sci.*, 1999, **55**, 563–572.
49. J. Habash, J. Raftery, R. Nuttall, H. J. Price, C. Wilkinson, A. J. Kalb (Gilboa) and J. R. Helliwell, *Acta Crystallogr., Sect. D: Biol. Crystallogr.*, 2000, **56**, 541–550.
50. T. Koritsanszky, *Acta Crystallogr., Sect. A: Found. Crystallogr.*, 1998, **54**, 686.
51. P. Munshi, A. Ø. Madsen, M. A. Spackman, S. Larsen and R. Destro, *Acta Crystallogr., Sect. A: Found. Crystallogr.*, 2008, **64**, 465–475.
52. P. Munshi, E. H. Snell, M. J. Van Der Woerd, R. A. Judge, D. A. A. Myles, Z. Ren and F. Meilleur, *Acta Crystallogr., Sect. D: Biol. Crystallogr.*, 2014, **70**, 414–420.
53. J.-P. Bacik, S. Mekasha, Z. Forsberg, A. Y. Kovalevsky, G. Vaaje-Kolstad, V. G. H. Eijsink, J. C. Nix, L. Coates, M. J. Cuneo, C. J. Unkeferand J. C.-H. Chen, *Biochemistry*, 2017, **56**, 2529–2532.

54. W. B. O'Dell, A. M. Bodenheimer and F. Meilleur, *Arch. Biochem. Biophys.*, 2016, **602**, 48–60.
55. M. Melzer, J. C. Chen, A. Heidenreich, J. Gab, M. Koller, K. Kehe and M. M. Blum, *J. Am. Chem. Soc.*, 2009, **131**, 17226–17232.
56. M. Haupt, M. P. Blakeley, S. J. Fisher, S. A. Mason, J. B. Cooper, E. P. Mitchell and V. T. Forsyth, *IUCrJ*, 2014, **1**, 429–438.
57. E. I. Howard, B. Guillot, M. P. Blakeley, M. Haertlein, M. Moulin, A. Mitschler, A. Cousido-Siah, F. Fadel, W. M. Valsecchi, T. Tomizaki, T. Petrova, J. Claudot and A. Podjarny, *IUCrJ*, 2016, **3**, 115–126.
58. S. Matsuoka, S. Sugiyama, D. Matsuoka, M. Hirose, S. Lethu, H. Ano, T. Hara, O. Ichihara, S. R. Kimura, S. Murakami, H. Ishida, E. Mizohata, T. Inoue and M. Murata, *Angew. Chem., Int. Ed.*, 2015, **54**, 1508–1511.
59. C. Lecomte, B. Guillot, C. Jelsch and A. Podjarny, *Int. J. Quantum Chem.*, 2005, **101**, 624–634.
60. N. Muzet, B. Guillot, C. Jelsch, E. Howard and C. Lecomte, *Proc. Natl. Acad. Sci. U. S. A.*, 2003, **100**, 8742–8747.
61. I. Mata, I. Alkorta, E. Molins and E. Espinosa, *Chem.–Eur. J.*, 2010, **16**, 2442–2452.
62. F. Fabiola, R. Bertram, A. Korostelev and M. S. Chapman, *Protein Sci.*, 2002, **11**, 1415–1423.
63. H. Choi, H. Kang and H. Park, *J. Comput. Chem.*, 2010, **31**, 897–903.
64. E. Espinosa, E. Molins and C. Lecomte, *Chem. Phys. Lett.*, 1998, **285**, 170–173.
65. Y. a. Abramov, *Acta Crystallogr., Sect. A: Found. Crystallogr.*, 1997, **53**, 264–272.
66. D. Liebschner, C. Jelsch, E. Espinosa, C. Lecomte, E. Chabrière and B. Guillot, *J. Phys. Chem. A*, 2011, **115**, 12895–12904.
67. P. R. Mallinson, G. T. Smith, C. C. Wilson, E. Grech and K. Wozniak, *J. Am. Chem. Soc.*, 2003, **125**, 4259–4270.
68. P. Munshi and T. N. G. Row, *CrystEngComm*, 2005, **7**, 608–611.

CHAPTER 7

Towards a Generalized Database of Atomic Polarizabilities

MICHELLE ERNST*[a], LEONARDO H. R. DOS SANTOS[b], ANNA KRAWCZUK[c] AND PIERO MACCHI*[a]

[a]University of Bern, Department of Chemistry and Biochemistry, Freiestrasse 3, Bern, 3012, Switzerland; [b]Universidade Federal de Minas Gerais, Departamento de Química, Av. Antônio Carlos 6627, Belo Horizonte, MG-31270-901, Brazil; [c]Jagiellonian University, Department of Crystal Chemistry and Crystal Physics, Gronostajowa 2, 30-387 Krakow, Poland
*E-mail: michelle.ernst@dcb.unibe.ch, piero.macchi@dcb.unibe.ch

7.1 Introduction

Present-day quantum chemical methods allow for precise calculations of molecular wavefunctions and, consequently, the electronic properties of molecules. The same holds true for crystals, for which periodic wavefunctions, atomic structures as well as electronic structures and properties can be quite accurately calculated. In this context, it could be expected that the interest in sub-molecular units, like functional groups, is small because the most obvious breakdown of a material may be molecules (for molecular materials), monomers (for polymers) or atoms (for extended covalent, metallic and single-atom ionic solids). However, when a materials scientist or a chemist aims at designing a new functionalized material, the basic constituents of

Monographs in Supramolecular Chemistry No. 26
Understanding Intermolecular Interactions in the Solid State: Approaches and Techniques
Edited by Deepak Chopra

Published by the Royal Society of Chemistry, www.rsc.org

molecules and crystals are still considered. As a matter of fact, the concept of functional groups is so deeply entangled in modern chemistry that it cannot easily be avoided. The characteristic behaviour of two or more functional groups justifies for instance the use of the arrow notation in synthetic chemistry, it allows for the explanation of reaction mechanisms and designing new synthetic routes. Functional groups are also important for the characterization of a material. For example, typical vibrations of atoms in a group give rise to absorption bands that allow their identification in an IR spectrum. Finally, the optical, electric, magnetic or mechanical properties of molecular materials can be analysed in terms of the contributions of the functional groups they consist of; ultimately, just one or a few functional groups might be enough to explain a molecular or crystal property, an important aspect for retro-designing new materials (reverse crystal engineering). The fact that functional groups approximately retain their identity and properties in different molecules goes along with the concept of *transferability*,[1–3] according to which the electron density and thereof derived properties depend, to a first approximation, on the local nuclear configuration and the bonding situation. Nevertheless, the properties of a functional group depend obviously also on the chemical environment, which perturbs its "isolated" electron density. These perturbations are of two kinds: through-bond and through-space. The first one is due to conjugative effects that may enable outer atoms to modify the electron distribution of a given functional group, even if they are quite distant. Of course, this perturbation decreases with the number of bonds that separate the perturbing atom and the functional group under analysis. The through-space perturbation, instead, is due to the polarization induced by the electric field of the perturbing group and it rapidly decreases with the distance. Thus, despite an overall similarity, if we compare a given property of a specific functional group in different molecules, a relatively large variance may be observed. The variance is smaller for functional groups of the same molecule embedded in different crystalline environments (for example in co-crystals or polymorphs). In fact, the electric field due to the surrounding molecules in a crystal, despite being of the order of magnitude of GV m^{-1}, is much smaller than an intra-molecular field. Moreover, through-bond effects are almost absent in inter-molecular interactions, apart from very strong hydrogen bonds. Therefore, changing some substituents in a molecule is much more perturbative than changing the surrounding of a molecule.

For these reasons, to use functional groups in chemistry and materials science, the concept of *similarities* and *dissimilarities* must be properly quantified, and it is quite useful to establish a classification that allows sorting them into groups (clusters) within which some *descriptors*, for example physical properties, are similar. This means that elements of a given cluster can be well represented by the average values of their descriptors, whereas functional groups belonging to different clusters are clearly distinguished. After this delicate procedure, one can build up a database of functional groups and their associated properties, to easily evaluate the role of a group of atoms in a molecular crystal. The cluster averages not only represent the molecules used to build up the database, but they are also representative (within a given confidence criterion) of the same functional groups in other molecules that were not employed for constructing the database. If this purpose is reached, the database entries would allow the rapid estimation of properties of large molecules or their aggregation in crystals, without the need of extensive and time-consuming calculations. Furthermore, the role of individual atoms or functional groups in the build-up of a molecular or crystal property can be evaluated, an important step towards the rational design of molecular-based materials.

Even though our approach is general, we will focus on the first-order polarizability tensor in this work. The electric polarizability is responsible for the optoelectronic properties of a material like refractive index or dielectric constant. Thus, a good comprehension of the key factors that affect the polarizability allows one to tune the optoelectronic properties and design new materials with targeted applications. For example, high refractive index (HRI) materials are desirable for lenses, optical waveguides,[4] or a high light extraction efficiency in LED components.[5] Anti-reflective coatings used in solar cells require a tuneable refractive index.[6] The ongoing downscaling of semiconducting devices demands for materials with continuously lower dielectric constants for the insulation of wires.[7] Moreover, fast market changes and the objective of producing more sophisticated optoelectronic devices has stimulated the development of more environment-friendly HRI materials. So far, aromatic rings, heteroatoms like sulphur, or halogen atoms have been used to enhance refractive indices.[8–11] However, due to environmental concerns, compounds with halogen substituents are not suitable. Moreover, a big challenge is to increase the refractive index of a material while keeping its low viscosity at ambient temperature. At present, a wide group of halogen-free acrylate derivatives is used for fabricating HRI optical devices but so far no organic compound with low birefringence, small optical dispersion and absence of visible spectra absorption has exhibited a refractive index above 1.7.[12–14] Inorganic compounds may exhibit higher refractive indices but they carry a number of disadvantages such as high density, low flexibility and highly intense visible absorption bands.[15] Therefore, there is still a demand for new and improved organic compounds with excellent optical properties.

In this context, the purpose of our research is to simplify the efforts necessary to select new promising materials (either organic, inorganic or metal–organic), before undergoing expensive tests and more sophisticated calculations, and before embarking on new and perhaps expensive fabrication procedures. For this purpose, a critical issue is being able to rapidly estimate an approximate value of the refractive indices of a molecular-based crystalline material in order to select the most promising candidates.

With modern computational resources, the calculation of static as well as frequency dependent polarizabilities of small to medium sized molecules is possible using quantum-mechanical methods at many levels of theory, thus allowing for acceptable precision. However, given the relatively high level of electronic correlation usually necessary for very accurate estimations of the optoelectronic properties, *ab initio* calculations for large molecules and condensed matter are often prohibitive.[16] Noteworthy is that a highly correlated level of theory is necessary to well represent local and non-local effects perturbing a functional group. This suggests that a partition of the material in functional groups may be preferable to compute the overall polarizability of the system. In fact, a database of functional groups may simultaneously guarantee the requested accuracy (depending on the level of theory used to pre-calculate the functional groups) and a very high rapidity (because new *ab initio* calculation is not necessary and the polarizability of the whole system is simply the sum of the polarizabilities of their constituent parts). The only concern is a proper spatial orientation of the database polarizability tensors, which can be solved with a proper treatment of local tensor axes, as deeply discussed in this chapter.

Once a database of functional group polarizabilities is compiled and at hand, the optical properties of molecular materials can be rapidly predicted. If the database also contains dipole moments of the functional groups, the inner crystal field can be approximated, and therefore the dielectric properties of a crystalline material as well, by also evaluating the mutual polarization of the molecules.

The road-map of our research is summarized in Figure 7.1, focusing on the build-up of a polarizability database, which is the subject of this chapter.

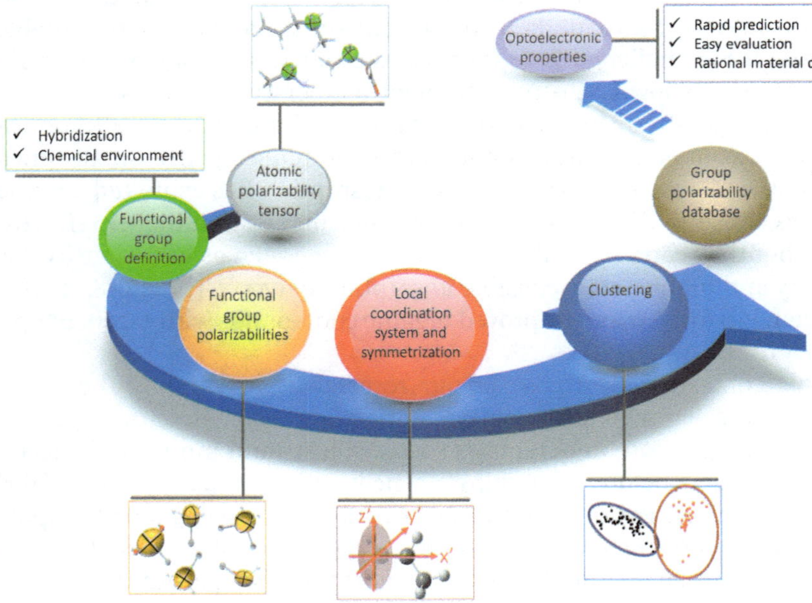

Figure 7.1 A step-by-step pathway for creating a generalized database of functional group polarizabilities.

7.2 Theoretical Background

7.2.1 Earlier Atomic Polarizability Databases and the Need for a New One

A molecule exposed to an electric field will respond to this perturbation by a polarization of its charge density (affecting both the electron and nuclear distributions). The strength and type of this response is described by the electric (hyper)polarizabilities, which are symmetric tensors relating the three-dimensional electric field vector $\boldsymbol{E}$ to the change in the dipole moment $\boldsymbol{\mu}$. If only the dipolar first order polarizability is considered, the molecular dipole moment will change linearly with the applied field and the induced dipole moment change is easily derived:[17]

$$\Delta\boldsymbol{\mu} = \boldsymbol{\alpha}\cdot\boldsymbol{E} \tag{7.1}$$

The first order polarizability $\boldsymbol{\alpha}$ is a 3 × 3 tensor whose individual components can hardly be determined experimentally. Alternatively, it is common to refer to the isotropic polarizability (the arithmetic average of the main diagonal tensor components), easily obtained from refractive measurements in solution or polycrystalline materials. However, knowing the individual tensor components would be essential as they inform on the orientation of the polarizability ellipsoid, which may be highly anisotropic. This is of relevance for rationally designing crystalline materials, because the anisotropy of the molecular polarizability may relate to the most important intra- and inter-molecular contacts that cause the anisotropy of the optoelectronic properties.

Vogel[18] was one of the first to evaluate and produce a database of isotropic polarizabilities for different atom and functional group types using additivity schemes. In this kind of approach, the group constituents of a chemical system are considered to not interact with each other, hence the "unperturbed" polarizabilities are simply summed together to produce the whole property. Based on Vogel's isotropic values, Applequist[19] set up a more advanced way to evaluate the polarizabilities of molecules *via* a non-additive model, taking into account the interaction of atomic polarizabilities *via* induced dipole moments. Therefore, the molecular polarizabilities, despite being based on isotropic atomic values, were transformed into anisotropic ones. Miller[20] used the so-called atomic hybrid components (AHCs), which were defined by the atomic number and the hybridization state of the atom, to construct isotropic molecular polarizabilities. Every AHC had an assigned polarizability that was obtained by fitting to experimental data. Kassimi *et al.*[21] calculated isotropic polarizabilities of amino acids as the sum of smaller fragments they consist of. The fragments had to be saturated with hydrogen atoms and this contribution was afterwards removed from the summation:

$$\boldsymbol{\alpha}(AB) = \boldsymbol{\alpha}(AH) + \boldsymbol{\alpha}(BH) - 2\boldsymbol{\alpha}(H) \tag{7.2}$$

The reference polarizability for hydrogen was obtained from the polarizabilities of methane and ethane in the following way:

$$\boldsymbol{\alpha}(H)=\boldsymbol{\alpha}(CH_4)-\frac{1}{2}\boldsymbol{\alpha}(C_2H_6) \tag{7.3}$$

The polarizabilities of the small fragments AH and BH can be calculated with much less effort or they can also come from a previously built database. Therefore, the authors pointed out the need for a database of such fragment polarizabilities.

Dykstra *et al.*[22] developed a database for different atom types including also anisotropic polarizabilities obtained by fitting their tensor components to the corresponding calculated molecular components. Unfortunately, no definition was given for the local coordinate system used to import the anisotropic tensors from the database, an issue that is not trivial to solve (as we will discuss in Section 7.3.2).

Alternatively, Stone[23] calculated distributed polarizabilities from the Hilbert-space based distributed multipole analysis previously developed by himself and co-workers.[24–26] After analysing many partition schemes, Stone advocated that the most transferable polarizabilities would be obtained from a real-space based partitioning, like the ones proposed by Bader or Hirshfeld. Ivanov-Smolenskii *et al.*[27] obtained polarizabilities of ions based on X-ray diffraction data. They examined crystals of the NaCl type where the Wigner–Seitz cell contains only one atom, and the separation into what they defined as pseudoatoms is very similar to Bader's scheme described in Section 7.2.2. From the number of electrons and their mean radius they estimated atomic polarizabilities. Whitten *et al.*,[28] instead, pointed out that the one-electron density, as for example obtained from a multipolar refinement, is not sufficient for determining polarizabilities. They proposed the X-ray constrained wavefunction approach to evaluate polarizabilities from X-ray measurements. Suponitsky *et al.*[29] calculated the intermolecular interaction energy of a molecule in a crystal as the energy difference of this molecule embedded in the crystal and in isolation. The induced dipole moment, necessary to calculate the electrostatic part of the energy, was derived from the difference of the electron density of the molecule in the crystal and in isolation. The electron density was described with a Hansen–Coppens multipolar model fitted on the one hand to X-ray data and on the other hand to Hartree–Fock calculations. They concluded that the uncertainty of the electron density obtained from a measurement, in their example of a non-centrosymmetric crystal of urea, is too large to allow this calculation, while it worked well for electron densities obtained from Hartree–Fock calculations.

Bader *et al.*[2] demonstrated the additivity of group polarizabilities for acyclic hydrocarbons and rationalized it based on the partitioning of the electron density with the quantum theory of atoms in molecules (QTAIM). From the polarizabilities of alkanes, Bader derived average polarizabilities for the methyl and methylene group. The QTAIM approach was further developed by Keith who mainly contributed to the solution of the origin dependence problem of dipole moment derivatives of non-neutral subunits of a molecule (like, in fact, atoms). This scheme was improved and implemented in the software *PolaBer*, which allows the calculation of full polarizability tensors for atoms and functional groups in isolated molecules or crystals.[30] Using this method, Dos Santos *et al.*[31] calculated the average polarizabilities of functional groups present in naturally occurring amino acids and evaluated the changes with respect to their aggregates, thus explaining the role of hydrogen bonds on perturbing isolated-molecule polarizabilities. On the one hand, this work paved the way for the database discussed here. On the other hand, that work also provided accurate analysis of the effect of intra- and intermolecular interactions on the transferable functional groups, thus showing how to use the database for accurate determination of polarizabilities of large molecular aggregations, and eventually also crystals.

In all the above studies, functional groups were defined *a priori* based on chemical intuition. Except for the QTAIM based studies, the polarizabilities of these atom types, bonds, or groups were always obtained by fitting them to a set of experimental or calculated molecular polarizabilities. As anticipated, our goal is to create a first principles and generally transferable database for functional group polarizabilities, obtained by a partitioning of the molecular electron density according to QTAIM (see Section 7.2.2). This method is self-consistent and does not require any assumptions. Moreover, the recognition of functional groups should also emerge from the statistical analysis of calculated polarizabilities on a number of representative molecules without assuming any *a priori* knowledge and adopting only unsupervised clustering algorithms.[32] Using a multivariate cluster analysis, one may in fact fulfil this condition. In addition, since the number of clusters may exceed that of functional groups, subtypes of functional groups may also be distinguished. Within one cluster, the polarizabilities are sufficiently similar and the average value of a particular cluster is the best representation of its constituent polarizabilities. Ideally, the cluster coincides with a functional group or a sub-type functional group.

Eventually, a database can be established where each entry contains information about the functional group type and a 3×3 (symmetric) polarizability tensor.

The following approximations are adopted: (a) the nuclear relaxations under the applied field and the vibrational contributions to the polarizability are ignored (b) only static (*i.e.* zero-frequency) polarizabilities are considered. By applying (a), the electric (hyper)polarizabilities are only determined by the electron density. The use of only the first order polarizability $\boldsymbol{\alpha}$ (a 3×3 symmetric tensor) is justified by the *quasi*-linear response of the dipole moments to electric fields of the magnitudes of those of most interacting radiations. Of course, this implies that non-linear effects (like for example second-harmonic generation) cannot be evaluated using the database. Even in the strong inner fields of molecular crystals the range of the response is in general sufficiently close to linearity, therefore the field induced dipole moments and inner fields are also accurately estimated.

The use of static fields is justified by the relative similarity between high frequency electric polarizabilities and static polarizabilities. In fact, at very high frequency, all the motions of the molecules and specific strong absorptions are quenched. A simple scaling could be, in this case, adopted to estimate the refractive index in the desired wavelength spectrum region.[33]

As described above, many scientists have proposed methods to estimate atomic $\boldsymbol{\alpha}$ tensors, and constructed similar databases, either empirically or semi-empirically, but lacking exportable local coordinate systems for the tensors. A major problem was also the definition of polarizabilities that avoid issues related to the origin-dependence of the framework. The novelty of our approach is to use the transferable Bader's zero-flux-bounded functional groups, as previously anticipated[30,31] to create a database that allows for easy transferability of properties, hence contributing to the rapid estimation of optoelectronic properties of rather complex materials such as proteins, polymers or crystals.

This kind of analysis is not restricted to the polarizability, but could be extended to other properties, like for example the atomic multipole moments. Good estimations of the electric moments (in particular the dipole moment) are important to evaluate the electrostatic potential "produced" by a molecule. As a matter of fact, in the field of charge density analysis, many databases of atomic electric multipoles have been developed in the past.[34–36] Together with the polarizability, they allow the approximation of the crystal field, which polarizes molecules in a crystal.

7.2.2 Distributed Atomic Polarizabilities

In order to derive atomic polarizabilities, it is necessary to define an atomic polarizability density. In the approach proposed here, we use the quantum theory of atoms in molecules, which provides a topological partition of the molecular electron density into atomic basins Ω, based on the gradient field. This enables defining atomic properties like the atomic dipole moments $\boldsymbol{\mu}(\Omega)$, consisting of two components, namely the atomic polarization $\boldsymbol{\mu}_p(\Omega)$ (*i.e.*, the charge shift within the atomic basin) and the charge translation $\boldsymbol{\mu}_C(\Omega)$ (*i.e.* the shift from one basin into another).

$$\boldsymbol{\mu}(\Omega) = \boldsymbol{\mu}_{\mathrm{p}}(\Omega) + \boldsymbol{\mu}_{\mathrm{C}}(\Omega) = -\int_{\Omega}(\boldsymbol{r} - \boldsymbol{R}_{\Omega})\rho(\boldsymbol{r})\mathrm{d}\mathbf{r} + (\boldsymbol{R}_{\Omega} - \boldsymbol{R}_0)q(\Omega) \quad (7.4)$$

In eqn (7.4), $\boldsymbol{R}_{\Omega}$ is the position vector of the nucleus Ω; $\boldsymbol{R}_0$ is the origin of the molecular coordinate system (arbitrarily chosen). If the atomic basin is not neutral, $\boldsymbol{\mu}_C(\Omega)$ does not vanish and is origin dependent. This is a serious issue for transferability. However, the procedure designed by Keith enables overcoming this dependency by re-writing the charge-translation component into a basin-to-basin summation:

$$\boldsymbol{\mu}(\Omega) = \boldsymbol{\mu}_{\mathrm{p}}(\Omega) + \boldsymbol{\mu}_{\mathrm{C}}(\Omega) = -\int_{\Omega}(\boldsymbol{r} - \boldsymbol{R}_{\Omega})\rho(\boldsymbol{r})\mathrm{d}\mathbf{r} + \Sigma(\boldsymbol{R}_{\Omega} - \boldsymbol{R}_{\mathrm{BCP}}(\Omega|\Omega'))Q(\Omega|\Omega') \quad (7.5)$$

where $Q(\Omega|\Omega')$ is the charge transferred to the atomic basin Ω from the bonded atom Ω' and $\boldsymbol{R}_{BCP}$ is the positional vector of the bond critical point between atoms Ω and Ω'. The scheme proposed by Keith has general validity beyond QTAIM itself. A different partitioning scheme may not only produce a different intrinsic polarization $\boldsymbol{\mu}_C(\Omega)$, but also a different connectivity between atomic basins, modifying quite substantially $\boldsymbol{\mu}_C(\Omega)$. Nevertheless, the sum of all atomic dipoles $\boldsymbol{\mu}_C(\Omega)$ and $\boldsymbol{\mu}_P(\Omega)$ must return the molecular dipole. A special case is that of non-neutral molecules. In this case, the molecular dipole is origin dependent. However, for any given partitioning, the procedure by Keith provides an origin independent solution (this basically transforms the origin dependence of the $\boldsymbol{\mu}_C(\Omega)$ terms into a "partition" dependence). This aspect is very important when considering transferability of functional group properties, given that a functional group in a molecule is not supposed to be neutral.

The polarizabilities are eventually obtained as the field derivatives of the total atomic dipoles ($\boldsymbol{\mu}_C(\Omega) + \boldsymbol{\mu}_p(\Omega)$), which could be computed analytically or numerically. Thus, atomic polarizabilities can be derived from the unperturbed and perturbed atomic dipole moments by numerical differentiation, provided that the electric field is sufficiently small to cause only a linear response:

$$\alpha(\Omega)_{ij} = \lim_{E\to 0} \frac{\mu_i^{E_j}(\Omega) - \mu_i^0(\Omega)}{E_j} \tag{7.6}$$

The resulting atomic polarizability is a 3 × 3 tensor, which is by its own definition not symmetrical, but for practical purposes it is symmetrized according to the classical scheme by Nye.[37] Otero *et al.*[38] have criticized the use of eqn (7.6), arguing that because the unperturbed, field-free electron density and the densities perturbed by the electric fields are independently calculated and partitioned, the atomic basins are not constrained to be constant in the process. According to the authors, this would be the ultimate reason as to why the polarizability tensors result in not being symmetrical. However, this criticism is not correct, as the numerical approach of eqn (7.6) gives the exact definition of a dipolar density derivative, which corresponds to the exact analytical definition in the limit of an infinitely small field. The reason why the tensors are not symmetrical lies in the atomic basins themselves, which are in fact by definition not symmetrical, which does not enable eqn (7.6) to be invariant under the exchange of *i* and *j*. This affects both $\boldsymbol{\mu}_P(\Omega)$ and $\boldsymbol{\mu}_C(\Omega)$ terms.

As QTAIM is an exact partitioning scheme, in the sense that all points in molecular or crystal space are assigned to one and only one atomic basin (except for the boundaries), the sum of all atomic polarizabilities in a functional group gives the polarizability of the functional group itself, and likewise for the molecule. The atomic, functional group and molecular polarizability tensors can be represented in position space as ellipsoids, centred at the centre of the group/molecule or at the atomic nuclei, with units of a volume.[30,31,39,40] It is also important to emphasize that, while by definition the total molecular polarizability must be a positive definite symmetric tensor, the individual atomic polarizability must not. However, this is very rarely observed and only for polarizabilities of hydrogen atoms that, in specific intermolecular contexts, could be negatively defined. This issue does not affect any of the molecules we used in the calculations reported here.

7.3 Constructing the Database

7.3.1 Computational Details

A sufficiently extended sampling is necessary to obtain a good estimation of the polarizability of a specific functional group, given that many different environments could be considered. However, because the number of possible molecules that contain a particular functional group is extremely large, even under the condition that the molecule contains a limited number of atoms, the chosen molecules for a given group were constructed using a preconceived approach. In particular, molecules containing only C, H, N, and O atoms were used. The 16 common functional groups or atoms shown in Table 7.1 were attached as first neighbours and as second neighbours to the functional group under analysis. Only one of these functional groups was attached at a time and all other positions were saturated with hydrogen or methyl groups. The cooperative effect of various functional groups is therefore not taken into account. The geometry of every molecule, in its more commonly observed conformation, was optimized with *GAUSSIAN09*[41] at the B3LYP level of theory using the cc-pVDZ basis set. An extensive conformational analysis, in search of more stable conformers, was not performed. Only for 1,3-propandiol was a full conformational analysis performed with the *Open Babel confab generator.*[42] The results confirm that using more

Table 7.1 Functional groups or atoms used in this work.

Added group	
-H	
$-CH_3$	Methyl
-OH	Hydroxyl
$-OCH_3$	Methoxy
-COOH	Carboxyl
-COH	Formyl
$-COCH_3$	Acetyl
$-COOCH_3$	Acetate ester
$-NH_2$	Primary amine
$-N(CH_3)_2$	Dimethylamine
$-CONH_2$	Primary amide
$-CHCOOHNH_2$	Amino acid group
$-CHCH_2$	Vinyl
$-C(CH_3)CH_2$	Isopropenyl
-CCH	Ethynyl
$-C_6H_5$	Phenyl

conformations for the same molecule would not increase the information, given the relatively lower variance. In fact, the standard deviation of the isotropic polarizability of the nine calculated conformers of 1,3-propandiol is 0.0026 compared to a standard deviation of 0.0056 for the entire dataset of the hydroxyl group.

After the geometry optimization of the selected conformation, calculations with and without a static electric field were carried out at the CAM-B3LYP/aug-cc-pVDZ level of theory to obtain the wavefunctions and molecular electron densities. Our previous work demonstrated that this level of theory gives quite accurate functional group polarizabilities for many amino acids.[31] The chosen electric field was 0.001 a.u. and it was applied in the $+x$, $-x$, $+y$, $-y$, $+z$ and $-z$ directions, to allow the application of eqn (7.6). *AIMAll*[43] was used to partition the seven obtained electron densities into atomic contributions.

7.3.2 The Local Coordinate System

The 3×3 tensors representing the atomic and functional group polarizabilities are expressed with respect to a coordinate system. For each molecule used to generate the database, this coordinate system is originally oriented with respect to the molecular inertia tensor. However, all the polarizabilities of atoms within a given functional group must be compared based on a common reference frame, independent from that of the molecules they belong to. Therefore, a local functional group coordinate system is necessary to define the functional group polarizabilities that become therefore transferable to any molecule containing that functional group. The definition of a local coordinate system is not unique and it is of course a rather crucial step to properly use the information stored in the database. Several definitions of local coordinate systems are for example found in scattering factor databases used for charge density studies, like the invariom[34] database, ELMAM2,[35] or UBDB.[36] In the invariom database, the axes are defined along bonds and the ranking of the bond lengths assigns the axes. ELMAM2 also uses bonds for the assignment; additionally, seven different axes systems are distinguished depending on the different bonding situations. At variance from our approach, these definitions hold for individual atoms, where the centre of the coordinate system must obviously coincide with the atomic nucleus and the axes orientation depends only on other atoms. In order to implement similar concepts for coordinate systems of functional groups, additional criteria are necessary, because the group has its own inertia system and intragroup bonds. While individual atoms (if assumed just as material points) are inherently spherically symmetric, functional groups are less symmetric and often non-symmetric. Nevertheless, they often have a topological symmetry, which could be larger than the actual symmetry of the molecule. In order to refer the local properties to a molecule-independent system, we defined a local coordinate system for functional groups. For those containing more than two atoms, an inertia-like tensor **I** was calculated using the position of the atoms in the functional group and, instead of the atomic masses, the nuclear charges Z. This is sometimes called the nuclear charge moment tensor.[44] The elements of the symmetric **I** tensor were obtained as follows:

$$\begin{aligned}
I_{xx} &= \sum_{i=1,N} \left(y_i^2 + z_i^2\right) Z_i \\
I_{yy} &= \sum_{i=1,N} \left(x_i^2 + z_i^2\right) Z_i \\
I_{zz} &= \sum_{i=1,N} \left(x_i^2 + y_i^2\right) Z_i \\
I_{xy} &= I_{yx} = -\sum_{i=1,N} x_i y_i Z_i \\
I_{yz} &= I_{zy} = -\sum_{i=1,N} y_i z_i Z_i \\
I_{xz} &= I_{zx} = -\sum_{i=1,N} x_i z_i Z_i
\end{aligned} \tag{7.7}$$

where N is the number of atoms in the functional group, x, y, z the atomic coordinates, and Z_i the atomic number.

The main axes of the tensor are taken as the local coordinate system with the origin located at the centre of nuclear charges. Mathematically, the eigenvectors from the diagonalization of the nuclear charge moment tensor correspond to the components of the rotation matrix, which rotates the global coordinate system to the local coordinate system, as shown in Figure 7.2a. The rotation matrix must be reordered according to the magnitude of its eigenvalues (while keeping the handedness of the system) in order to obtain consistent results for the same functional groups in different molecules. Furthermore, the same orientation of a given functional group in different molecules must be assured by choosing axes along particular directions specified by given atoms in the functional group. The advantage of this "nuclear charge moment tensor" definition is that it is less dependent on small changes of the molecular geometry compared to other definitions. For functional groups consisting of only two atoms, the local axes system was defined along the bond direction using a procedure very similar to the one developed for ELMAM2.[35] The first axis z' was defined along the bond connecting the two atoms, the second and third axes, x' and y', were simply defined as perpendicular to the first one. For an unambiguous definition, an additional criterion was obviously needed, therefore, a supporting vector s', shown in Figure 7.2b, was defined along the bond to a neighbouring atom and y' was chosen to be perpendicular to the plane spanned by z' and s'. In both cases the vectors were normalized, giving rise to a locally defined right-handed coordinate system.

The rotation matrix Q corresponds to the matrix rotating the global coordinate system into the local one. This matrix must be applied on the originally calculated polarizability tensor as well as on the molecular coordinates to express them with respect to the local coordinate system.

$$\boldsymbol{Q}^T \boldsymbol{\alpha} \boldsymbol{Q} = \boldsymbol{\alpha}^{FG} \tag{7.8}$$

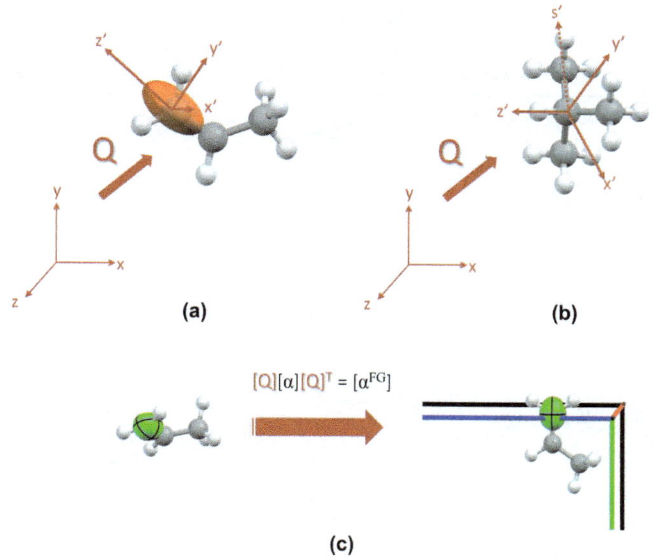

Figure 7.2 (a) Definition of the local coordinate system based on the nuclear charge moment tensor of the CH_2 functional group. (b) Definition of the local coordinate system along bonds for a functional group with just two atoms (C–H). s' is a supporting vector to a neighbouring atom needed for a unique definition. (c) The matrix **Q** rotates properties from the external to the local coordinate system.

After the polarizability tensors were rotated to the local coordinate system, they were symmetrized with respect to the topological symmetry of the functional group given in Table 7.2. Due to the symmetrization of the polarizability tensor of the functional group, a small fraction of the information about the orientation of the tensor is lost. However, the isotropic polarizability is not influenced by this procedure.

In principle, the database can not only contain polarizability tensors for functional groups but also tensors for each atom in a given group. The atomic polarizabilities can enter the database without further symmetrization besides that required when applying eqn (7.6). In order to use them, they would be averaged over all entries in the database and eventually symmetrized.

7.3.3 Multivariate Data Analysis and Clustering

Once a representative amount of data is available, more insight comes from statistical analysis of the variables that describe the objects. The tools we adopted for the analysis of this multivariate data are the principal component analysis and the unsupervised cluster analysis.[32]

For a given dataset of calculated polarizabilities the mean $\langle\alpha_{ij}\rangle$ and the standard deviation σ_{ij} were calculated for every element of the polarizability

Table 7.2 Symmetry constraints applied to the functional groups used in the database. Due to the inherent symmetry of a second order tensor, the symmetry of the polarizability is always higher than that of the functional group.

Functional group	Hybridization	Functional group polarizability	Atomic polarizabilities	Topological symmetry of fragment
CH_3	C sp^3			C_{3v}
CH_2	C sp^2		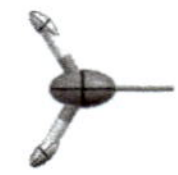	C_{2v}
	C sp^3			C_{2v}
CH	C sp^3			C_∞
	C sp^2			C_{2v}
	C sp			$C_{\infty v}$
OH				C_s
NH_2				C_s
NH	N sp^3		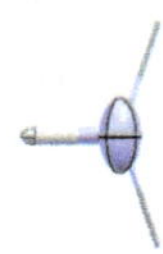	C_s
	N sp^2			C_s
COOH		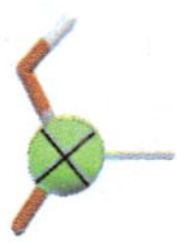		C_s
$CONH_2$				C_s
$CHCOOHNH_2$			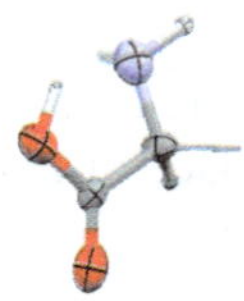	C_1

tensor (α_{xx}, α_{yy}, α_{zz}, α_{xy}, α_{xz}, α_{yz}) individually. Afterwards, the data can be standardized according to eqn (7.9). This procedure corresponds to setting the mean to 0 and the standard deviation to 1 for all variables, thus bringing them to a comparable scale. In the case of polarizabilities, this is relevant given the large discrepancy between diagonal (large) and out of diagonal components (smaller, often close or equal to zero). In this way, all the elements of the polarizability tensor have the same weight.[45]

$$z_{ij} = \frac{\alpha_{ij} - \alpha_{ij}}{\sigma_{ij}} \tag{7.9}$$

The next step is the clustering, which means recognizing that in some molecules a given functional group assumes values that are more similar than in others, implying a sub-classification of groups. For this purpose, one can adopt a cluster analysis. The average of all polarizabilities within the same cluster will eventually be the database entry for a functional group in a specific context. It is expected that this context coincides with a specific chemical environment. For example, the functional group CH_2 may behave differently according to the hybridization of the carbon atom and the nature of its bond(s). Therefore, one may expect to obtain a cluster for C(sp^2), one for C(sp^3) and several sub-clusters depending on the groups attached to the C atom. However, the multivariate analysis is unbiased and therefore no *a priori* information is given on expected different CH_2 groups.

For a good representability of the average values and a more appropriate multivariate analysis, the ensemble of tested molecules should be sufficiently large. The purpose of the database is dual: on the one hand, it should represent as many bonding situations as possible for each functional group in order to represent well the "chemical variance"; on the other hand, it must synthesize the information in a not too large number of clearly distinguishable cases. A database would be useless if it is insufficiently representative (too few test molecules) or if it is too much diversified (it contains too many clusters). It is important to note that different clusters can contain different numbers of molecules, meaning that the clustering is only based on the variables and there are no external restraints or interventions.

In order to perform a cluster analysis, it is necessary to evaluate the proximity of two polarizability tensors. For this purpose, we use the Euclidean distance:

$$d_{AB} = \sqrt{\sum_{K=1} \left(P_K(A) - P_K(B)\right)^2} \tag{7.10}$$

where K is the dimension of the property P. Thus, for polarizabilities the summation runs from 1 to 6. A and B are different data points, thus representing a functional group in two different molecules.

Clustering was performed using a genetic algorithm, a procedure that is often adopted for finding the best set of parameters fitting a very large number of observations. Genetic algorithms use methods that mimic evolutionary processes in nature. A given state can be encoded in a string of numbers (or also matrices, *etc.*), so called chromosomes. Initially, a set of chromosomes representing different starting states is created giving rise to a set of parent chromosomes. Each of them is attributed a fitness according to the chromosome's ability to minimize or maximize a given function. Subsequently, two chromosomes are chosen out of the parental set and combined to give a new one. The probability of picking chromosomes depends on their respective fitness; the higher the fitness, the higher the probability to use them in the recombination step. Pairing of two chromosomes out of the parental set is repeated until a set of offspring chromosomes is generated with the same size as the parental set. Mutation of offspring can happen, which means that a chromosome is changed at an arbitrary number of random positions. After recombination and mutation, a new set of chromosomes is obtained with potentially higher fitness. While cycles of selection, recombination, and mutation are repeated, the general fitness should increase. The overall fittest chromosome corresponds to the best solution. When the genetic algorithm is applied to our case of grouping polarizabilities, the problem lies in finding an optimum clustering that minimizes a global function F defining the average distance between the points in a cluster to their cluster average:[46]

$$F = \sum_{i}^{\text{clusters}} \sum_{\boldsymbol{\alpha}_j \in C_i} \| \langle \boldsymbol{\alpha} \rangle_{C_i} - \boldsymbol{\alpha}_j \| \tag{7.11}$$

where $\langle \boldsymbol{\alpha} \rangle_{C_i}$ is the average polarizability of cluster i and the sum runs over all polarizabilities in all clusters. The optimization is done by expressing the cluster averages as chromosomes that are subject to evolutionary processes. After each cycle of fitness calculation, selection, recombination and mutation all polarizabilities are assigned to the closest cluster average. Thereafter, a new cluster average is calculated based on all the points in this cluster. Then the next generation is produced and this cycle is repeated many times.

The symmetrized functional group polarizability tensors can be grouped into clusters based on either standardized or unstandardized data. The only bias in the procedure is that the number of clusters is not a "result", but it is chosen *a priori* based on certain criteria. There is some degree of arbitrariness in selecting these indicators. A criterion similar to that of Popelier was adopted here, namely the *intercluster ratio* (IR).[47] A Gaussian distribution has a given mean and standard deviation. Using the standard deviation, a confidence interval around the mean, *e.g.* 2σ, can be specified within which a data point is located with a given probability. The IR is a similar criterion to judge the separability of two clusters. An IR with a specific number signifies, in analogy to a single distribution, that the points in the two clusters are correctly classified with a given probability. It is calculated for each diagonal component of the polarizability tensor individually. For two clusters *A* and *B* the *IR* is defined as the ratio of the difference of their means to the sum of their respective standard deviations:

$$IR = \frac{\langle \alpha_A \rangle - \langle \alpha_B \rangle}{\sigma_A + \sigma_B} \tag{7.12}$$

The number of distinct clusters was chosen such that each pair of clusters has at least one component with an IR higher than a given number, which we chose to be 2. This means that for both clusters the chance that a data point is correctly classified is at least 95.4%.

7.3.4 Recognizing a Functional Group

The clustering into subcategories that we have presented is based on the polarizabilities of functional groups. Of course this information is not available when only the molecular geometry is known and an assignment is needed in order to use the database. Therefore, additional criteria are needed that are based solely on structural features. Calculations for CH_2 show that a clustering based on the inertia tensor of the functional group results in a very similar clustering as that based on the functional group polarizabilities. Selecting two clusters, for example, leads in both cases to a separation of sp^3 and sp^2 hybridized carbon atoms.

This correspondence is essential in order to classify the functional groups based solely on their geometry and therefore to make use of the more detailed information stored in the database. This enables one to make the functional group recognition and polarizability calculation quite automatic, following this simple sequence: (a) define the functional groups within a molecule (based on the molecular topology); (b) calculate the inertia tensor of each functional group; (c) compare the inertia tensors with those calculated for the functional groups stored in the database and assign to the functional group the mean polarizability of the selected cluster; (e) calculate the molecular polarizability as sum of the functional group polarizabilities.

7.4 Results

7.4.1 Clustering the CH_2 Polarizabilities

As introduced above, the polarizabilities of a given functional group depend on the type and connectivity of the atoms they consist of, perturbed by the chemical environment (mainly due to *through-bond* conjugative effects along a given connectivity path or *through-space* interactions). The clustering procedure described in the previous section is a way to identify and classify these perturbations.

We illustrate here functional group clustering using the CH_2 group as a working example. Obviously, the polarizability of this group depends strongly on the hybridization state of the carbon atom. An sp^3 hybridized carbon atom will form two additional bonds besides those to the hydrogen atoms, while an sp^2 hybridized carbon atom will form only one additional (double) bond. Therefore, we first expect to find a clear clustering of functional group polarizabilities based on the carbon hybridization. The effect of groups bonded to the carbon atom is less easy to predict, although one may anticipate different effects depending on the electronegativity of the bonded atoms, the presence of electron resonance forms and the proximity of other non-bonded groups (that may bear strong charges and therefore produce a strongly polarizing field). Due to the tensor symmetrization in C_{2v}, the principal axes of the CH_2 polarizability tensors coincide with the axes of the local coordinate system. Therefore, the off-diagonal elements are zero and the polarizabilities can be plotted with the values of the diagonal only (Figure 7.3a). In the case of functional groups with lower symmetry, the off-diagonal elements are not constrained to zero. In this case, representation of the six independent variables is difficult to visualize, and the correlation is not easily understood. For this reason, a principal component analysis is necessary. This allows the identification of the combinations of variables (principal components, PC) that enable explaining the largest variance in the dataset.

For CH_2, the scores of PC1 and PC2 and the local coordinate system are reported in Figure 7.3b. PC1 is composed of $-0.56\alpha_{xx}$, $0.57\alpha_{yy}$, $0.60\alpha_{zz}$ and PC2 is composed of $0.74\alpha_{xx}$, $0.67\alpha_{yy}$, $0.06\alpha_{zz}$. Noteworthy is that PC1, which expresses the negative correlation between polarizability in the directions x and y, explains more than 80% of the variance.

The molecules used for the statistical analysis are listed in Table 7.3. The colours in Figure 7.3 indicate the hybridization of the carbon atom (red-orange for sp^2; black-blue for sp^3) as well as bonds to heteroatoms (blue or orange). As expected, the polarizability is affected by the hybridization (PC1) more than by the direct neighbour (PC2). The clustering depicted in Figure 7.3b is simply driven by chemical intuition. By applying the genetic algorithm clustering, we can check whether this intuition reflects the data.

In Figure 7.4, results are reported for calculation with 2, 4, 6 and 8 clusters. In keeping with the principal component analysis, the effect of the carbon hybridization overwhelms the presence of a heteroatom. In fact, the result of the clustering with two clusters corresponds almost entirely to the separation of sp^2 and sp^3 hybridized carbons. In the case where four clusters were chosen, the groups with different hybridizations remain separated, but very similar polarizabilities are distributed on three clusters (blue); a result which does not represent a chemically intuitive separation because it is not strictly related to the presence of heteroatoms. When going to eight clusters, the cases where the carbon atom is bound to a heteroatom are in a distinct cluster.

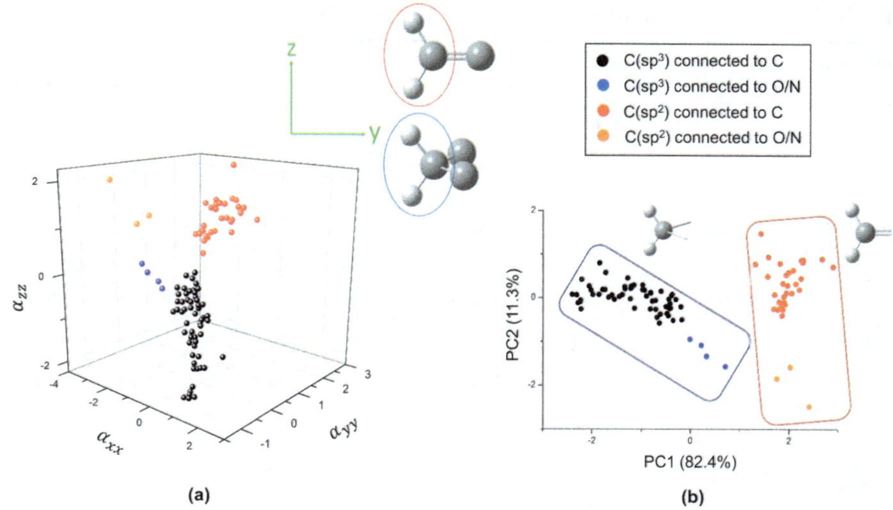

Figure 7.3 (a) Representation of the standardized CH_2 polarizabilities in three-dimensional space and (b) projected on the plane of the two principal components. The percentage gives the amount of variance explained by each principal component. The blue rectangle highlights functional groups with an sp^3-hybridized carbon atom and the red rectangle highlights functional groups with an sp^2-hybridized carbon atom. Furthermore, the orientation of the axes is shown with the green vectors y and z.

7.4.2 Clustering all Functional Groups

Figure 7.5 shows the 456 standardized polarizabilities of all the calculated functional groups plotted in principal component space. Figure 7.5a shows the different types of functional groups, highlighted with colours. The hybridization is indicated with the size of the squares. From this picture, one can see that the functional groups characterized by the highest polarizabilites, namely amino acids, amides and carboxylic acids, present the highest variability as well. Figure 7.5b shows the outcome of a clustering including the polarizabilities of all different functional groups.

Figure 7.6 shows the 13 clusters originating from the clustering. The size of the symbol is proportional to the number of entries from a given functional group in this cluster. Since the polarizabilities of the amino acid functional group are rather high, the variability is high as well and they go into different clusters. By contrast, the NH(sp^2) and CH(sp^2) polarizabilities go into the same cluster.

7.4.3 Using the Database to Compute Polarizabilities

The polarizabilities of some amino acids were calculated and results are summarized in Figures 7.7–7.10. The molecular tensors are computed in four different ways: (1) analytical double derivation of the molecular electronic

Table 7.3 Molecules created for the CH_2 functional group.

Molecule	Label	Molecule	Label	Molecule	Label	Molecule	Label
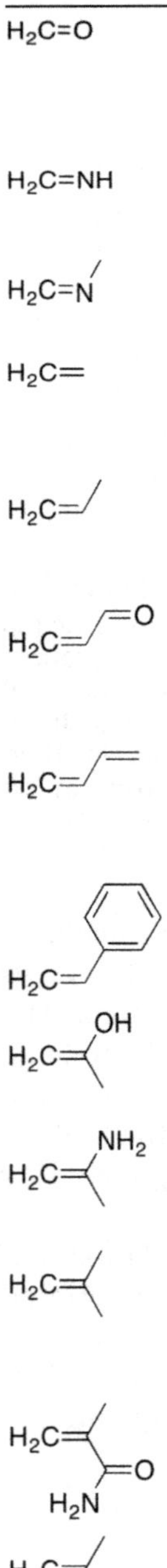	**1**	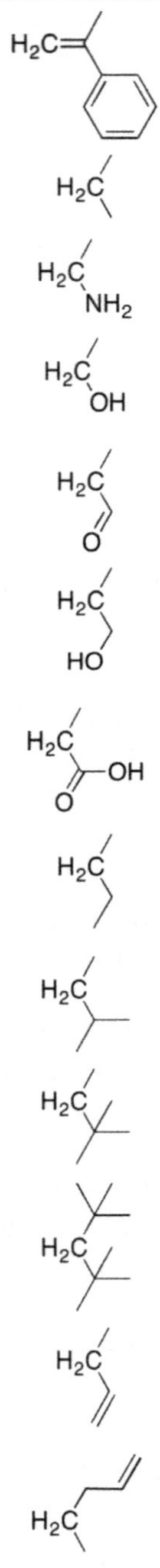	**10h**	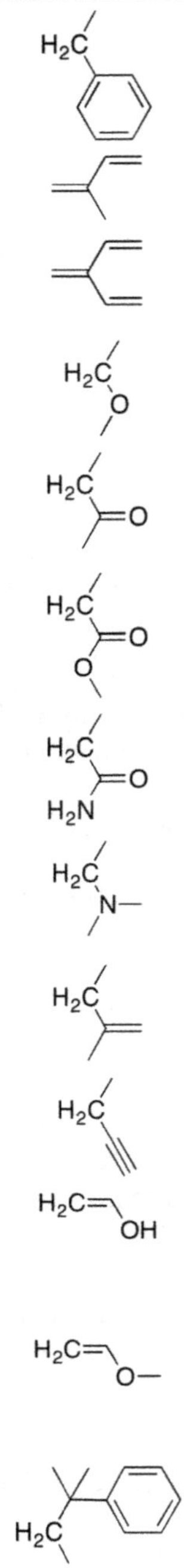	**28**	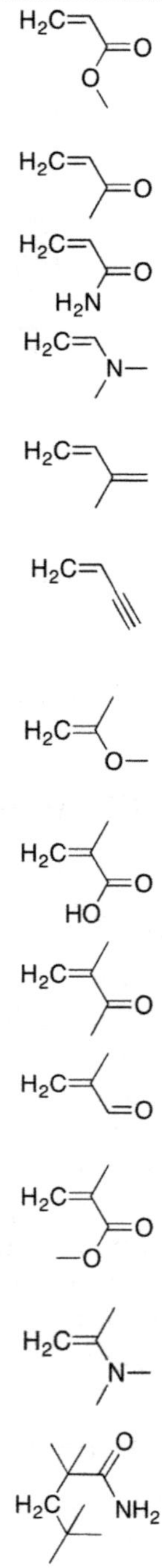	**51**
	2a1		**14a**		**40**		**52**
	5		**14f**		**41**		**53**
	6		**14g**		**42**		**54**
	7a		**14h**		**43**		**55**
	7e		**16b**		**44**		**56**
	7i		**16c**		**45**		**57**
	7j		**16d**		**46**		**58**
	8a		**16e**		**47**		**59**
	9a		**14b**		**48**		**60**
	10a		**14c**		**49**		**61**
	10g		**21b**		**50**		**62**
	63		75		87		99

(continued)

Table 7.3 (*continued*)

Molecule	Label
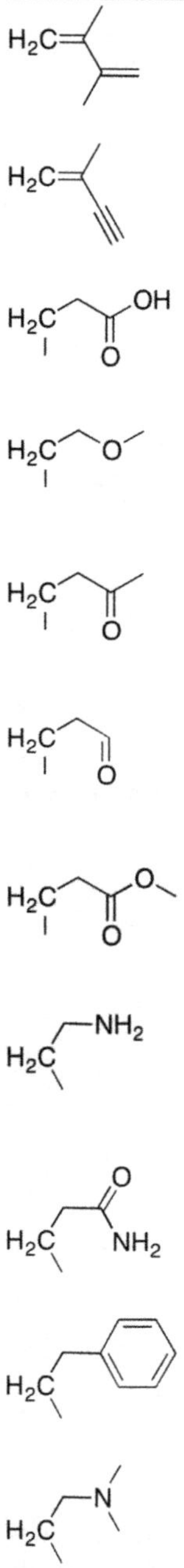	64, 65, 66, 67, 68, 69, 70, 71, 72, 73, 74
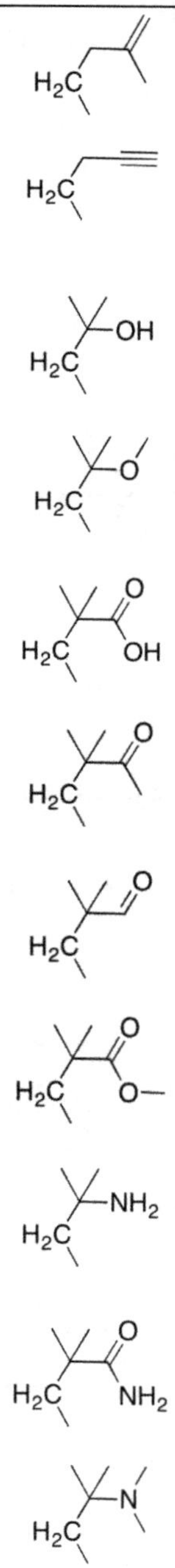	76, 77, 78, 79, 80, 81, 82, 83, 84, 85, 86
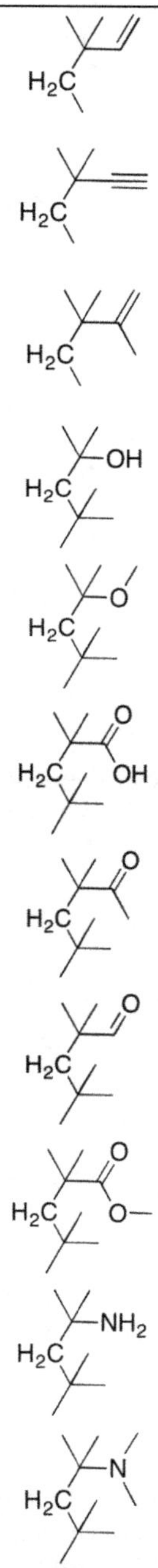	88, 89, 90, 91, 92, 93, 94, 95, 96, 97, 98
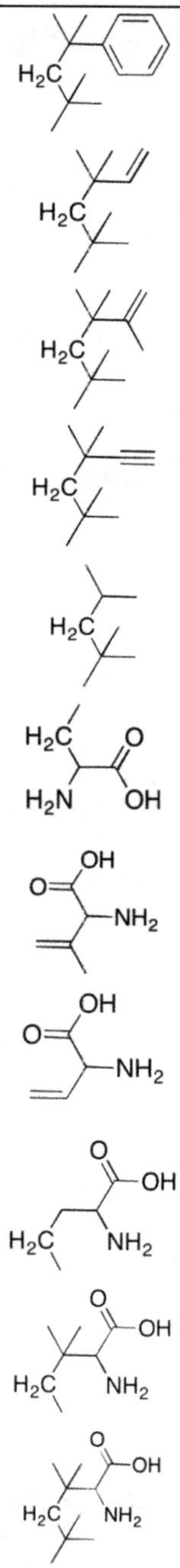	100, 101, 102, 103, 104, 105, 106, 107, 108, 109, 110

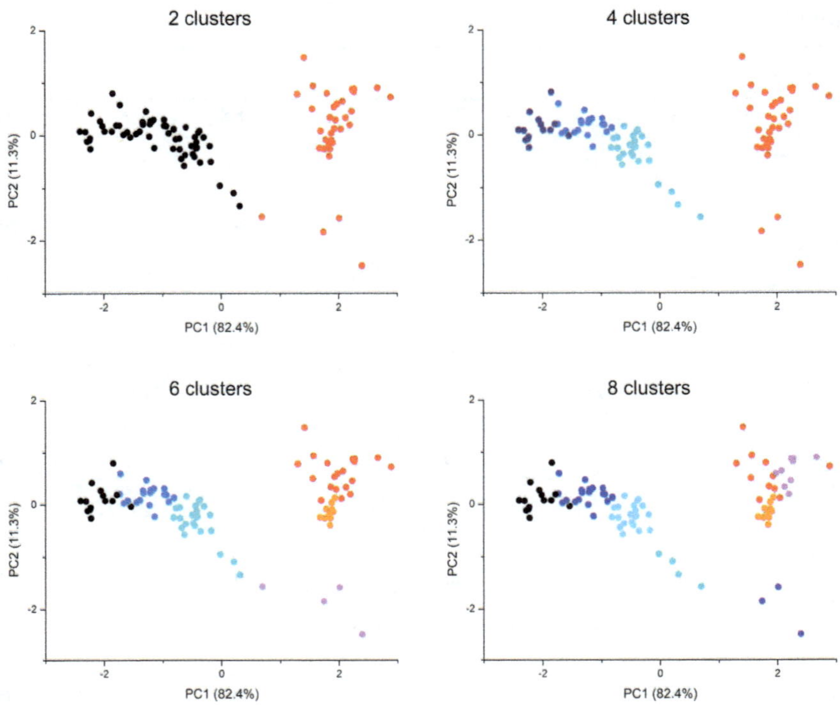

Figure 7.4 Outcome of the genetic algorithm clustering with 2, 4, 6, and 8 clusters indicated with different colours.

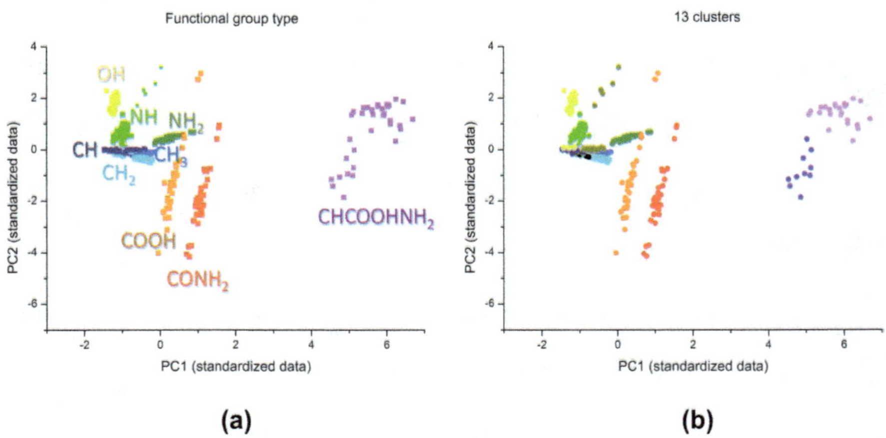

Figure 7.5 (a) Polarizabilities of eight different functional groups. For NH, CH_2 and CH the size of the squares corresponds to the hybridization of the carbon atom. (The smallest is sp, the middle one sp^2 and the largest squares are sp^3.) This makes the total number of functional groups equal 13. (b) Outcome of the genetic algorithm clustering with 13 clusters shown with different colours.

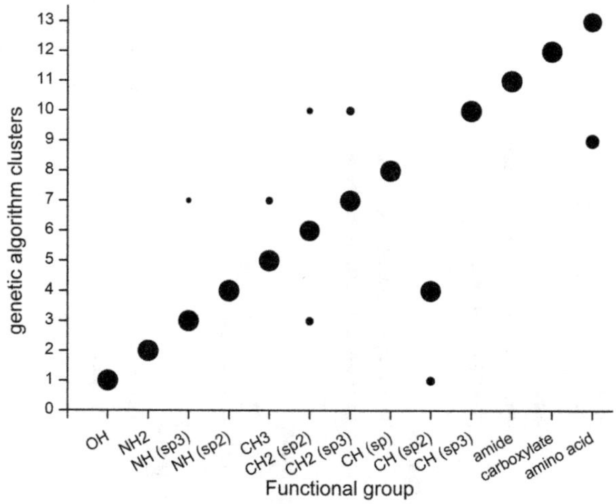

Figure 7.6 13 different types of functional groups for which polarizabilities were calculated are shown on the *x*-axis and the 13 clusters of the genetic algorithm clustering on the *y*-axis. The size of the circles is proportional to the percentage of the functional group type given on the *x*-axis in a particular cluster.

energy with respect to the field; (2) sum of the QTAIM partitioned atomic polarizabilities obtained from the calculated molecular electron density and finite field approximation; (3) sum of average functional group polarizabilities, from our database entries without clustering; (4) sum of functional group polarizabilities, from our database entries after the genetic algorithm clustering procedure.

Both methods 1 and 2 use couple-perturbed Kohn–Sham theory to calculate the molecular wave function under perturbation of a field. From our previous studies,[39] we already know that the QTAIM distributed polarizabilities (method 2) reconstruct rather accurately the molecular polarizability calculated with method 1, in keeping with the electron density partitioning being an "exact" one. Slight differences are due to the numerical derivatives of the atomic dipoles and the numerical integrations of the density in the atomic basins. Because the purpose of the present work is testing the clustering and database procedures, we use as a benchmark the QTAIM distributed atomic polarizabilities (method 2) and do not further discuss in terms of the exact polarizabilities. Moreover, the benchmark polarizabilities are calculated at the same level of theory as the database entries, in order to avoid issues due to the choice of functionals and basis sets (albeit of course important). In previous work, we have shown that the level of these calculations was sufficient to reproduce the experimentally measured molecular polarizabilities of the natural α-amino acids.

The number of clusters per functional group was determined by the intercluster ratio, chosen in such a way that no clusters have an intercluster ratio lower than 2.

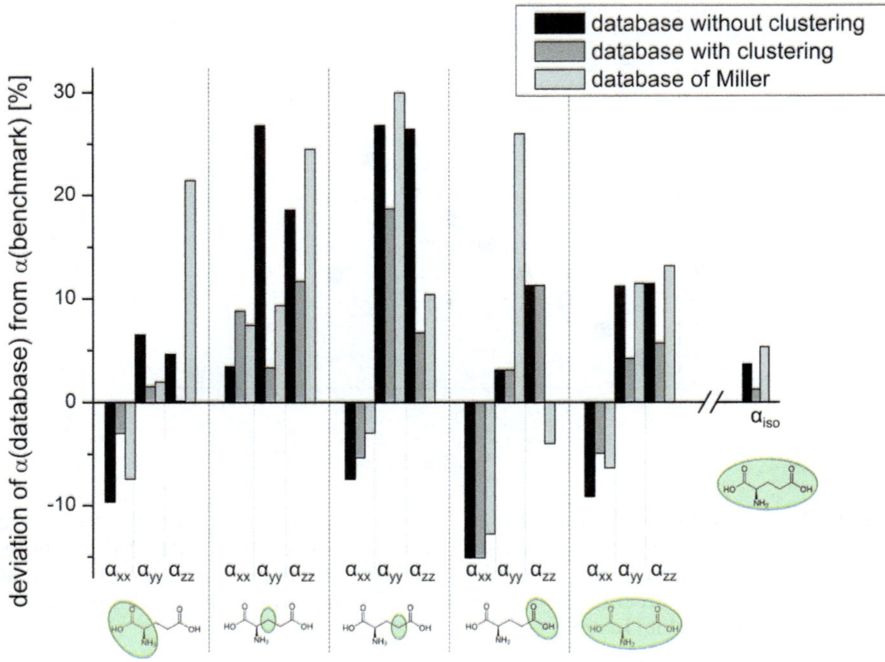

Figure 7.7 Percentage deviation of the functional group polarizabilities of glutamic acid calculated with different databases from the benchmark values. The diagonal components are shown.

The purpose of the comparison is to assess whether the database is able to predict with sufficient precision the molecular polarizability, compared to a fully *ab initio* calculation on the same molecule, given the enormous difference in computational costs. Moreover, it is interesting to check if using a single entry per functional group is significantly worse than using clustered and thus more specific entries.

Figure 7.7 shows the percentage deviation of the database polarizabilities from the benchmark ones for the functional groups of glutamic acid for α_{xx}, α_{yy}, α_{zz}. Additionally, the functional group polarizabilities were calculated using Miller's[20] isotropic atomic polarizabilities. The deviation of the so-obtained polarizabilities to the benchmark ones is also given. As all calculations refer to gas phase molecular geometries, the configuration that is considered is the neutral, instead of the zwitterionic (stable in the solid state) one. We already calculated a set of polarizabilities for zwitterionic forms, by fixing the molecular geometries to neutron diffraction values to mimic the actual behaviour of amino acids in solid state.[31] Figure 7.7 shows that the group most difficult to model with a database is in fact CH_2, although it plays a smaller role for the molecular polarizability because it carries fewer electrons. Noteworthy is the improvement with the clustering procedure. In fact, the simple database averages produce largest errors up to 25% for the CH_2 group, which are drastically reduced with the clustering (apart for α_{yy} of one CH_2 group and α_{xx} of the COOH group); all other components are estimated with 10% discrepancy from the benchmark. The use of Miller's isotropic polarizability leads in general to a worse approximation of the benchmark polarizability when compared to the database with clustered entries.

The sum of the various functional group polarizabilities tends to compensate some errors so that the molecular polarizabilities are estimated within 11% error on individual components without clustering and within 6% with clustering. The error of the molecular isotropic polarizability is further reduced (because of error compensation) to 3.66% with the general database polarizabilities and to 1.24% with the clustered values. Given that we are interested in predicting the optical properties of crystals consisting of these molecules, it is important to calculate how this error would propagate, for example on the refractive index. If approximated with the gas phase molar volume, the error on the polarizability corresponds to an error on the refractive index of 0.011, when calculated without clustering, or 0.004 with clustering. In both cases, the deviation is well within the uncertainty of a typical refractive index measurement. This means that calculating an *ab initio* molecular polarizability for the evaluation of a crystal refractive index is no more accurate than simply using the database entries (especially after clustering). On the other hand, the estimation of the anisotropy and therefore the birefringence is worse.

As already discussed in previous work,[31] an accurate evaluation of the crystal refractive index should of course also account for the mutual polarization of molecules. Noteworthy is that the functional group approach has an intrinsic advantage with respect to the molecular approach, because it allows for a more detailed evaluation of the mutual polarizations.[31] Therefore, the database functional group approach may guarantee a more accurate evaluation of mutual polarizations with respect to the exact, but global, molecular polarizabilities (unless of course they are partitioned into distributed atomic or functional group polarizabilities).

Figure 7.8 illustrates the difference between the benchmark QTAIM polarizability tensors and the same tensors from the database with clustering for glutamic acid drawn with the software *peanut*.[48]

A comparison between the benchmark isotropic polarizability calculated with *PolaBer* and the isotropic polarizability calculated from database entries was made for ten amino acids that were not used to construct the database. Figure 7.9 shows the percentage deviation of the molecular isotropic polarizabilities obtained with database entries from the benchmark values. The deviations between the reference values and the values obtained without clustering are shown in black, whereas the deviations between the reference values and those obtained after clustering are in red. The average error of the isotropic polarizability among all examined amino acids calculated without clustering is 2.32% while it is 1.43% when calculated

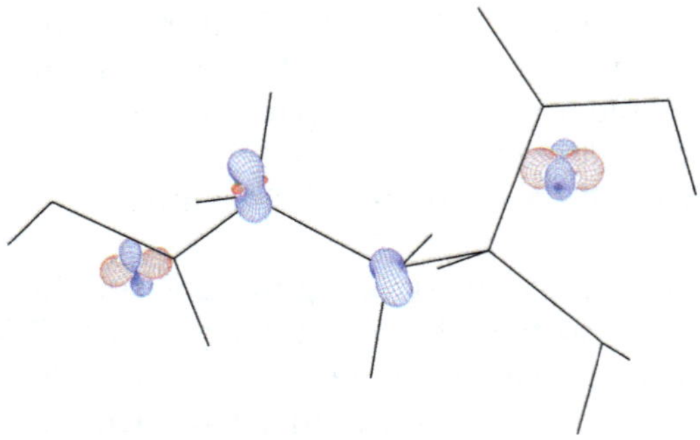

(a)

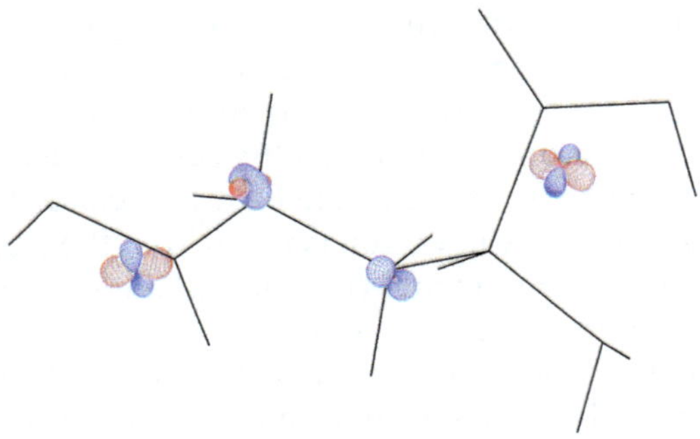

(b)

Figure 7.8 Difference between the benchmark polarizability tensor and the polarizability tensor from the database for glutamic acid drawn at the center of charge. Red colour indicates that the benchmark polarizability is higher and blue that the database polarizability is higher. (a) The database without clustering and (b) the database with clustering.

with clustering. It is interesting to note that almost all the values are positive, meaning that the database polarizability is always higher (Figure 7.10). This is in contrast to many models discussed in Section 7.2.1 where the polarizability from the database is usually lower because the mutual induction of polarizability between different fragments is not explicitly taken into account.

The database was also tested for β- and γ-amino acids (Figure 7.11), which reveal similar features as the natural amino acids. The deviation of the polarizability obtained with the database from the benchmark polarizability is shown for the α_{xx}, α_{yy}, and α_{zz} components. The results from the database with clustering are better than when no clustering is performed. The deviation of the isotropic polarizability is less than 3% in both cases, while it is larger for the individual components.

These analyses enable us to conclude that the performances of the database are very promising, because they allow us to estimate the isotropic

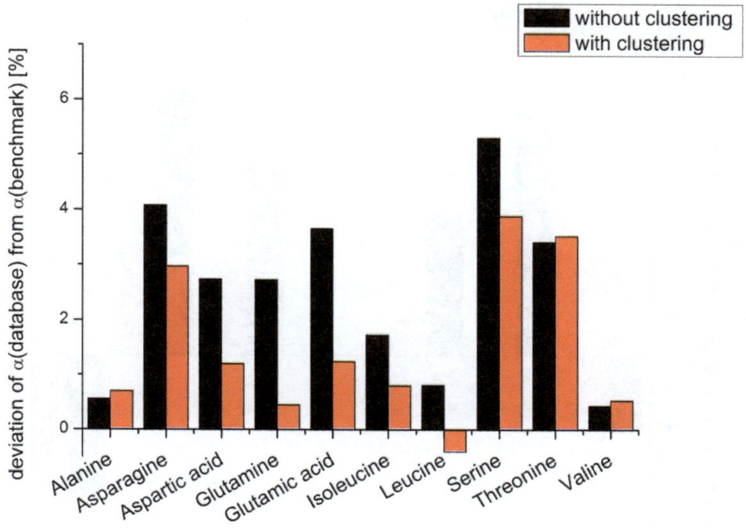

Figure 7.9 Plot representing the percentage deviation of the database polarizabilities from the benchmark polarizabilities for ten amino acids. Black columns refer to a database without clustering and red ones to a database with clustering.

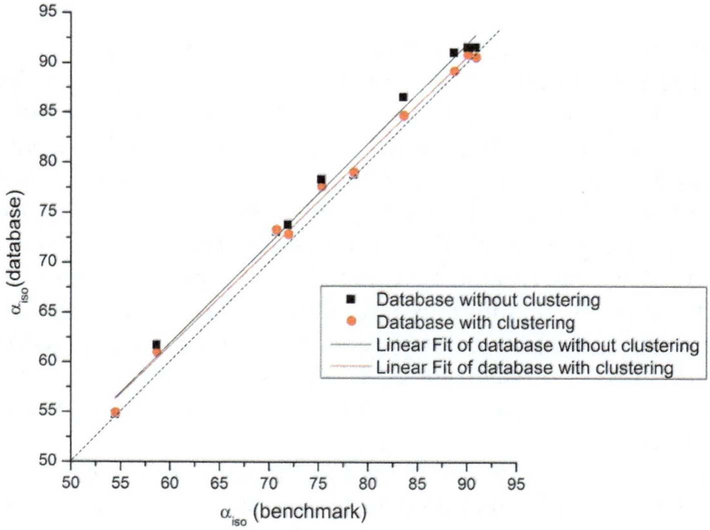

Figure 7.10 Comparison of the isotropic benchmark polarizability to the database polarizabilities for the ten calculated amino acids. The dotted line represents perfect coincidence.

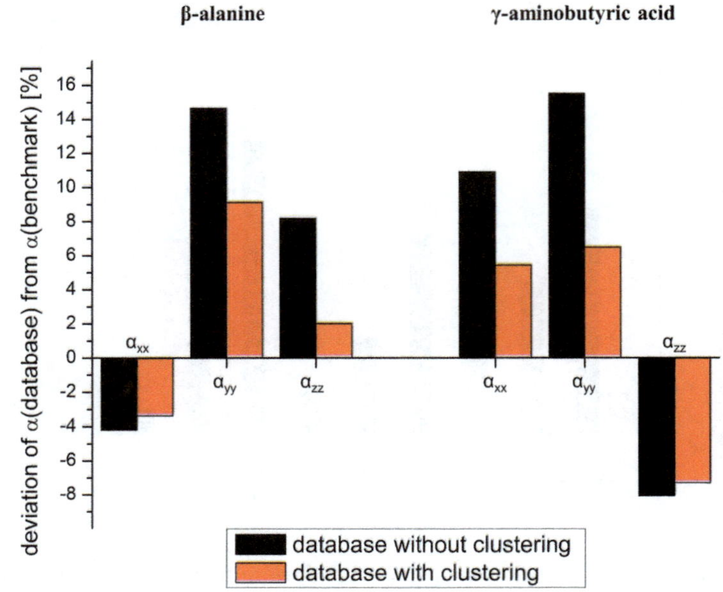

Figure 7.11 Percentage deviation of the polarizabilities of β-alanine and γ-aminobutyric acid calculated with the database (with and without clustering) from the benchmark values. The diagonal components are shown.

polarizabilities for molecules that were not used to build the database with an accuracy of *ca.* 2%. For anisotropic components the discrepancies are higher, but still acceptable. Therefore, we may conclude that the necessary condition of transferability is demonstrated. Of course, the success of the database also depends on the quality of the calculations on the set of test molecules. In this respect, there is room for improvement, for example using functionals and basis sets that are able to approach even better the experimental results. For this purpose, more tests are necessary as well as more experimental information. Noteworthy is that the computational costs to produce a similar database with a more sophisticated level of theory are not so high. Nevertheless, the current database already seems to be suitable for the purposes mentioned in the introduction.

The database will be included in the forthcoming new version of the software *PolaBer*. For each functional group, the averages of atomic and group polarizability components are stored (in Bohr^3), together with the averages within each clusters. The database molecules used for calculations are listed for each cluster.

In the future, the program *PolaBer* will be modified in order to recognize a functional group based only on geometrical features and to automatically propose adequate polarizabilities (average group polarizabilities or specific cluster). The program will allow users to quickly estimate the polarizability of any given molecule for which they just need to input the geometry. Considering also the molecular volume, which may be given by the user or also stored in the database, dielectric properties can directly be calculated.

7.5 Conclusions

In this chapter, we presented a methodology to derive a functional group database of fully exportable polarizabilities. We defined a local coordinate system that enables the transferability of functional group polarizabilities. These polarizabilities were symmetrized and clustered into subcategories using a genetic algorithm. The database entries thereby obtained were used to estimate the polarizabilities of different molecules.

The work has many purposes: (1) a multivariate analysis is not only a way to produce representative values for given quantities, but also a method to gain insight into the machinery of electron polarizability effects; (2) the database provides a means to rapidly estimate polarizabilities in larger molecules or aggregates consisting of the building blocks present in the database; (3) the derived molecular polarizabilities can be used as such in force fields for the evaluation of polarization and dispersion interactions in molecular recognition processes; (4) the polarizability can be used to rapidly estimate dielectric constants in the solid state as well as derived quantities such as refractive indices.

One of the goals of our research is in fact connected with point (3); the availability of a polarizability database, especially if extended to all main group atoms and to transition metals as well, would enable a very rapid screening of molecular/polymeric materials for optoelectronic applications, for example HRI materials,[4–6,8–15,49] low dielectric constant porous materials,[7] *etc.*

Moreover, our purpose is that of giving the guidelines for extending the database, so that interested researchers could contribute and expand it by adding functional groups interesting for their own specific purposes.

Our future research, apart from further development of the database by adding more descriptors (*e.g.* dipole moments, hyperpolarizabilities, *etc.*) and improving the statistics with more test molecules, will be dedicated to improve the automatism of the calculations, using adequate routines that will enable recognition of the functional groups. Given the speed of database calculations of polarizabilities and the subsequent derivation of static dielectric constants and refractive indices, an automatic procedure of structure screening would be extremely rapid. These structures could be actually observed and already characterized structures (for example present in structural databases) as well as structures simulated during processes of crystal structure prediction. The implications for material design are then rather straightforward.

Acknowledgements

We thank the Swiss National Science Foundation for financial support (project 160157).

References

1. L. P. Hammett, *J. Am. Chem. Soc.*, 1937, **59**, 96–103.
2. R. F. W. Bader, T. A. Keith, K. M. Gough and K. E. Laidig, *Mol. Phys.*, 1992, **75**, 1167–1189.
3. R. F. W. Bader, A. Larouche, C. Gatti, M. T. Carroll, P. J. MacDougall and K. B. Wiberg, *J. Chem. Phys.*, 1987, **87**, 1142–1152.
4. T. Matsuda, Y. Funae, M. Yoshida, T. Yamamoto and T. Takaya, *J. Appl. Polym. Sci.*, 2000, **76**, 45–49.
5. Y. Ju, G. Almuneau, T. Kim and B. Lee, *Jpn. J. Appl. Phys.*, 2006, **45**, 2546–2549.
6. K. C. Krogman, T. Druffel and M. K. Sunkara, *Nanotechnology*, 2005, **16**, 338–343.
7. B. D. Hatton, K. Landskron, W. J. Hunks, M. R. Bennett, D. Shukaris, D. D. Perovic and G. A. Ozin, *Mater. Today*, 2006, **9**, 22–31.
8. J. Liu and M. Ueda, *J. Mater. Chem.*, 2009, **19**, 8907.
9. J. Wang, M. Hutchins, K. Woo, T. Konish and M. J. Idacavage, *RadTech Rep.*, 2010, 11–17.
10. T. Higashihara and M. Ueda, *Macromolecules*, 2015, **48**, 1915–1929.
11. E. K. Macdonald and M. P. Shaver, *Polym. Int.*, 2015, **64**, 6–14.
12. R. A. Gaudiana and R. A. Minns, Rogers H.G. U.S. Patent No. 5,132,430, 1992.
13. A. Stiegman, U.S. Patent No. 8,470,948, 2013.
14. C. L. Jones, B. U. Kolb, D. B. Olson, E. S. Goenner, S. D. Hart, and J. T. Cowher, U. S. Patent No. 7622164 B2, 2010.
15. W. Caseri, *Macromol. Rapid Commun.*, 2000, **21**, 705–722.
16. B. Champagne and D. M. Bishop, *Adv. Chem. Phys.*, 2003, **126**, 41–92.
17. D. J. Griffiths, *Introduction to Electrodynamics*, Pearson Benjamin Cummings, San Francisco, 3rd edn, 2008, p. 161.
18. A. Vogel, *J. Chem. Soc.*, 1948, 1833–1855.
19. J. Applequist, *Acc. Chem. Res.*, 1976, **10**, 79–85.
20. K. J. Miller, *J. Am. Chem. Soc.*, 1990, **112**, 8533–8542.
21. N. E. B. Kassimi and A. J. Thakkar, *Chem. Phys. Lett.*, 2009, **472**, 232–236.
22. J. M. Stout and C. E. Dykstra, *J. Am. Chem. Soc.*, 1995, **117**, 5127–5132.
23. A. J. Stone, *Mol. Phys.*, 1985, **56**, 1065–1082.
24. A. J. Stone, *Chem. Phys. Lett.*, 1981, **83**, 233–239.
25. A. J. Stone and M. Alderton, *Mol. Phys.*, 1985, **56**, 1047–1064.
26. C. R. Le Sueur and A. J. Stone, *Mol. Phys.*, 1994, **83**, 293–307.
27. G. A. Ivanov-Smolenskii, V. G. Tsirel'son and R. P. Ozerov, *Acta Crystallogr., Sect. A: Found. Crystallogr.*, 1983, **39**, 411–415.

28. A. E. Whitten, D. Jayatilaka and M. A. Spackman, *J. Chem. Phys.*, 2006, **125**, 174505.
29. K. Y. Suponitsky, V. G. Tsirelson and D. Feil, *Acta Crystallogr., Sect. A: Found. Crystallogr.*, 1999, **55**, 821–827.
30. A. Krawczuk, D. Pérez and P. Macchi, *J. Appl. Crystallogr.*, 2014, **47**, 1452–1458.
31. L. H. R. Dos Santos, A. Krawczuk and P. Macchi, *J. Phys. Chem. A*, 2015, **119**, 3285–3298.
32. A. K. Jain, *Pattern Recogn. Lett.*, 2010, **31**, 651–666.
33. T. Seidler, A. Krawczuk, B. Champagne and K. Stadnicka, *J. Phys. Chem. C*, 2016, **120**, 4481–4494.
34. B. Dittrich, C. B. Hübschle, K. Pröpper, F. Dietrich, T. Stolper and J. J. Holstein, *Acta Crystallogr., Sect. B: Struct. Sci., Cryst. Eng. Mater.*, 2013, **69**, 91–104.
35. S. Domagała and C. Jelsch, *J. Appl. Crystallogr.*, 2008, **41**, 1140–1149.
36. A. Volkov, X. Li, T. Koritsanszky and P. Coppens, *J. Phys. Chem. A*, 2004, **108**, 4283–4300.
37. J. F. Nye, *Physical Properties of Crystals: Their Representation by Tensors and Matrices*, Oxford University Press, Oxford, 1985, p. 15.
38. N. Otero, C. Van Alsenoy, C. Pouchan and P. Karamanis, *J. Comput. Chem.*, 2015, **36**, 1831–1843.
39. A. Krawczuk-Pantula, D. Pérez, K. Stadnicka and P. Macchi, *Trans. Am. Crystallogr. Assoc.*, 2011, **42**, 1–25.
40. T. A. Keith, *The Quantum Theory of Atoms in Molecules: From Solid State to DNA and Drug Design*, ed. C. F. Matta and R. J. Boyd, Wiley-VCH, Weinheim, 2007, pp. 61–94.
41. M. J. Frisch, G. W. Trucks, H. B. Schlegel, G. E. Scuseria, M. A. Robb, J. R. Cheeseman, G. Scalmani, V. Barone, B. Mennucci, G. A. Petersson, H. Nakatsuji, M. Caricato, X. Li, H. P. Hratchian, A. F. Izmaylov, J. Bloino, G. Zheng, J. L. Sonnenberg, M. Hada, M. Ehara, K. Toyota, R. Fukuda, J. Hasegawa, M. Ishida, T. Nakajima, Y. Honda, O. Kitao, H. Nakai, T. Vreven, J. A. Montgomery, Jr., J. E. Peralta, F. Ogliaro, M. Bearpark, J. J. Heyd, E. Brothers, K. N. Kudin, V. N. Staroverov, R. Kobayashi, J. Normand, K. Raghavachari, A. Rendell, J. C. Burant, S. S. Iyengar, J. Tomasi, M. Cossi, N. Rega, J. M. Millam, M. Klene, J. E. Knox, J. B. Cross, V. Bakken, C. Adamo, J. Jaramillo, R. Gomperts, R. E. Stratmann, O. Yazyev, A. J. Austin, R. Cammi, C. Pomelli, J. W. Ochterski, R. L. Martin, K. Morokuma, V. G. Zakrzewski, G. A. Voth, P. Salvador, J. J. Dannenberg, S. Dapprich, A. D. Daniels, O. Farkas, J. B. Foresman, J. V. Ortiz, J. Cioslowski, and D. J. Fox, *Gaussian 09, Revision A.02*, Gaussian, Inc., Wallingford CT, 2009.
42. N. M. O'Boyle, T. Vandermeersch, C. J. Flynn, A. R. Maguire and G. R. Hutchison, *J. Cheminf.*, 2011, **3**, 8.
43. T. A. Keith, *AIMAll (Version 14.11.23)*, TK Gristmill Software, Overland Park KS, USA, 2017.
44. G. Zhou, S. Zhang and C. Guan, *Prog. Electromagn. Res. Symp.*, Beijing, 2009, pp. 997–1002.

45. P. Murray-Rust and S. Motherwell, *Acta Crystallogr., Sect. B: Struct. Crystallogr. Cryst. Chem.*, 1978, **34**, 2518–2526.
46. U. Maulik and S. Bandyopadhyay, *Pattern Recogn.*, 2000, **33**, 1455–1465.
47. P. L. A. Popelier and F. M. Aicken, *J. Am. Chem. Soc.*, 2003, **125**, 1284–1292.
48. W. Hummel, J. Hauser and H. B. Bürgi, *J. Mol. Graphics*, 1990, **8**, 214–220.
49. M. Ernst, L. H. R. Dos Santos and P. Macchi, *CrystEngComm*, 2016, **18**, 7339–7346.

CHAPTER 8

Solid-state NMR in the Study of Intermolecular Interactions

F. ROSSI[a], P. CERREIA VIOGLIO[b], M. R. CHIEROTTI[a] AND R. GOBETTO*[a]

[a]University of Torino, Department of Chemistry and NIS Centre, Via P. Giuria 7, Torino 10125, Italy; [b]Aix-Marseille Université, CNRS, ICR (UMR 7273), Marseille 13013, France
*E-mail: roberto.gobetto@unito.it

8.1 Introduction

Over the last few decades, solid-state NMR (SSNMR) has developed into an instrumental tool to investigate chemical structures of both crystalline and amorphous solids.[1] The main advantages of this technique are that the solid sample can be in powder form and it does not need long-range order or homogeneity, as diffraction techniques do. Indeed, SSNMR spectra reflect the local environment of the nuclei under investigation, resulting in a significant contribution to all those cases in which diffraction methods encounter difficulties. However, it is worth noting that the concurrent application of both SSNMR and diffraction methods can benefit from their complementarity, ultimately leading to a robust and detailed picture of a broad variety of materials, including for instance enzymes,[2] drugs,[3] and porous materials.[4]

The microscopic environment at each individual site in a solid sample does influence the isotropic chemical shifts of the nuclei, but also many other NMR measurables, including chemical shift anisotropy (CSA) and

Monographs in Supramolecular Chemistry No. 26
Understanding Intermolecular Interactions in the Solid State: Approaches and Techniques
Edited by Deepak Chopra

Published by the Royal Society of Chemistry, www.rsc.org

internuclear couplings both through space (dipolar and quadrupolar couplings) or through bonds (*J* couplings). Indeed, these local parameters can provide substantial information for the complete characterization of the material in terms of connectivity and spatial relationships among atoms. Rapid developments in SSNMR, such as new pulse sequences, more sensitive probes and rotor spinning speeds up to 110 kHz, have promoted this technique into a powerful tool for investigating subtle changes in the local structure, the crystal packing and different types of intermolecular interactions. SSNMR together with quantum mechanical computations of magnetic shielding is extremely useful to obtain pivotal information in the field of supramolecular chemistry and in the study of polymorphs, co-crystals, stoichiometric and non-stoichiometric solvates, as well as hydrogen bond arrangements.

The purpose of this chapter is to describe and highlight the potentialities of SSNMR and to offer an overview of the state-of-the-art of this technique by introducing some interesting examples. After discussing the basic principles of the technique, attention will be mainly focused on the determination and characterization of hydrogen and halogen bonds (from now on HB and XB), as they represent paradigmatic cases of intermolecular interactions. The HB is certainly a unique phenomenon in chemistry and of paramount importance in nature;[5] the XB has instead garnered much attention across the scientific community in the last twenty years due to its promising application in various areas of science.[6] The chapter is intended not only to provide valuable information but also to encourage further advances in the supramolecular field.

8.2 Essential Techniques and Parameters in Solid-state NMR

In order to understand how useful information is extracted from SSNMR spectra, essential techniques and fundamental parameters are presented in this section. Some introductory textbooks on SSNMR are nevertheless available;[7,8] the interested reader is referred to them for further details.

8.2.1 Magic-angle Spinning, High-power Proton Decoupling and Cross Polarization

In solution NMR experiments, all orientation-dependent properties are rapidly averaged to isotropic values, due to the fast molecular tumbling and translational motions. By contrast, in SSNMR, powder samples consist of many crystallites with random orientation, each giving rise to a different spectral frequency. Hence, not all spatial-dependent interactions, namely CSA, dipolar coupling and quadrupolar interaction can be ignored, causing substantial line broadening and loss of spectra resolution. High-resolution

SSNMR spectra of diluted spins, *i.e.* ^{13}C and ^{15}N, can be achieved by combining three techniques: (i) magic-angle spinning (MAS), (ii) cross-polarization (CP) from abundant spins (typically ^{1}H or ^{19}F) and (iii) high-power heteronuclear decoupling. MAS is performed by fast spinning macroscopic samples about an axis inclined at the so-called magic angle (54.7°) with respect to the external applied magnetic field (B_0). The rapid rotation mimics the Brownian motions in solution and thus removes CSA effects due to their dependence from the geometrical factor $3\cos^2\theta - 1$ present in the Hamiltonian. This results in averaging the chemical shift tensor of a particular nucleus to its isotropic value. In practice, if MAS rates are smaller than the chemical shift anisotropies, a set of so-called *spinning sidebands* (ssbs) for each isotropic peak will appear in the spectrum, spaced at integer multiples of the MAS rate. Ssbs present both problems and opportunities. They may hamper the assignment of the resonances as well as causing a loss in sensitivity since the intensity of the isotropic peak generally diminishes with increasing the chemical shift anisotropy; on the other hand, they do contain precious structural information because they describe the shape of the chemical shift tensor which, in turn, is a representation of the electron distribution around a nucleus, *i.e.* its chemical environment. Thus, by varying the MAS rate is possible to either reduce or increase the number of ssbs, depending on the information needed. Typically, moderate spinning rates (12–30 kHz) are sufficient for dilute nuclei such as ^{13}C or ^{15}N. Higher spinning speeds of up to 110 kHz are instead used to exploit the capability of MAS for removing the effects of homonuclear dipolar coupling. Ultrafast MAS is a key to the successful acquisition of proton SSNMR spectra, wherein ^{1}H–^{1}H dipolar couplings severely broaden the line shapes.[9] Moreover, when observing a dilute spin with abundant spins nearby such as proton, heteronuclear dipolar coupling causes additional broadening to the spectrum. This effect can be removed by high-power proton decoupling (HPPD), *i.e.*, the application of radio frequency fields of the order of at least 60–100 kHz at the ^{1}H Larmor frequency. Standard HPPD in the solid state involves continuous-wave irradiation and leads to a substantial narrowing of the peaks. Another widely used decoupling scheme is the two-pulse phase-modulated scheme,[10] which is characterized by a series of phase alternated pulses applied to the ^{1}H channel and provides enhancement in both sensitivity and resolution of the dilute-spin spectrum. Finally, the benefit of the cross-polarization technique is two-fold: firstly, allowing the transfer of magnetization from abundant nuclei (typically ^{1}H) to rare nuclei such as ^{13}C and ^{15}N, CP results in a signal enhancement given by γ_I/γ_S, where γ_I and γ_S are the gyromagnetic ratios of the abundant spin I and the dilute spin S, respectively. Secondly, CP enables improved signal-to-noise ratios with faster experimental times than direct-excitation experiments, as relaxation delays are calculated from the T_1 of the I-spin, which is usually shorter than the T_1 of the S-spin.

Nowadays, the great majority of SSNMR spectra are acquired by combining CP and MAS techniques, including also the HPPD.

8.2.2 Chemical Shift

Chemical shift (CS) values obtained from CPMAS or MAS SSNMR spectra enable valuable information to be gained on solid compounds. In particular, CS provides a wealth of structural data about the local environment at each site, *i.e.*, surrounding substituents, atom hybridization, crystal packing, intermolecular interactions, *etc.* Since, in general, electrons are not spherically distributed around any given nucleus, and therefore the shielding field is anisotropic, the chemical shift results in a second-rank tensor defined by a 3×3 matrix in the x, y and z coordinates:

$$\delta = \begin{pmatrix} \delta_{xx} & \delta_{xy} & \delta_{xz} \\ \delta_{yx} & \delta_{yy} & \delta_{yz} \\ \delta_{zx} & \delta_{zy} & \delta_{zz} \end{pmatrix}. \tag{8.1}$$

If a particular reference frame (*principal axis system*) fixed on the nucleus in a molecule and defined by the X, Y and Z axes is chosen, the tensor can be diagonalized and thus defined by the three principal components δ_{XX}, δ_{YY}, δ_{ZZ}. In this way, the isotropic part δ_{iso} can be easily obtained by

$$\delta_{iso} = (\delta_{XX} + \delta_{YY} + \delta_{ZZ})/3. \tag{8.2}$$

In the literature, three main notations for describing the CS tensor are generally accepted and used. In the Mehring notation, the three principal components of the tensor are labelled as:

$$\delta_{33} \le \delta_{22} \le \delta_{11} \tag{8.3}$$

where δ_{11} is the highest frequency value and δ_{33} is the lowest frequency value.

In the Haeberlen notation, the most widely used when reporting experimental measurements, the three components are defined by:

$$|\delta_{ZZ} - \delta_{iso}| \ge |\delta_{XX} - \delta_{iso}| \ge |\delta_{YY} - \delta_{iso}| \tag{8.4}$$

where δ_{ZZ} is the farthest from and δ_{YY} is the closest to δ_{iso}. The anisotropy is defined by:

$$\zeta = \delta_{ZZ} - (\delta_{XX} + \delta_{YY})/2 = 3\delta/2 \tag{8.5}$$

whereas the shielding asymmetry is:

$$\eta = (\delta_{YY} + \delta_{XX})/\delta;\ (0 \le \eta \le +1). \tag{8.6}$$

In the Herzfeld–Berger notation the anisotropy and asymmetry are replaced with the span (Ω) and the skew (κ) parameters, defined respectively as:

$$\Omega = \delta_{11} - \delta_{33};\ (\Omega \ge 0) \tag{8.7}$$

$$\kappa = 3(\delta_{22} - \delta_{iso})/\Omega;\ (-1 \le \kappa \le +1). \tag{8.8}$$

CS tensors strongly depend on the nucleus under investigation, the local environment, the nature and mobility of the functional group and the intermolecular interactions. The CS tensor can be determined through different approaches: (i) SSNMR measurements of single crystals giving full access to both the chemical shift tensor and its orientation in the sample frame;[11] (ii) analysis of static powder pattern spectra[12] or (iii) study of the spinning sidebands in slowly rotating samples,[13] both providing just the principal components (which is the most useful part). The last method can be also performed as 2D experiments in which the chemical shift anisotropy is recoupled in some form in one dimension by applying strong π pulses at specified time delays under one rotor period, whilst isotropic signals are retained in the other dimension.[14–16] CS tensor analysis is now attainable also for fast-spinning MAS NMR samples, by the use of some particular multidimensional experiments for recovering CSA information such that, at a high spinning rate, a well-defined spectrum is achieved in the direct dimension.[17–20]

The access to the full chemical shift tensor offers valuable information for elucidating the three-dimensional structure of molecular crystals and other materials; for instance, by means of proton CSA as a sensitive probe for both structure and hydrogen bonding, intermolecular interactions in supramolecular patterns can be determined. Important structural information can also be achieved from CS parameters of other nuclei such as ^{13}C, ^{15}N or ^{11}B.[21–23] It is noteworthy that this information cannot be provided by solution NMR techniques, since the CS tensor is reduced to a single, averaged isotropic value because of the molecular tumbling.

Solid-state NMR analysis is increasingly often supported by computational methods able to relate chemical shift values to the crystallographic site of atoms in the unit cell, thereby helping in the assignment of resonances.[11,24] First-principles quantum-mechanical methods have been largely employed in the study of small isolated molecules, although more advanced techniques are needed when evaluating periodic systems. Theoretical calculations for solid-state systems are mainly achieved by the well-known DFT (density functional theory)-based GIPAW (gauge including projector augmented waves) method.[25,26] The use of pseudopotentials and a plane wave basis set makes it possible to take into account the periodicity of the whole crystal, and thus today magnetic resonance parameters can be measured for solids with an excellent balance between accuracy and computational costs.

When considering crystal packings wherein intermolecular interactions are dominant, namely molecular crystals, π–π, CH···π and host–guest complexes, dispersion interactions must be taken into account. This is one of the greatest challenges the DFT approach is called to face today, as the commonly utilized functionals yield a poor description of intra- and intermolecular interactions.[27] The effects of neglecting dispersion interactions generally translate into an expansion of the unit cell and unrealistic geometries. The magnitude of these changes depends upon the functional used and the type of crystal structure being considered.[28] A common strategy is to constrain the unit cell dimensions to the experimental ones obtained by diffraction; however, a more rigorous treatment involves semi-empirical dispersion correction schemes within the DFT framework: the so-called DFT-D method. The efficacy of the DFT-D approach has been proven, for instance, in the elucidation of the crystal structure of galactose, for which diffraction data were not able to identify the correct hydrogen bond network.[29] Another successful application of the DFT-D method involves a pharmaceutical material such as indomethacin; this represents a paradigmatic case in which the HB network and the π–π interactions have been probed separately to better understand co-crystal packing arrangements.[30]

It is important to stress that the combination between experimental SSNMR and computational approaches allows for accurate structure determination and refinement. In fact, this procedure is at the basis of "NMR crystallography", which has added even more significance to SSNMR. "NMR crystallography" is now a well-established discipline that exploits the power of SSNMR to give crystallographic information to support and facilitate the analysis of single-crystal and powder diffraction patterns. Indeed, SSNMR can and does address, as it always did since the earliest days of the technique, crystallographic matters that diffraction methods struggle with. In this sense, X-ray diffraction and NMR are extremely complementary since the former relies on long-range-order whereas NMR reflects the local environment at each site. The combination of the two techniques is especially powerful, in particular when combined also with DFT computational data to help and support NMR spectra interpretation and assignments. This combined approach can now be used to refine diffraction data and, in favourable cases, to solve crystal structures with poor[31–33] or even no diffraction data.[34] Comprehensive guides to the practice of NMR crystallography are available: the reader is referred to the works of Harris *et al.*[35,36] and of Ashbrook and McKay.[28]

8.2.3 Dipolar Interaction

Dipolar interaction represents a key parameter in NMR crystallography thanks to its high crystallographic content, which allows measuring internuclear distances and evaluating atom connectivities. Indeed, owing to its dependence on the atom distance ($D \propto r^{-3}$), dipolar interaction has been widely used for geometrical and structural elucidation, in particular in H-bonded samples such as co-crystals and polymorphs.[37]

Dipolar couplings arise due to direct interaction between the magnetic dipoles of two nuclear spins close in space. It is proportional to the inverse third power of the distance between the two spins I and S and to the product of their gyromagnetic ratios (γ), according to:

$$D = \frac{\mu_D}{4\pi}\frac{h\gamma_I\gamma_S}{r_{IS}^{-3}}. \tag{8.9}$$

Since dipolar coupling depends on the geometric factor ($3\cos^2\theta - 1$), in high-resolution spectra it is averaged out under MAS conditions, leading to the loss of essential information. However, many recoupling sequences have been developed to reintroduce it in the spectra and thus recover internuclear distances.[38] For instance, the rotational-echo double resonance (REDOR) technique prevents the complete refocusing of dipolar coupling and thus provides internuclear distances between atoms such as ^{13}C and ^{15}N (up to about 5 Å) or ^{13}C and ^{31}P (up to about 6 Å).[39] Longer distances and thus weaker dipolar coupling measurements are intrinsically restricted by the decay of the transverse magnetization of nuclei, yet many different MAS SSNMR approaches have been developed to measure distances up to 15–20 Å, taking advantage of the high gyromagnetic ratios of ^1H and ^{19}F, and their spin diffusion that probes distances in the nanometer range.[40]

We refer the reader to the literature for information on quantitative measurement of the dipolar coupling with symmetry-based sequences,[41] and other popular schemes such as DRAMA,[42] TEDOR[43] and RFDR.[44,45] Dipolar coupling measurements involving quadrupolar nuclei have also been reported (TRAPDOR[46] and REAPDOR[47]).

Double quantum (DQ) spectroscopy under MAS NMR is usually considered as the technique of choice to study ^1H–^1H dipolar couplings in the solid state. There is a wide range of DQ experiments allowing the measurement of proton–proton proximities.[48] In rotor-synchronized two-dimensional ^1H DQ MAS experiment acquired at fast MAS (>30 kHz), DQ coherences (DQCs) can be generated between neighbouring hydrogen atoms, thus providing specific information about intra- and intermolecular proton–proton spatial proximities.[49] Finally, significant line narrowing in the ^1H solid-state spectra can be achieved by the combined rotation and multiple pulse sequences (CRAMPS) technique, characterized by the combination of physical rotation under MAS and the application of RF pulses, such as PMLG[50] or DUMBO.[51,52]

8.2.4 Quadrupolar Interaction

Nuclei with spins greater than 1/2 are called *quadrupolar* and are characterized by a non-spherical charge distribution, quantified by the nuclear electric quadrupole moment (Q). For such nuclei, the spectrum is dominated by the quadrupolar interaction (QI), arising from the coupling between Q and the surrounding electric field gradient (EFG).[53]

QI is defined using the quadrupolar coupling constant (C_Q) and asymmetry parameter (η_Q):

$$C_Q = eV_{33}Q/h \tag{8.10}$$

$$\eta_Q = (V_{11} - V_{22})/V_{33} \tag{8.11}$$

where $|V_{33}| \geq |V_{22}| \geq |V_{11}|$ are the principal components of the EFG second-rank tensor.

For small quadrupolar interactions, only first-order perturbation to the Zeeman levels can be considered and MAS may be enough to average the anisotropy to zero. For large quadrupolar interactions, where also second-order effects are significant, MAS is unable to remove the anisotropic effects and very broad and distorted line shapes appear in the spectrum (see Figure 8.13 for an example of second order line shapes). Thus, high-resolution spectra are needed to achieve great insight into the local and electronic structure and symmetry. This can be obtained by using some special experimental techniques for the study of quadrupolar nuclei with half-integer spins. We refer the reader to the literature for information on measurements of quadrupolar nuclei by means of techniques requiring special probes (DOR and DAS),[54,55] 2D techniques relying on single or multiple quantum evolution such as STMAS[56] and MQMAS,[57] respectively, or recent echo sequences developed to improve experimental sensitivity such as QCPMG.[58]

Concerning quadrupolar nuclei, ^{2}H NMR spectroscopy is of particular interest, even though a significant obstacle to its routine application is the very low natural abundance of ^{2}H with the consequent problem of inherently low sensitivity. Because of its relatively small quadrupole moment (0.286 fm^{2}) its lines are generally narrow compared to those of other common quadrupolar nuclei and methods analogous to those used for half-integer nuclei generally suffice. Its acquisition provides a method for measuring ^{1}H NMR chemical shifts in solids with several advantages over the direct measurement of high-resolution solid-state ^{1}H NMR spectra. In fact, it requires only the application of routine MAS and routine ^{1}H decoupling methods instead of ultrafast MAS or complex pulse sequences for homonuclear ^{1}H decoupling.[59] In this sense, it is a convenient and often-used tool for studying H-bonding and water interactions in solids.[60,61] For instance, well-defined correlations between the magnitude of C_Q and D···O distance have been established.[62,63] On the other hand, it is also very useful in probing dynamics in solids since the static band shape or the spinning sideband patterns are extremely affected by local molecular-level mobility.[8] For instance, it has been used to investigate structural arrangements and dynamics of several solvates and hydrates.[64,65]

8.3 SSNMR and Hydrogen Bond

In the supramolecular field, the HB has been and is still being the most important and widely exploited non-covalent interaction.[66] Its strength, high selectivity and directionality make it the most powerful organizing force in molecular assembly,[67] determining both intra- and intermolecular arrangements and generating new molecular architectures.[68] Due to intrinsic limitations, X-ray diffraction techniques often fail in characterizing HB networks, particularly when dealing with hydrogen positions or when long-range crystal order is missing. Among the different techniques used for HB detection and characterization, SSNMR plays a central role. Indeed, NMR parameters are strongly sensitive to the local electronic environment and, moreover, the SSNMR multinuclear approach (carbon, nitrogen or even directly hydrogen atoms) makes it possible to achieve information on several types of contacts according to their strength and geometry. Strong, moderate and weak

HBs, as classified from Jeffrey (Scheme 8.1),[69] have been easily detected and characterized.[70]

The main SSNMR parameters that can provide useful insights into HB interactions are (i) the CS and CSA, (ii) the dipolar interaction and (iii) the relaxation parameters. Each NMR active nucleus involved in the weak interaction, such as ^{1}H, ^{13}C, ^{15}N or ^{19}F, can be investigated to obtain the aforementioned parameters. The influence of HBs on these nuclei has recently been reviewed.[71]

8.3.1 Hydrogen Bond and Chemical Shift/Chemical Shift Anisotropy

In standard experiments, ^{1}H SSNMR spectra of rigid solids appear strongly broadened and thus characterized by severe peak overlaps, mainly because of strong ^{1}H–^{1}H homonuclear dipolar couplings and restricted ^{1}H CS ranges. However, an increasing number of proton-based SSNMR experiments are being performed, owing to the significant enhancement in spectral resolution and sensitivity afforded by the noteworthy advances in ultrafast MAS (up to 110–120 kHz) and higher magnetic fields available (up to 1 GHz). An exhaustive review of high-resolution ^{1}H SSNMR applications has recently been published.[72] Concerning H-bonded supramolecular systems, most ^{1}H studies are focused on the direct chemical shift measurement of protons in HB. In particular, signals related to the H-bonded protons are increasingly shifted toward higher frequencies (around 10–20 ppm), far from the aliphatic and aromatic regions (Scheme 8.2). The magnitude of the shift is directly correlated with the strength and the length of the HB.[69] Moreover, the ^{1}H CS is very sensitive to the location of the hydrogen atom along the interaction and thus direct relationships between $\delta(^{1}\text{H})$ and HB strength, and between $\delta(^{1}\text{H})$ and X–H distance for different classes of compounds have been observed.[73,74]

Similarly, the ^{13}C and ^{15}N chemical shifts are also correlated to the presence and strength of HBs (Scheme 8.2). In the case of ^{13}C C=O signals, the shift due to HB formation is of about 3–7 ppm and always toward higher frequencies, whether carboxylic or carbonyl groups. In cases of both carboxylic and carbonyl groups there are approximately linear relationships between δ and X···O distances (X = N or O). If a proton transfer occurs across the HB, a charge-assisted N^{+}–H···O^{-} HB and a carboxylate group are formed and the high-frequency shift is a maximum (6–7 ppm).[75] Very strong HBs, such as those formed in the cyclic dimerization of the COOH functions, usually present CSs similar to that of carboxylate groups.[76]

On the other hand, ^{15}N chemical shifts are very sensitive to the protonation state of the nitrogen involved in the HB due to the wider range of CS with respect to ^{13}C. HB formation produces high- or low-frequency shifts according to type of nitrogen atom and synthons involved (Scheme 8.2). The protonation-induced shifts are of the order of about 25 ppm towards high frequencies for aliphatic amines and of 100 ppm towards lower frequencies for aromatic amines.[77]

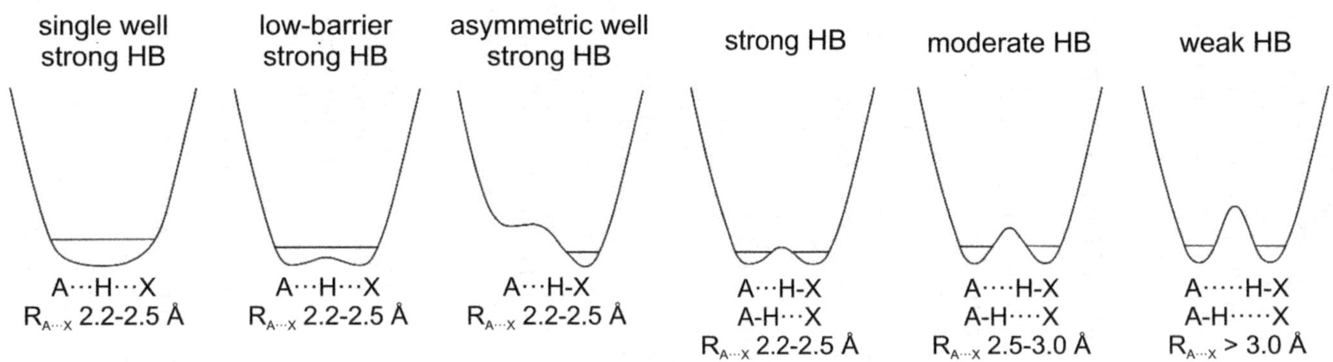

Scheme 8.1 Different types of HB according to their strength. A···X distances are according to Jeffrey's classification.[69]

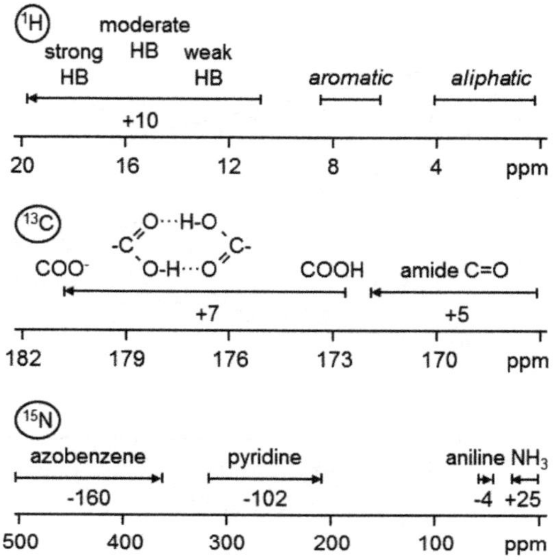

Scheme 8.2 ^{1}H, ^{13}C (carboxylic and amide carbonyl only) and ^{15}N SSNMR shifts due to HB.

Thus, ^{1}H, ^{13}C and ^{15}N SSNMR shift data can be used to evaluate the presence and the strength of HBs and to discriminate between the formation of neutral hydrogen bonds (*e.g.* O–H···N) and charge-assisted HBs involving proton transfer (*e.g.* $^{-}$O··· H–N^{+}).

As already stated above, the chemical shift is a second-rank tensor whose isotropic value (δ_{iso}) represents an average of three components δ_{11}, δ_{22} and δ_{33}. Hydrogen bond formation can cause the tensor values to change dramatically with respect to the same system without HB interactions.[78] Since tensor shift variations are larger in magnitude than ones involving isotropic values, individual principal components are usually more sensitive indicators of changes in the HB environment. Moreover, hydrogen bonding has important effects on only certain principal components, usually those perpendicular to the X–H···Y HB direction. It is worth noting that the CS tensors can be obtained for every nucleus involved in HB interaction: ^{1}H, ^{13}C, ^{15}N, or even ^{17}O. However, measurements of the ^{1}H shift tensor in solids are challenging, mainly because of low resolution and broadened spectra due to strong ^{1}H homonuclear dipolar couplings. Some progress in structural studies relying on ^{1}H CSA has now been made and in general the ^{1}H shift tensor can be extracted through 2D NMR experiments. These methodologies, performed either at slow or very fast MAS, correlate the anisotropic and isotropic parts of the CS interaction.[79] By using rotor-synchronous pulse sequences, different ^{1}H sites are resolved in the detection dimension *via* their isotropic shifts and ^{1}H CSA is reintroduced or "recoupled" during the evolution time in the indirect dimension.[80–82]

As confirmed by *ab initio* calculations, the main origin of the proton high-frequency shift upon HB formation is the deshielding of the principal

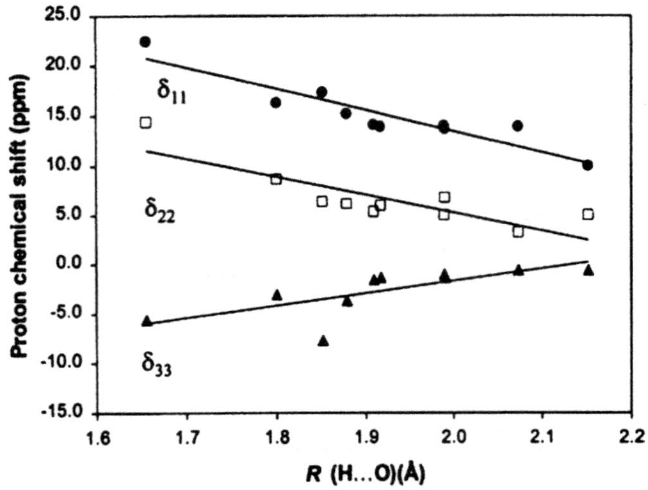

Figure 8.1 Individual tensor components *versus* hydrogen bond length for the flipping water molecule in hydrates. The correlation coefficient for δ_{11}, δ_{22} and δ_{33} is 0.9256, 0.83952 and 0.7254 respectively. Reprinted with permission from G. Wu, C. J. Freure and E. Verdurand, *J. Am. Chem. Soc.*, 1998, **120**, 13187–13193, Copyright 1998 American Chemical Society.

component of the ^{1}H CSA tensor perpendicular to the HB that occurs as the bond length decreases.[83] Actually, clear linear correlation between calculated ^{1}H CSA values and O–H distance in α-D-galactose have been established.[29] Similarly, Wu *et al.* found clear correlation between each individual tensor component and HB length in crystalline hydrates.[84] In this example, as the HB length decreased, δ_{11} and δ_{22} increased, but δ_{33} changed in the opposite direction (Figure 8.1) and as a result, the span ($\Omega = \delta_{11} - \delta_{33}$) of the ^{1}H CS tensor is highly correlated to both HB length and geometry (O–H···O angle).

These approaches are generally suited for recording proton CSA parameters even in more complex biological and organic systems, including protein assemblies and nucleic acids. For instance, accurate ^{1}H CS tensor parameters have been extracted for an 89-residue protein, U–^{13}C, ^{15}N–CAP-Gly domain of dynactin.[85] The CSA parameters correlate with the HB distances (Figure 8.2) and the trends are in excellent agreement with the prior solution NMR results.[86,87]

In a similar way, ^{13}C CS tensor values and orientation also provide insights into HB interaction properties. They can be easily obtained from the sideband (ssb) analysis of low spinning speed MAS NMR spectra.[13] However, when a strong ssb overlap is observed, the preferred choice is the 2D PASS (phase adjustment spinning sideband) technique. It can separate the isotropic chemical shift in the direct dimension from the anisotropic pattern (ssb) related to each different site in the indirect dimension.[16] Most ^{13}C CSA studies concern amide carbonyl and carboxylic groups, in which tensors possess specific orientations (Scheme 8.3)

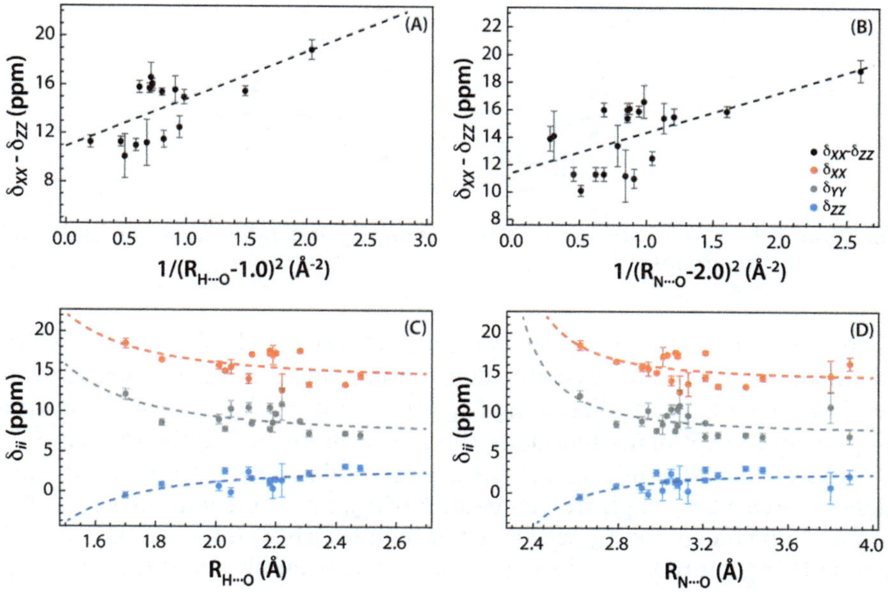

Figure 8.2 Correlation between principal components of ^{1}H CSA tensors and HB length in the U–^{13}C,^{15}N–CAP-Gly domain of dynactin. (A) and (B) show the correlation between the span, δ_{XX}–δ_{ZZ} (black), and the H···O ($R_{H\cdots O}$) and N···O ($R_{N\cdots O}$) distances, respectively. In (C) and (D), correlations of the principal components δ_{XX} (red), δ_{YY} (grey) and δ_{ZZ} (blue) are shown with the H···O and N···O distances, respectively. The H···O and N···O distances are extracted from the MAS NMR structure of CAP-Gly (PDB code 2m02). Reprinted with permission from G. Hou, S. Paramasivam, S. Yan, T. Polenova and A. J. Vega, *J. Am. Chem. Soc.*, 2013, **135**, 1358–1368, Copyright 2013 American Chemical Society.

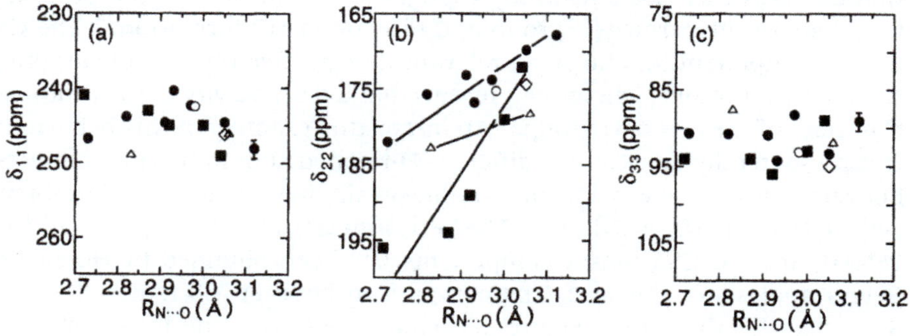

Figure 8.3 Plots of the observed ^{13}C chemical shift tensor components for δ_{11} (a), δ_{22} (b) and δ_{33} (c) for the amide carbonyl carbon in Gly (●), Ala (■), Val (◇), Leu (Δ) and Asp (□) residues in peptides against the N···O hydrogen bond length ($R_{N\cdots O}$). Reprinted from *Journal of Molecular Structure*, **384**, T. Kameda, N. Takeda, S. Kuroki, H. Kurosu, S. Ando, I. Ando, A. Shoji and T. Ozaki, Hydrogen-bonded structure and 13C NMR chemical shift tensor of amino acid residue carbonyl carbons of peptides and polypeptides in the crystalline state. Part I, 17–23, Copyright 1996, with permission from Elsevier.

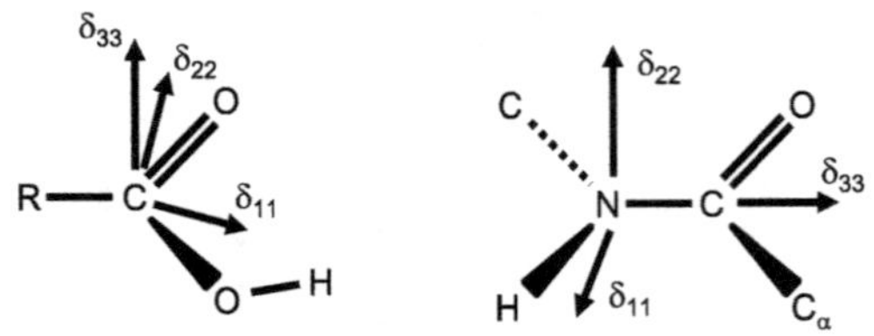

Scheme 8.3 Orientation of ^{13}C CS tensors (δ_{11}, δ_{22} and δ_{33}) in carboxylic and amide groups.

In the case of amide carbonyl groups, different studies came up with the same observation: the most shielded component δ_{33} is perpendicular to the carbonyl sp^2 plane, the direction of δ_{22} is nearly along the amide CO bond and δ_{11} lies in the sp^2 plane but along the direction normal to the CO bond.[88,89] Moreover, SSNMR studies on single crystals and powder samples of peptides have shown a large high-frequency shift of δ_{22}, a slight low-frequency shift of δ_{11} and no change for δ_{33} with a decrease in the N···O distance ($R_{N\cdots O}$) in N–H···O HBs (Figure 8.3).[90,91] In addition, it has been observed that the slope and intercept of the variation of the δ_{22} against $R_{N\cdots O}$ depends on the amino acid residue, whereas δ_{11} and δ_{33} remain almost the same when changing amino acid.

It has also been reported that the principal values of the ^{13}C shielding tensor depend on the secondary structure of polypeptides.[92,93] In particular, the δ_{22} tensor element results in the most sensitive to the distance of hydrogen bonding partner, which in turn varies among secondary structural motifs.[94,95]

Concerning the carboxylic groups, δ_{11} and δ_{22} lie in the carboxylic plane while consequently δ_{33} is perpendicular to the plane (Scheme 8.3). Different studies on carboxyl groups in peptides and amino acids have demonstrated that the ^{13}C CS tensor anisotropy changes significantly with the protonation state of the carboxylic group and thus upon HB formation.[96] The CS tensor component δ_{33} shows no relevant dependence on the surrounding structural variations, whereas δ_{22} reveals marked sensitivity to the protonation state of the carboxyl group.[97] An interesting example of this behaviour is represented by the study of different H-bonded supramolecular adducts between dicarboxylic acids of variable chain length and 1,4-diazabicyclo[2.2.2]octane (DABCO).[75,98,99] The series investigated by ^{1}H, ^{13}C and ^{15}N SSNMR and ^{13}C CSA tensor components has been obtained by Herzfeld–Berger analysis of the ssbs. The comparison between CS tensor and the X-ray data for the series (Figure 8.4) shows that δ_{22} is the most sensitive parameter to the localization of the HB, since it varies from 155 ± 20 ppm for the protonated form to 177 ± 10 ppm for the deprotonated one.

The isotropic CS δ_{iso} is sometimes enough to discriminate the protonation state as, for instance, in the DABCO-glutaric acid adduct in which δ_{iso} is 176.5 ppm for the COOH and 181.8 ppm for the COO– group. However, it is worth noting that often the isotropic chemical shift shows only small increases in shielding upon protonation since the information is

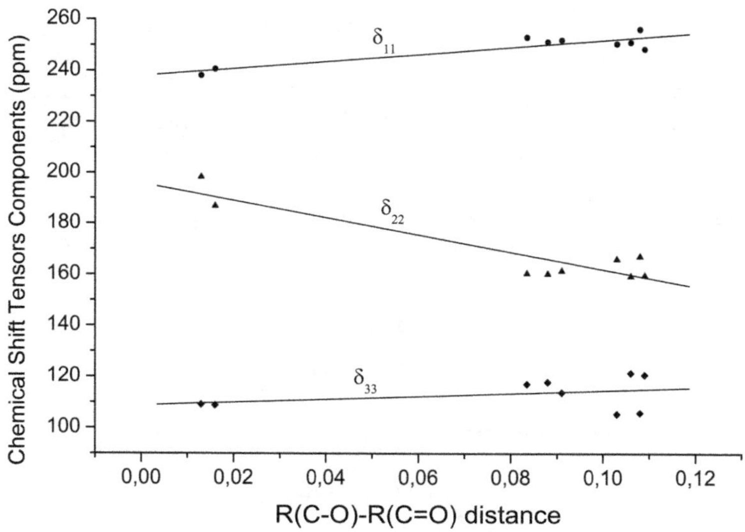

Figure 8.4 Chemical shift tensor components (δ_{11}, δ_{22} and δ_{33}) of DABCO-dicarboxylic acid compounds plotted as function of the difference between the C–O and C=O bond lengths from crystallographic data. Reprinted and adapted with permission from D. Braga, L. Maini, G. de Sanctis, K. Rubini, F. Grepioni, M. R. Chierotti and R. Gobetto, *Chem.-Eur. J.*, 2003, **9**, 5538–5548, Copyright © 2003 WILEY-VCH Verlag GmbH & Co. KGaA, Weinheim.

intrinsically limited by the fact that δ_{11} and δ_{22} change their values in the opposite direction. Anisotropic interactions, such as CSA and dipolar interactions, can provide detailed information regarding packing structures,[100] conformations,[101] orientations,[102] intermolecular interactions[103] and molecular dynamics of polymers at the atomic level.[104] This is the case, for example, of the secondary structure of Samia Cynthia ricini silk that was determined by measurements of the relative orientation of the CSA tensors of the carbonyl groups in the polymer backbone.[105]

Advanced 2D solid-state NMR techniques have been employed to obtain structural constraints for determining the backbone conformation of poly(L-isocyanoalanyl-D-alanine) methyl ester (L,D-PIAA).[106] A 2D-separated local field (2D-SLF) experiment was used to determine the orientation of the $^{13}C{=}^{15}N$ dipolar vector in the principal axis frame of the ^{13}C CSA tensor, while a 2D double-quantum single-quantum spectroscopy (DOQSY) experiment was used to determine the relative ^{13}C–^{13}C CSA tensor orientation of neighbouring carbons. The structural constraints obtained from the DOQSY and 2D-SLF data were used to build the L,D-PIAA model. This polymer model was subsequently subjected to molecular dynamics simulations in order to obtain the energy minimized structure.

Compared to ^{13}C CSA, dealing with ^{15}N CSA tensors is more challenging because of the very poor sensitivity of the ^{15}N resonance, due to the small γ value and the low natural abundance (~0.4%). However, interest in ^{15}N CSA

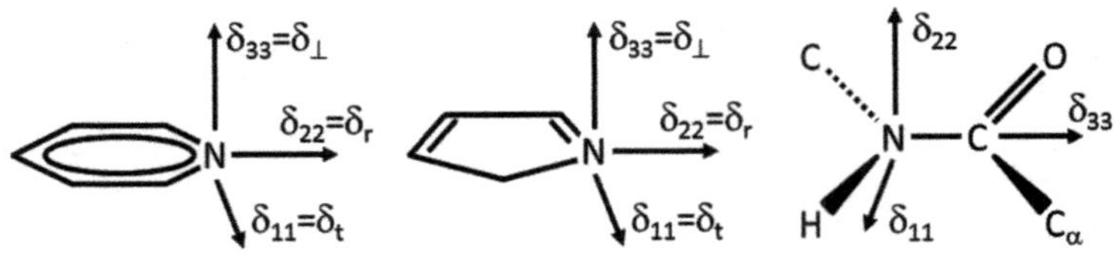

Scheme 8.4 Orientation of ^{15}N chemical shift tensors (δ_{11}, δ_{22} and δ_{33}) in pyridine, and pyrrole/pyrazole/imidazole derivatives and amide groups.

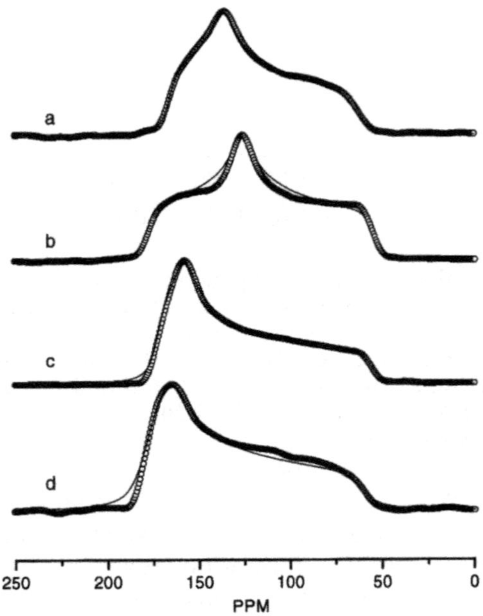

Figure 8.5 ^{15}N CSA spectra of (a) phtalimide, (b) *N*-methylphtalimide, (c) *N*-stearylimide and (d) *N*-benzylphtalimide. Experimental data are represented by open circles, spectral simulation by a solid line. Reprinted from *Journal of Magnetic Resonance, Series A*, **116**, W. L. Jarrett, G. C. Johnson and L. J. Mathias, Solid-state NMR characterization of ^{13}C- and ^{15}N-labeled phthalimides as model compounds for studying polyimides, 156–160, Copyright 1995, with permission from Elsevier.

has been growing since it appears to be more sensitive than ^{13}C to HB, producing larger shift changes upon bonding.

The principal values and the orientation of the ^{15}N CS tensors (Scheme 8.4) have been derived for different classes of nitrogen-containing compounds such as amine, imine, pyridine-like and pyrazole/imidazole nitrogen atoms.

The tensor values are usually extracted from ^{15}N line shape simulation of static powder spectra, as reported for instance by Jarrett *et al.* for four model imides (Figure 8.5).[107]

Studies on ^{15}N tensor values and hydrogen bonding have primarily focused on nitrogen heterocycles such as nucleic acid bases.[108,109] In these compounds, for protonated nitrogen atoms (N–H) the δ_{22} value that lies in the plane of the molecule and is almost perpendicular to the N–H bond (δ_{11} direction) is more sensitive and moves to higher frequencies upon HB formation. On the other hand, in non-protonated nitrogen atoms δ_{11} and δ_{22} have reverse orientation, such as δ_{11} component results are more influenced and become more shielded upon HB formation.

Concerning amide groups, an intriguing study on crystalline histidine and histidine-containing peptides have been published.[110] Systematic trends in the values of imidazole CSA with changes in HBs have been observed, in particular between the ^{15}N δ_{11} tensor value and the HB length for cationic species; as the HB distance decreases, the δ_{22} tensor value shifts downfield.

To summarize, knowledge of the chemical shift tensor values and orientations is fundamental in the characterization of HB systems, since for each functional group these are rather conservative parameters and can be fixed relative to a molecular reference frame for a small molecular fragment. In particular for ^{1}H, ^{13}C and ^{15}N nuclei involved in HB, general trends have been observed. In these cases (be aware that ^{15}N applies to NH groups only), δ_{22} is perpendicular to the HB axes and is generally the most sensitive parameter of the HB features, mainly HB length and hydrogen atom position along the HB axes, *i.e.*, proton transfer. Its value usually increases upon decreasing the HB length. In this context, a key role has been played by extensive computational work commonly carried out within the well-established density functional theory using the plane waves approach as implemented in the GIPAW method (see above). Indeed, computation definitely allowed the chemical shift tensor components to be related to the crystallographic location of relevant atoms in the unit cell, with the geometry of the chemical bond adding significantly to the analysis of weak interactions and thus to the practice of NMR crystallography.

8.3.2 Hydrogen Bond and Dipolar Interaction

Many applications of SSNMR concerning structure determination rely upon the dependence of the dipolar coupling D on the inverse cubed power of the internuclear distance r_{IS} *i.e.* r_{IS}^{-3}, where I and S are the two involved spins. Indeed, as already stated above, the dipolar coupling is the NMR parameter with the highest crystallographic content.[37] For this reason, dipolar coupling is the most exploited interaction to measure heteronuclear distances and thus to get information on the geometric and electronic structure of solids. Moreover, with respect to hydrogen bonds, the analysis of dipolar interaction can provide pivotal insights concerning the localization of hydrogen atoms along HB axes. As already commented in Section 8.2.3, dipolar coupling is averaged out by MAS but numerous recoupling techniques are available to reintroduce it in the spectrum.

For instance, ^{1}H DQ MAS spectroscopy has been extensively applied to probe proton–proton proximities and to quantitatively determine ^{1}H–^{1}H dipolar couplings. This powerful method relies on the generation of DQCs

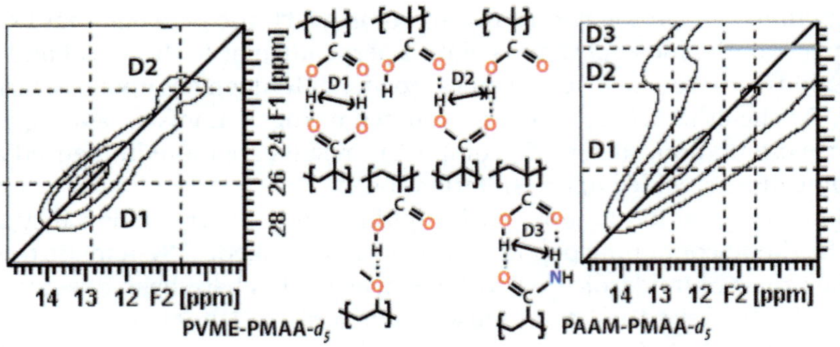

Figure 8.6 HB regions, SQ/DQ, 9–15 ppm/18–30 ppm, of the 2D ^{1}H DQ MAS of the (left) poly(vinyl methylether)-poly(methacrylic acid) dimer, (right) poly(acrylamide)-poly(methacrylic acid) complexes. The labels D1 and D2 identify the auto-peaks of cyclic and open dimers of poly(methacrylic acid) respectively, while the D3 cross-peak indicates the poly(acrylamide)-poly(methacrylic acid) hetero dimer. Reprinted with permission from B. Fortier-McGill, V. Toader and L. Reven, *Macromolecules*, 2012, **45**, 6015–6026, Copyright 2012 American Chemical Society.

between proximal ^{1}H spins within 3.5 Å for a typical organic solid, thus probing the spatial arrangement of protons according to the strength of the through-space dipolar coupling.[111,112] To efficiently excite the DQC, the use of a recoupling pulse sequence is required to restore the dipolar couplings averaged out by MAS.[113,114] Despite several recoupling methods available and used, to date the technique of choice for such experiments at fast MAS rates (>30 kHz) has been the BABA (BAck-to-BAck) sequence.[115] The applications of the ^{1}H DQ MAS technique include distinguishing between two chain-like structures of benzoxazine oligomers,[116] the study of structure and dynamics in H-bonded polymers,[117,118] proton-conducting materials,[119,120] the investigation of keto–enol tautomerism in a series of barbituric and thiobarbituric acid polymorphs,[121,121] the study of supramolecular host–guest complexes,[122,123] π–π stacked polycyclic aromatic systems,[124] and surface organometallic species.[125]

In the polymer field, a detailed NMR study on the presence, relative strengths and network of HBs has been published for poly(methacrylic acid) complexes with complementary polymers.[103] Three possible HB arrangements within PMAA complexes have been deducted from ^{1}H DQ MAS spectra (Figure 8.6) and proton–proton distances were estimated from the analysis of ^{1}H DQ spinning sidebands patterns (Figure 8.7).

As an example of an important and general pharmaceutical application concerning pseudopolymorphism (the formation of hydrates/solvates), Brown and collaborators demonstrated that the presence of a specific pseudopolymorph of an API in a tablet formulation can be revealed using ^{1}H DQ CRAMPS.[126] They collected ^{1}H DQ spectra of the pure anhydrous and monohydrate form of an API with about 20 carbon atoms (Figure 8.8) to obtain detailed spectroscopic fingerprints that were compared with the ^{1}H DQ CRAMPS spectrum of

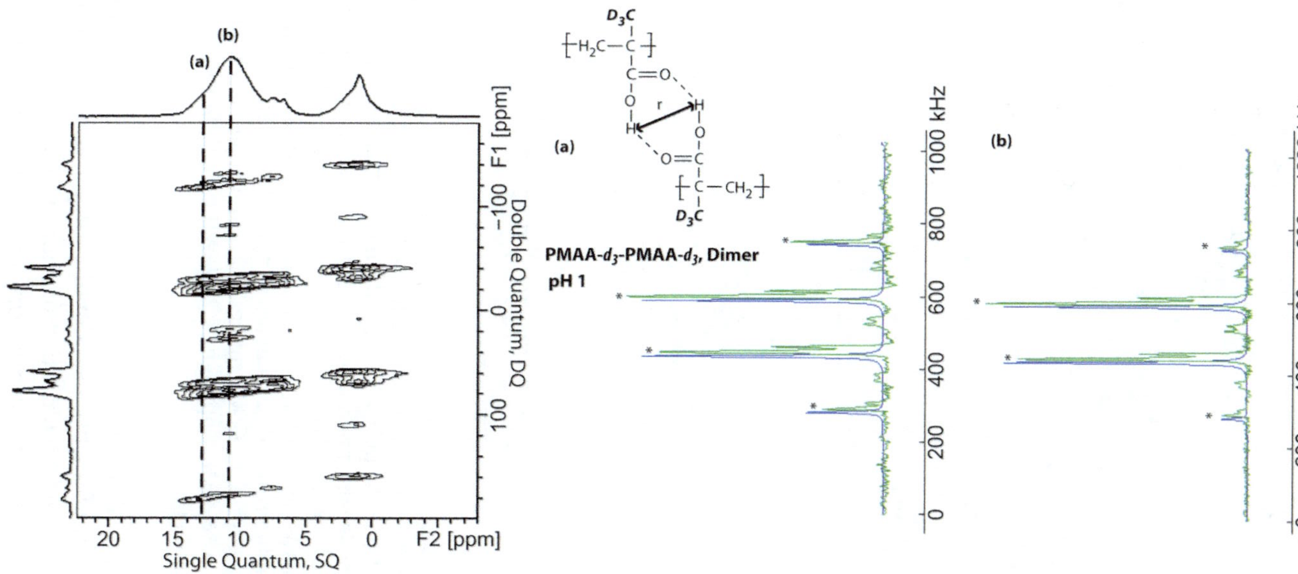

Figure 8.7 Left: 2D ^{1}H DQ MAS NMR spectrum of poly(methacrylic acid)-d$_3$, using an $N = 4$ cycle excitation/reconversion BABA sequence at $\nu_r = 30$ kHz. Right: experimental (green) and simulated (blue) 1D ^{1}H DQ MAS NMR sideband spectra for the slices taken at 12.5 ppm (a) and 10.5 ppm (b). The asterisk denotes the DQ spinning sidebands of interest as the unlabelled peaks are the result of the overlapping DQC signal between the H-bonded COOH and aliphatic protons. Reprinted with permission from B. Fortier-McGill, V. Toader and L. Reven, *Macromolecules*, 2012, **45**, 6015–6026, Copyright 2012 American Chemical Society.

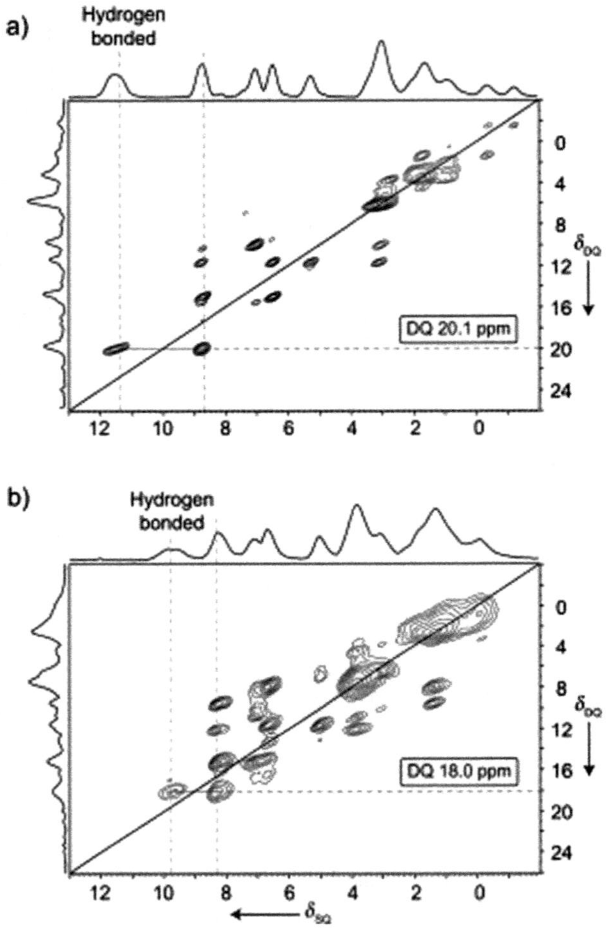

Figure 8.8 600 MHz ^{1}H DQ CRAMPS NMR spectra together with skyline projections of (a) the anhydrous and (b) the monohydrate form of the API under consideration. The pair of DQ signals in both spectra corresponding to the intramolecular proximity of hydrogen-bonded protons with high-ppm resonance signals to the same nearby proton are highlighted. Each spectrum was recorded in 105 min. Base contours are shown at 11% of maximum intensity. All axes in are in ppm. Reprinted with permission from J. M. Griffin, D. R. Martin and S. P. Brown, *Angew. Chem., Int. Ed.*, 2007, **46**, 8036–8038, Copyright © 2007 WILEY-VCH Verlag GmbH & Co. KGaA, Weinheim.

a crushed tablet. This method allowed the clear identification of resonances characteristic of only one of the two pseudopolymorphs, and was proposed as a routine procedure in the characterization of pharmaceuticals.

In a similar way, ^{1}H DQ MAS spectroscopy has been used to unravel the complex HB network in the biologically important molecule bilirubin and its dimethyl ester.[127] In particular, the really rare carboxylic acid to amide hydrogen-bonded arrangement has been probed and the proton–proton distances

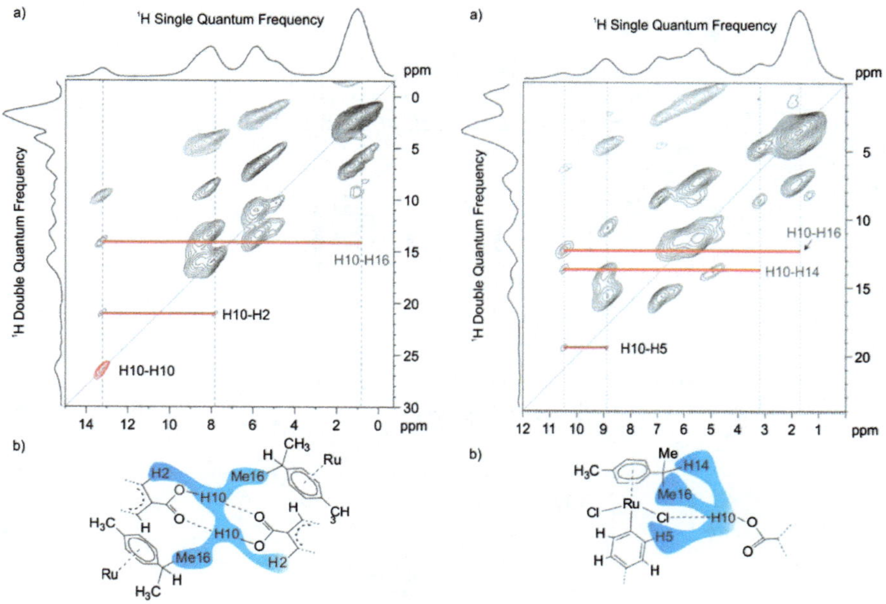

Figure 8.9 (a) ^{1}H DQ CRAMPS (PMLG5–POSTC7–*w*PMLG5) spectra of (left) [(*p*-cymene)Ru(κ*N*-A4AB)Cl$_2$] and (right) [(*p*-cymene)Ru(κ*N*-INA)Cl$_2$], together with skyline projections recorded at 12.5 kHz MAS. Solid red horizontal bars indicate specific DQCs between OH (H10) and nearby protons. The OH diagonal peak indicating the cyclic dimerization of the COOH groups is highlighted in red. (b) Representation of the crystal structures showing the main inter- and intramolecular proton–proton proximities. Reprinted with permission from M. R. Chierotti, R. Gobetto, C. Nervi, A. Bacchi, P. Pelagatti, V. Colombo and A. Sironi, *Inorg. Chem.*, 2014, **53**, 139–146, Copyright 2013 American Chemical Society.

have been measured through an analysis of ^{1}H DQ MAS spinning sideband patterns.

In the organometallic field, ^{1}H DQ CRAMPS spectra have been fundamental in determining the HB network and, thus, the self-assembly of several organometallic Ru(II) complexes.[76] The hydrogen bond networks in all systems have been investigated in term of supramolecular synthons. The presence or absence of peculiar DQ auto-peaks in the ^{1}H DQ CRAMPS spectra (Figure 8.9) allowed determining that only one complex showed the expected supramolecular cyclic dimerization of the carboxylic functions of the ligand.

Beside the possibility of probing proton–proton proximities *via* homonuclear dipolar coupling, SSNMR has provided tools for measuring the heteronuclear dipolar coupling and thus achieving the average lengths of the X–H bonds (X usually = N).

In a recent study, the tautomeric state of the C.I. Pigment Yellow 138 has been investigated by means of ^{1}H–^{13}C on- and off-resonance FSLG HETCOR and ^{14}N–^{1}H J-HMQC.[128] The ^{1}H–^{13}C on-resonance, performed with a short

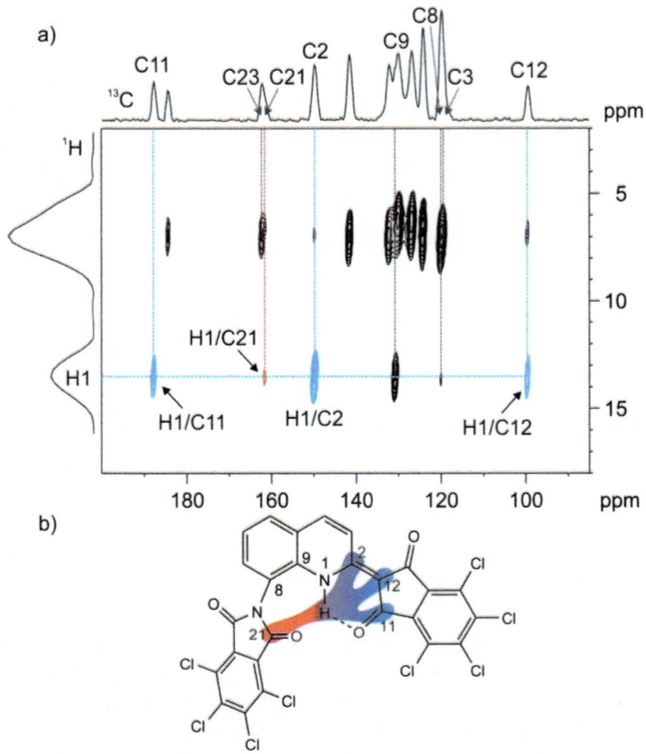

Figure 8.10 (a) ^{1}H–^{13}C FSLG off-resonance HETCOR spectrum of P. Y. 138, recorded with a contact time of 2 ms and a spinning speed of 12 kHz. (b) Representation of the crystal structure of P. Y. 138 showing the main ^{1}H–^{13}C proximities involving the NH moiety. Reprinted from *Dyes and Pigments*, **131**, S. D. Gumbert, M. Koerbitzer, E. Alig, M. U. Schmidt, M. R. Chierotti, R. Gobetto, X. Li and J. van de Streek, Crystal structure and tautomerism of Pigment Yellow 138 determined by X-ray powder diffraction and solid-state NMR, 364–372, Copyright 2016, with permission from Elsevier.

mixing time to give correlations only between directly bonded C–H, was fundamental for the complete assignment of the CH resonances. The off-resonance experiment, performed with long mixing time but with the CP under Lee–Goldburg conditions to prevent proton–proton spin diffusion, provided real long-range spatial proximities (*i.e.* correlations between C and H within a distance up to 5 Å) necessary to assist and confirm the structure solved from XRPD. Moreover, it was pivotal to detecting the presence of an intramolecular N–H···O hydrogen bond with formation of a "six-membered aromatic ring", since polarization transfer was observed from the NH hydrogen (H1) to six different carbon atoms (Figure 8.10). The presence of a N–H bond rather than a O–H bond was directly provided by the rotor-synchronized ^{14}N–^{1}H J-HMQC experiment acquired at 70 kHz through the indirect detection

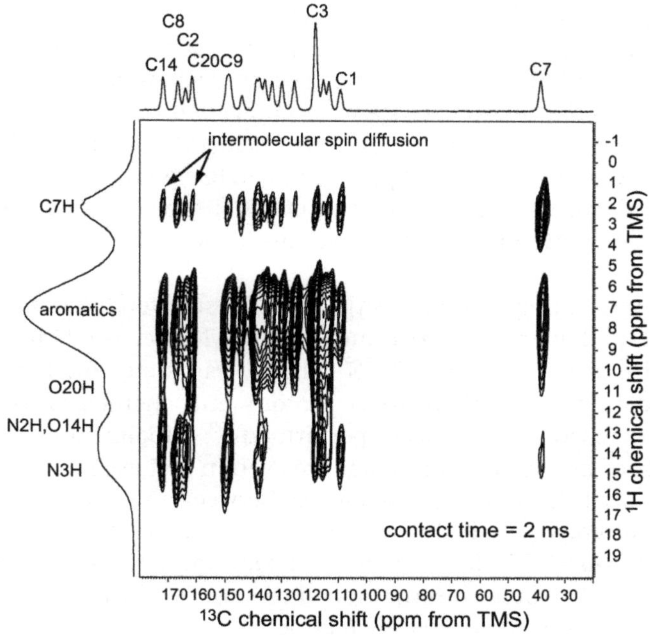

Figure 8.11 2D ^{1}H–^{13}C CP-HETCOR spectrum (ν_r = 12.5 kHz) of tenoxicam saccharin obtained at a contact time of 2 ms. The ^{1}H MAS (ν_r = 35 kHz) and ^{13}C CP–TOSS (ν_r = 8 kHz) spectra are plotted along the F1 (vertical) and F2 (horizontal) dimensions, respectively. Reprinted from *International Journal of Pharmaceutics*, **436**, J. R. Patel, R. A. Carlton, T. E. Needham, C. O. Chichester and F. G. Vogt, Preparation, structural analysis, and properties of tenoxicam cocrystals, 685–706, Copyright 2012, with permission from Elsevier.

scheme. This 2D experiment takes advantage of the direct detection of the highly sensitive ^{1}H nucleus and indirect detection of the ^{14}N to provide information on covalently bonded ^{1}H–^{14}N pairs, since the mixing time relies only on the evolution of the ^{1}H–^{14}N *J*-coupling while the other interactions are effectively averaged by very fast MAS.[129] Thus, the presence of a correlation peak reveals the existence of a covalently bonded N–H group. This is fundamental when one needs to evaluate the position of a hydrogen atom along N···H···X (X = O, N,...) interactions.

In a series of nine tenoxicam (non-steroidal anti-inflammatory drug) co-crystals the analysis of ^{1}H–^{13}C proximities achieved from ^{1}H–^{13}C short- and long-range CP-HETCOR combined with data from ^{1}H DQ MAS, allowed molecular association to be demonstrated.[130] For instance, in the ^{1}H–^{13}C long-range CP-HETCOR spectrum of the co-crystal, intermolecular correlations between one carbon atom of saccharin and the methyl protons of tenoxicam due to spin-diffusion showed direct evidence of the molecular association between the two coformers as a co-crystal (Figure 8.11).

In the field of proteins, an intriguing example of the application of 2D heteronuclear correlation experiments has been reported for the influenza M2 protein.[131] Attention was focused on assessing whether a critical histidine residue (His37) only interacts with water molecules or whether a low-barrier hydrogen bond (LBHB) also exists between histidines to stabilize charges. 2D ^{15}N–^{1}H HETCOR experiment probed that at low temperature, His37 forms conventional hydrogen bonds instead of LBHBs, whereas at ambient temperature it gives HBs only to water and low-barrier H-bonded dimers are not formed.

Another interesting approach is the use of perdeuterated microcrystalline proteins that allows the observation of hydroxyl protons by MAS solid-state NMR. In the case of a [U-^{2}H,^{13}C,^{15}N]-labelled sample of the R-spectrin SH3 domain, identification of hydroxyl protons and characterization of their exchange behaviour have been performed.[132] Dipolar couplings enable assignment of hydroxyl protons that are even in fast exchange with the solvent. The quantification of the distances between the hydroxyl proton and the carbon atoms in the hydrogen-bonding donor as well as acceptor group was achieved *via* a REDOR experiment. In combination with X-ray data, availability of ^{1}H–^{13}C distance information allows the localization of the position of the hydroxyl proton in a hydrogen bond.

Recently, a rare phenomenon related to tautomerism called desmotropy has been investigated for two different forms of the antiparasitic Albendazole (ABZ) (Scheme 8.5).[133]

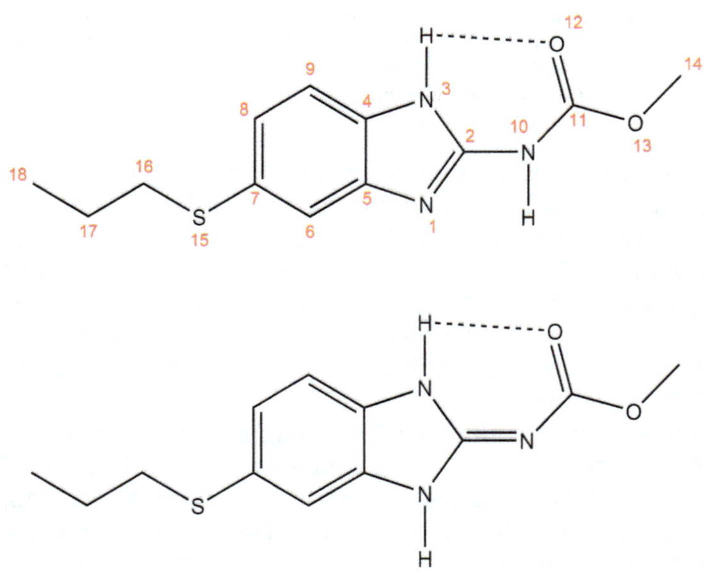

Scheme 8.5 ABZ tautomers corresponding to desmotropic forms I (top) and II (bottom). Reprinted with permission from A. K. Chattah, R. Zhang, K. H. Mroue, L. Y. Pfund, M. R. Longhi, A. Ramamoorthy and C. Garnero, *Mol. Pharm.*, 2015, **12**, 731–741, Copyright 2015 American Chemical Society.

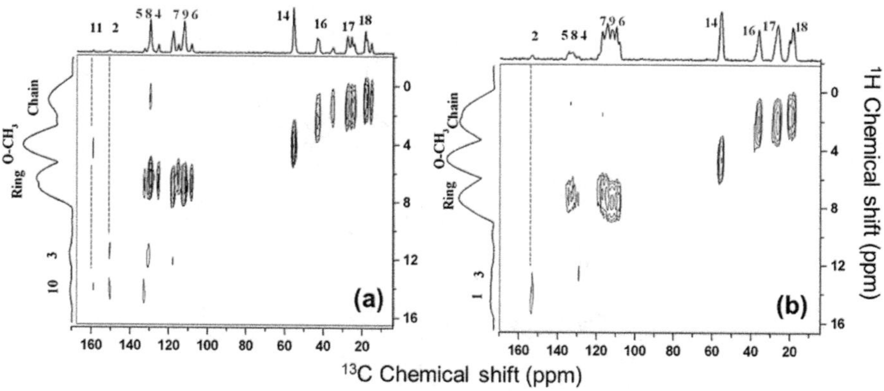

Figure 8.12 Two-dimensional ^{1}H–^{13}C HETCOR spectra of ABZ I (a) and II (b). ^{13}C and ^{1}H projections are displayed in the direct and indirect frequency dimensions, respectively. Some of the carbon–proton correlation peaks are marked with dashed lines. Reprinted with permission from A. K. Chattah, R. Zhang, K. H. Mroue, L. Y. Pfund, M. R. Longhi, A. Ramamoorthy and C. Garnero, *Mol. Pharm.*, 2015, **12**, 731–741, Copyright 2015 American Chemical Society.

The ^{1}H–^{13}C HETCOR spectrum together with one dimensional experiments and data from ^{1}H DQ MAS have been fundamental to decipher the intra- and intermolecular HB interactions present in both forms, thus enabling the unequivocal identification of the tautomers of each desmotrope. In particular, by comparing the correlations between NH protons and their neighbouring carbon atoms it was possible to observe that a NH proton in form I showed two correlations with C11 and C2, whereas in form II it correlates with C2 but not with C11, in agreement with the two different molecular structures (Figure 8.12).

8.3.3 Hydrogen Bond and Quadrupolar Interaction

A powerful tool for studying hydrogen bonding through quadrupolar nuclei is ^{17}O solid-state NMR, since it gives direct access to one of the HB partners in X–H···O (X = O, N or C) bonds. Moreover, ^{17}O chemical shift and quadrupolar coupling have been proved to be very sensitive to the strength and geometry of HBs.[134]

Solid-state ^{17}O data have been obtained for a ^{17}O-labelled diflunisal pyrazinamide co-crystal.[135] By comparing the experimental and theoretical results, both chemical shift and quadrupolar constant values, for the co-crystal and the form I of diflunisal, several insights into the HB network have been elucidated. The δ_{iso} for the O2 nucleus increases upon formation of the co-crystal, suggesting a strong hydrogen bond involving the O–H donor. Similarly, the δ_{iso} value obtained for the O1 site in the co-crystal is significantly increased with respect to form I, which suggests disruption of the diflunisal dimers in favour of a longer intramolecular HB concerning the C=O group (Scheme 8.6 for molecule numbering).

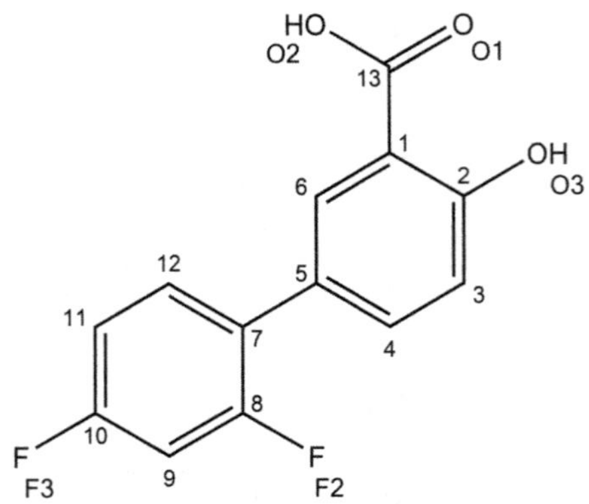

Scheme 8.6 Chemical structure and numbering scheme for diflunisal.

In addition, an interesting ^{1}H–^{17}O CP-HETCOR experiment has also been employed to further investigate the HB interactions. The strongest correlation in Figure 8.13 is consistent with the covalent interaction between O2 and O2H, whereas the weaker correlation is assigned to the intramolecular HB involving O1 and phenol proton O3H. Hence, the ^{1}H–^{17}O CP-HETCOR spectrum also provided resolution for the O2H and O3H resonances, otherwise they are not available.

Concerning the field of proteins, an interesting ^{17}O NMR characterization of three unstable enzymatic intermediates has been reported.[136] In particular, attention has been focused on probing the HB interactions between the carbonyl (C=^{17}O) functional group of the substrate acyl moiety and the two NH groups from the protein backbone (Ser195 and Gly193), in the so-called oxyanion hole of the serine protease chymotrypsin. Experimental data together with computations have proved that ^{17}O NMR parameters strongly vary according to the different hydrogen bonding environment in the oxyanion hole and that HB interaction is responsible for the deacylation kinetics of acyl-enzymes.

As a conclusive remark, it should be pointed out that despite the clear attractiveness of ^{17}O SSNMR as a tool for investigating HBs, the extremely low natural abundance of ^{17}O makes it necessary to enrich the sample isotopically. Unfortunately, ^{17}O labelling is often not a realistic option, as it requires custom synthesis with associated high manufacturing costs.

8.4 SSNMR and Halogen Bonds

Halogen bonds (XBs) constitute an important class of non-covalent interaction that can be investigated by SSNMR. This interaction typically occurs between a region of positive, electrostatic potential of a polarized halogen (X = I, Br, Cl, F) in a molecule R–X, and a nucleophilic region in another or the same molecular entity (Y).[137,138] Hence, the XB is usually depicted as in Scheme 8.7.

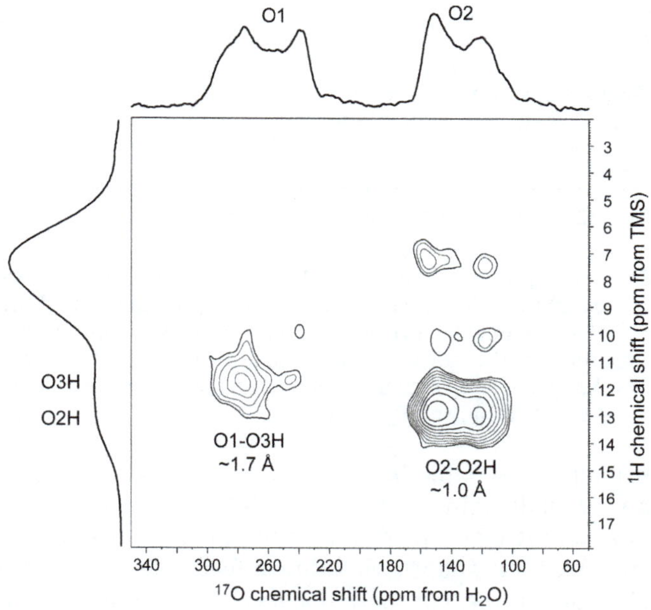

Figure 8.13 2D ^{1}H–^{17}O CP-HETCOR spectrum (ν_r = 25 kHz) of the ^{17}O-diflunisal pyrazinamide co-crystal. The spectrum was obtained using a contact time of 1 ms. The ^{17}O MAS echo spectrum (ν_r = 30 kHz) is plotted along the horizontal (F2) axis and the ^{1}H DP-MAS spectrum (ν_r = 30 kHz) is plotted along the vertical (F1) axis. All spectra were obtained at 16.4 T and 283 K. Distances given are approximate and are based on values taken from the form I crystal structure. Reprinted with permission from A. K. Chattah, R. Zhang, K. H. Mroue, L. Y. Pfund, M. R. Longhi, A. Ramamoorthy and C. Garnero, *Mol. Pharm.*, 2015, **12**, 731–741, Copyright 2015 American Chemical Society.

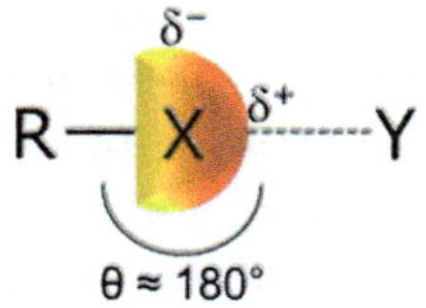

Scheme 8.7 Typical halogen-bond geometry, where R = C, N, O; halogen, X = F, Cl, Br, I; and Y = N, O, S, Se, I$^-$, Br$^-$, Cl$^-$, F$^-$,... The red-coloured area is a graphical representation of the σ-hole.

It typifies the building block exploited in modern supramolecular chemistry to assemble a number of advanced functional materials, such as porous systems, conducting and magnetic materials and biomimetic materials.[6,139] The main NMR observables for investigating the XB are chemical shifts, dipolar interaction, *J*-coupling and quadrupolar interaction. XB formation leads to substantial changes in the isotropic CSs of atoms directly involved in, or in close proximity to, the non-covalent interaction. All three fragments of the

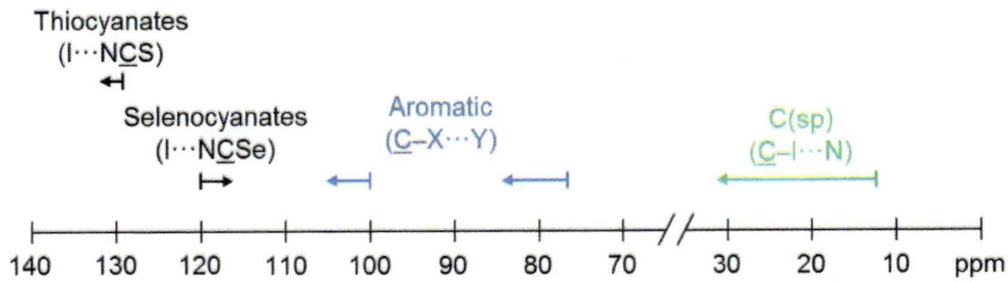

Scheme 8.8 Schematic representation of XB-induced changes in ^{13}C SSNMR chemical shifts. The difference in the shift direction noted for the selenocyanates may be attributed to an interplay of competing factors that influence the isotropic CS, which are difficult to define univocally. The changes are referred to carbon atoms of either the acceptor (black), or the donor (blue and green).

XB are in principle amenable to be studied by SSNMR; however, halogens represent a very challenging task for the NMR spectroscopist and SSNMR itself, as they exhibit large quadrupolar interactions that cause extremely broad and distorted line shapes with almost featureless spectra, and technically difficult acquisitions. For these reasons, the analysis of XBs by SSNMR has traditionally resorted to probing the atoms in the R and Y moieties: ^{13}C and ^{15}N are the most investigated nuclei, and their acquisition and interpretation is relatively straightforward. We begin the following discussion by presenting some paradigmatic examples and trends involving XB detection and characterization by several carbon and nitrogen SSNMR techniques. In fact, in the vast majority of X-bonded molecular solids the donor bears a carbon atom covalently bonded to the halogen, whereas nitrogen compounds are the most used acceptors, followed by oxygen and sulphur compounds.[140] This section will conclude with an overview of the strategies to address the observation of quadrupolar nuclei.

When the interaction involves as a XB donor a halogen atom directly bonded to a carbon atom, and as a XB acceptor another carbon-rich molecule, ^{13}C represents a convenient probe of the interaction. However, while the changes in the $\delta_{iso}(^{13}C)$ of several acceptors have been reported to be in the range of 1–3 ppm,[141] the extent of the shift is much wider from the donor point of view (Scheme 8.8).[142]

Several studies have provided evidence that ^{13}C CSs are sensitive probes of the XB occurrence and a nice correlation between $\delta_{iso}(^{13}C)$ values and the C–I distance for iodoperfluoroalkanes and iodoperfluoro benzene co-crystals with ammonium or phosphonium halide salts has been reported.[142–144] ^{13}C chemical shift change upon XB formation is directly and linearly related to the C–I distance, *i.e.*, an increase in the carbon–iodine bond length corresponds to an increase in the CS (Figure 8.14a).

It is worth noting that MAS does not entirely average the dipolar coupling between the spin-1/2 carbon, and the spin-5/2 iodine, giving rise to residual dipolar broadening of the ^{13}C signals. Furthermore, their ^{13}C CSs are markedly shielded due to a well-known spin–orbit-induced heavy atom effect of iodine;[145,146] this relativistic effect has been revealed to be very useful for an unambiguous assignment of the CSs of such carbon atoms.

Other studies have employed XB donors with sp-hybridized carbons, and have found additional indications that the ^{13}C SSNMR CSs usually move toward high frequencies upon XB formation. Some of us have recently reported high field shift upon XB formation for the C(sp)–I resonance in two pharmaceutical compounds, X-bonded with 4,4′-bipyridine (4,4′-bipy), 1,2-bis-4(pyridyl)ethane (bipyet) and alkyl ammonium halide salts.[147,148] An increase in the $\delta_{iso}(^{13}C)$ values upon XB formation, ranging from about 7 to 15 ppm, has been observed. ^{13}C spectral editing experiments, in particular the non-quaternary suppression (NQS), which retains signals of quaternary carbon atoms only while suppressing CH and CH_2 and strongly reducing CH_3, was crucial to unambiguously assigning the CSs of the overlapped C(sp)–I resonances, in particular for one of the two pharmaceutical compounds (namely 3-Iodo-2-propynyl-*N*-butylcarbamate, IPBC) and its co-crystals (Figure 8.15).

^{15}N chemical shifts are also sensitive to the XB occurrence; results from systematic XB investigations by ^{15}N SSNMR are outlined in Scheme 8.9. It is of interest to note that this ^{15}N peculiarity had already been observed in the case of HB formation,[98,99] and also upon metal coordination.[149–152]

Recently, a complete screening of different XB donors in two series of X-bonded supramolecular structures, constructed by assembling dipyridyl derivatives and either halobenzenes or haloalkanes by means of ^{15}N SSNMR, FTIR, SCXRD, and a computational approach have been reported.[97] In these cases solid-state ^{15}N chemical shifts generally decrease upon XB formation. A good linear correlation is found (Figure 8.14b) when the change in ^{15}N chemical shifts is plotted against the corresponding normalized contacts (R_{XB}) calculated as the ratio $D_{XN}/(r_X + r_N)$, where D_{XN} is the experimental distance between the halogen atoms (X) and the XB-acceptor nitrogen, and r_X and r_N are the corresponding vdW radii. The use of RXB allows contacts of chemically different interacting sites to be directly compared.

This experimental trend has been also reproduced by GIPAW DFT calculations; hence, natural abundance ^{15}N SSNMR spectroscopy is a diagnostic of the geometry of the XB, and the chemical shift change upon XB occurrence is generally greater than the change of ^{13}C chemical shifts in a large set of X-bonded systems.

Weingarth *et al.* proposed a method based on the determination of the dipolar interaction (D) between ^{15}N and ^{127}I for evaluating the heteronuclear N⋯I distance in a ^{15}N-enriched benzyl-di(4-iodobenzyl)-amine adduct.[153] The numerical value of D, as well as the N⋯I distance related to the XB bond, were obtained by fitting the rotary resonance lineshapes with simulated ones, which in turn were obtained by experimental CSA parameters. This method is particularly useful when the quality of single crystals is not suitable for XRD analysis.

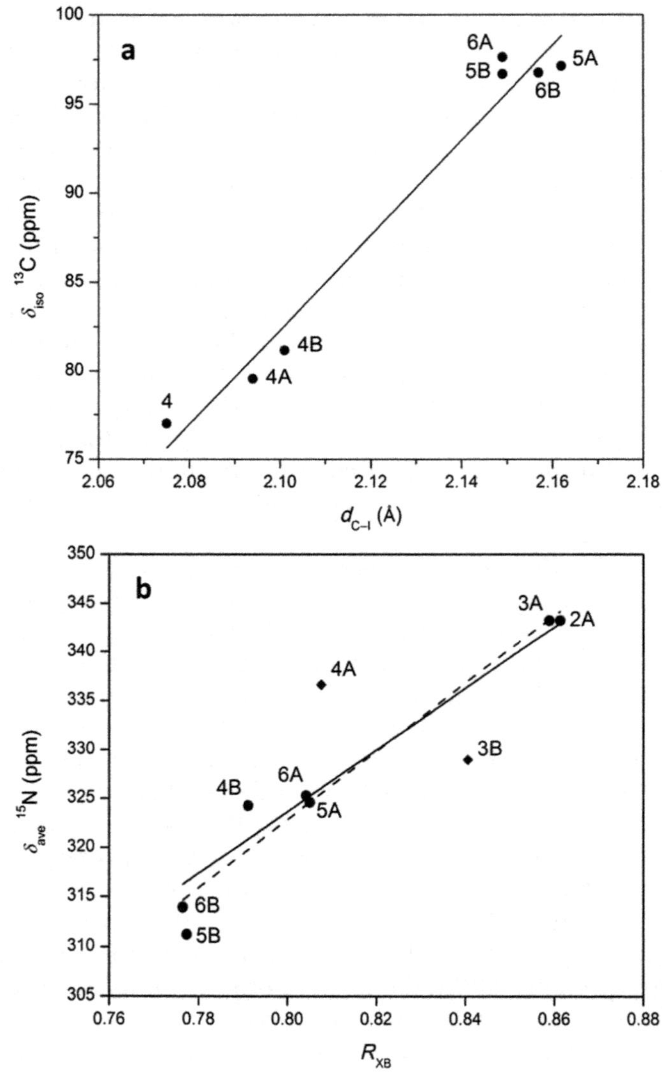

Figure 8.14 (a) Plot of the experimental values of ^{13}C chemical shift of carbons covalently bonded to iodine for 1,4-diiodotetrafluorobenzene (4), [(4,4′-bipyridine)·(1,4-diiodotetrafluorobenzene)] (4A), [(4,4′-bipyridine)·(1,8-diiodohexadecafluorooctane)] (5A), [(4,4′-bipyridine)·(1,6-diiodododecafluorohexane)] (6A), [(1,2-bis(4-pyridyl)ethane)·(1,4-diiodotetrafluorobenzene)] (4B), [(1,2-bis(4-pyridyl)ethane)·(1,8-diiodohexadecafluorooctane)], (5B), [(1,2-bis(4-pyridyl)ethane)·(1,6-diiodododecafluorohexane)] (6B). The best fit is represented by a linear function: $\delta_{iso}{}^{13}\text{C} = 266.5\delta_{\text{C-I}} - 477.3$, $R^2 = 0.9664$. (b) Plot of the experimental values of ^{15}N chemical shifts as a function of the corresponding normalized distance parameter for [(4,4′-bipyridine)·(1,4-dibromotetrafluorobenzene)], (2A), [(4,4′-bipyridine)·(1,4-diodobenzene)] (3A), [(4,4′-bipyridine)·(1,4-diiodotetrafluorobenzene)] (4A), [(4,4′-bipyridine)·(1,8-diiodohexadecafluorooctane)] (5A), [(4,4′-bipyridine)·

The NMR spectra of $^{35/37}$Cl, $^{79/81}$Br and ^{127}I are typically dominated by QI. SSNMR techniques on such nuclei are therefore extremely challenging since QI can be almost equal in magnitude to Zeeman interaction (10^3–10^7 Hz), thus severely enlarging the breadth of the spectrum. Hence, the vast majority of SSNMR studies of chlorine, bromine and iodine has been focused on observing the more accessible anionic forms of these elements, rather than covalently bonded halogens.[154] Szell *et al.* were able to obtain direct observation of ^{35}Cl SSNMR spectra of a covalently bonded chlorine engaged in XB in several small organic molecules using a WURST-QCPMG sequence at 21.1 T, corroborating the numerical values of C_Q and η_Q with ^{35}Cl NQR spectroscopy.[155] The method is sensitive to small changes in the electric field gradient (EFG) caused by XB, *i.e.*, the magnitude of the ^{35}Cl quadrupolar coupling constant (C_Q) was found to increase, and the quadrupolar asymmetry parameter (η) was found to decrease upon halogen bonding. (Figure 8.16).

Bryce and co-workers investigated a series of haloanilinium halides using ^{35}Cl, ^{81}Br and ^{127}I SSNMR, performing a full analysis of the chemical shift, CSA, and quadrupolar parameters. They were able to demonstrate the importance of the EFG in describing the XB environment: in particular, the C_Q of a chloride anion may be used to differentiate between two distinct ion coordination environments.[156]

Nevertheless the examination of ^{35}Cl, ^{81}Br and ^{127}I nuclei through SSNMR experiments remains a formidable task. A possible alternative approach is the use of nuclear quadrupole resonance (NQR) spectroscopy that enables direct detection of quadrupolar nuclei without requiring any external magnetic field. Hence, the absence of a Zeeman interaction ensures that only QI is present, whereas the other interactions are generally too weak to be considered. Pure NQR measurements provide the quadrupolar frequency ν_Q, and the full quadrupole coupling tensor; however, in the case of spin-3/2 nuclei, only a single quadrupolar frequency can be obtained, which is insufficient to determine the values of C_Q and η_Q. Techniques are available to overcome the problem, such as the addition of a weak magnetic field, or 2D nutation experiments.[157,158] NQR studies have shown changes in the quadrupolar frequency upon XB formation.[159–162]

(1,6-diiodododecafluorohexane)] (6A), [(1,2-bis(4-pyridyl)ethane)·(1,4-dibromotetrafluorobenzene)] (2B), [(1,2-bis(4-pyridyl)ethane)·(1,4-diodobenzene)] (3B), [(1,2-bis(4-pyridyl)ethane)·(1,4-diiodotetrafluorobenzene)] (4B), [(1,2-bis(4-pyridyl)ethane)·(1,8-diiodohexadecafluorooctane)], (5B), [(1,2-bis(4-pyridyl)ethane)·(1,6-diiodododecafluorohexane)] (6B). The diamonds represent the two X-bonded co-crystals, the space groups of which are different from the others. The solid line represents the best fit including the latter: $\delta_{iso}{}^{15}N = 233.4R_{XB} + 89.6$, $R^2 = 0.8273$. The dashed line represents the best fit excluding 3 B and 4 A data points: $\delta_{iso}{}^{15}N = 251.9R_{XB} + 74.2$, $R^2 = 0.9650$. Reprinted with permission from P. C. Vioglio, L. Catalano, V. Vasylyeva, C. Nervi, M. R. Chierotti, G. Resnati, R. Gobetto and P. Metrangolo, *Chem.–Eur. J.*, 2016, **22**, 16817–16826, Copyright © 2016 WILEY-VCH Verlag GmbH & Co. KGaA, Weinheim.

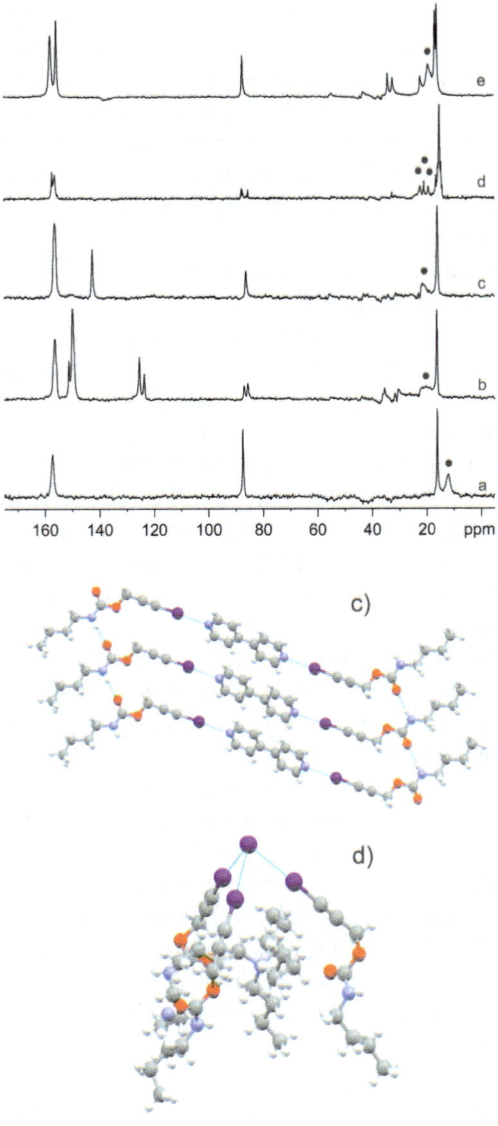

Figure 8.15 Top: ^{13}C (100 MHz) non-quaternary suppression (NQS) spectra of IPBC (a), and its X-bonded adducts with 1,2-bis-4(pyridyl)ethane, (b) 4,4′-bipyridine (c), and tetrabutylammonium iodide (d). C(sp)–I signals are marked with a black circle. Due to the NQS technique the spectra show signals related to quaternary and methyl carbon atoms only. Bottom: ball and stick representation of the crystal packing of (c) and (d). Colour code: grey, C; red, O; blue, N; purple, I; white, H. Reprinted and adapted with permission from M. Baldrighi, G. Cavallo, M. R. Chierotti, R. Gobetto, P. Metrangolo, T. Pilati, G. Resnati and G. Terraneo, *Mol. Pharm.*, 2013, **10**, 1760–1772, Copyright 2013 American Chemical Society.

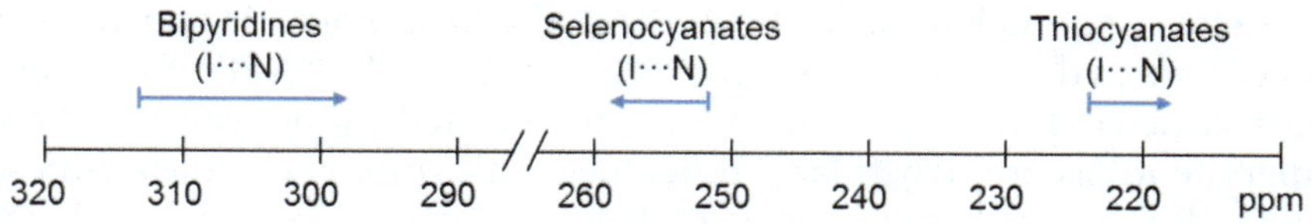

Scheme 8.9 Schematic representation of XB-induced changes in ^{15}N isotropic CSs. The scale is referenced to liquid NH_3 (*i.e.*, $\delta(^{15}N, NH_3) = 0$ ppm).

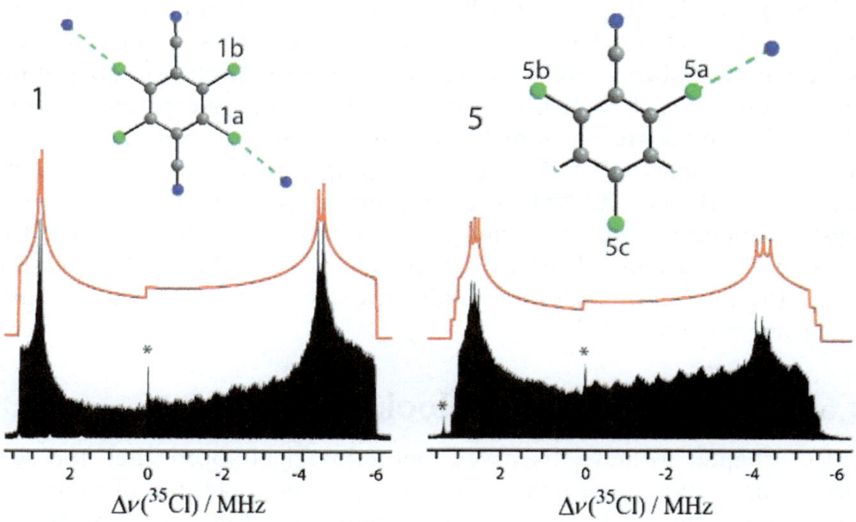

Figure 8.16 ^{35}Cl SSNMR spectra (21.1 T) of tetrachloroterephtalonitrile (1), and 2,4,6-trichlorobenzonitrile (5). Red traces are spectral simulations. The asterisks denote interference from local radio stations and trace NaCl, used as a CS reference. Numbers on atoms denote non-equivalent crystallographic chlorines. Reprinted and adapted with permission from P. M. J. Szell and D. L. Bryce, *J. Phys. Chem. C*, 2016, **120**, 11121–11130, Copyright 2016 American Chemical Society.

8.5 SSNMR and π–π Stacking

In the last two decades π-stacking interactions between aromatic systems have been extensively studied. Indeed, although much less than HB, aromatic π–π interactions and ring-current effects associated with the presence of aromatic moieties represent another important source of structural information. This interaction has been proposed to be a pivotal assembly force in many important supramolecular systems. including proteins and enzyme substrates,[163] nucleic acids[164] and molecular catalysts.[165] Intra- or intermolecular ring currents due to nearby through-space aromatic moieties produce variation in the chemical shift in the "exposed" part of the molecule; in particular a more pronounced effect can be visualized on the small CS range of a proton nucleus. A clear example of ring-current effects in ^1H MAS SSNMR has been reported by Brown *et al.* in the study of the host–guest interaction between a naphthalene-spaced tweezer as the host molecule and 1,4-dicyanobenzene as the aromatic, electron-deficient guest.[166] SSNMR investigation has revealed that the guest remains complexed on the time scale of the NMR experiment. Accordingly, a large difference in the proton chemical shift was detected in the rotor synchronized ^{1}H DQ MAS NMR spectrum. Such splitting, which is not present in solution, is a strong indication of ring-current effects.

Another interesting example is provided by the polycyclic aromatic molecule hexa-*n*-dodecylhexa dodecylhexa-perihexabenzocoronene.[167] By means of a ^{1}H MAS spectrum, it has been demonstrated that this compound forms a columnar mesophase with a very high one-dimensional charge carrier mobility. The observation of distinct aromatic resonances is explained in terms of the differing degrees to which the aromatic protons experience the ring current of adjacent layers. Moreover, the proton–proton proximities identified by the ^{1}H DQ MAS NMR spectrum are in agreement with the known crystal structure of unsubstituted hexabenzocoronenes (HBCs). 2D ^{1}H DQ MAS spectra have also been presented for various other HBC derivatives, whose DQ correlation peaks were indicative of different 3D solid-state packing arrangements.[124,168–170]

8.6 Conclusion and Outlook

In this chapter we have discussed methodological approaches and some recent applications that demonstrate the growing potential of SSNMR for investigating local structure, crystal packing and different types of intermolecular interactions. The coverage presented here provides only a few snapshots of recent developments. The availability of high-field instrumentation and expanding methodological collections of new pulse sequences have increased the potential of SSNMR to also infer structural information about supramolecular arrangements in complex molecular environments. Chemical shifts (both isotropic and anisotropic), and dipolar and quadrupolar couplings are extremely sensitive probes of the chemical microenvironment and can distinguish, for example, direct protonation of an ionizable group or the change in hybridization of a reacting substrate. Combination of X-ray diffraction and NMR methods, often paired with *ab initio* computational analysis, is also able to provide information in microcrystalline or partially ordered samples, which are all rather difficult to investigate *via* the standard diffraction approach. Thanks to a multi-nuclear and multi-parametric approach, SSNMR can really speed up and increase the reliability of the XRPD structure determinations. This approach has been proved successful for a broad variety of materials as well as for the study of active sites in biomolecules. Recently, particular attention has been devoted to the formation of polymorphs, hydrates, solvates, salts and co-crystals where SSNMR unravels the number of independent molecules and their symmetry, proximities and distances, connections and orientation relations; namely all information on local and intermediate length scales. Advancements in signal enhancement methods, such as dynamic nuclear polarization for applications to complex molecular systems, will further increase the potential of SSNMR spectroscopy for opening up a deeper knowledge of the dynamic and often transient nature of molecular interactions in organic, inorganic and biological materials.

References

1. K. J. D. MacKenzie and M. E. Smith, in *Multinuclear Solid-state NMR of Inorganic Materials*, ed. R. W. Cahn, Pergamon, Oxford, 2002.
2. L. J. Mueller and M. F. Dunn, *Acc. Chem. Res.*, 2013, **46**, 2008–2017.
3. C. Martineau, J. Senker and F. Taulelle, *Annu. Rep. NMR Spectrosc.*, 2014, **82**, 1–57.
4. H. C. Hoffmann, M. Debowski, P. Müller, S. Paasch, I. Senkovska, S. Kaskel and E. Brunner, *Materials*, 2012, **5**, 2537–2572.
5. G. Gilli and P. Gilli, *The Nature of the Hydrogen Bond: Outline of a Comprehensive Hydrogen Bond Theory*, Oxford University Press, Oxford, New York, 2009.
6. G. Cavallo, P. Metrangolo, R. Milani, T. Pilati, A. Priimagi, G. Resnati and G. Terraneo, *Chem. Rev.*, 2016, **116**, 2478–2601.
7. M. J. Duer, *Introduction to Solid-state NMR Spectroscopy*, Blackwell Publishing Ltd., Oxford (UK), 2004.
8. D. C. Apperley, R. K. Harris and P. Hodgkinson, *Solid-state NMR: Basic Principles and Practice*, Momentum Press LLC, New York, 2012.
9. Y. Nishiyama, *Solid State Nucl. Magn. Reson.*, 2016, **78**, 24–36.
10. A. E. Bennett, C. M. Rienstra, M. Auger, K. V. Lakshmi and R. G. Griffin, *J. Chem. Phys.*, 1995, **103**, 6951–6958.
11. R. K. Harris, *Solid State Sci.*, 2004, **6**, 1025–1037.
12. W. S. Veeman, *Prog. Nucl. Magn. Reson. Spectrosc.*, 1984, **16**, 193–235.
13. J. Herzfeld and A. E. Berger, *J. Chem. Phys.*, 1980, **73**, 6021–6030.
14. S. M. De Paul, K. Saalwächter, R. Graf and H. W. Spiess, *J. Magn. Reson.*, 2000, **146**, 140–156.
15. Y. Wei, D.-K. Lee, A. E. McDermott and A. Ramamoorthy, *J. Magn. Reson.*, 2002, **158**, 23–35.
16. O. N. Antzutkin, S. C. Shekar and M. H. Levitt, *J. Magn. Reson., Ser. A*, 1995, **115**, 7–19.
17. R. M. Orr and M. J. Duer, *Solid State Nucl. Magn. Reson.*, 2006, **30**, 1–8.
18. R. Witter, S. Hesse and U. Sternberg, *J. Magn. Reson.*, 2003, **161**, 35–42.
19. S.-F. Liu, J.-D. Mao and K. Schmidt-Rohr, *J. Magn. Reson.*, 2002, **155**, 15–28.
20. R. M. Orr, M. J. Duer and S. E. Ashbrook, *J. Magn. Reson.*, 2005, **174**, 301–309.
21. D. V. Dudenko, P. A. Williams, C. E. Hughes, O. N. Antzutkin, S. P. Velaga, S. P. Brown and K. D. M. Harris, *J. Phys. Chem. C*, 2013, **117**, 12258–12265.

22. E. D. L. Smith, R. B. Hammond, M. J. Jones, K. J. Roberts, J. B. O. Mitchell, S. L. Price, R. K. Harris, D. C. Apperley, J. C. Cherryman and R. Docherty, *J. Phys. Chem. B*, 2001, **105**, 5818–5826.
23. F. Franco, M. Baricco, M. R. Chierotti, R. Gobetto and C. Nervi, *J. Phys. Chem. C*, 2013, **117**, 9991–9998.
24. K. D. M. Harris and M. Xu, in *NMR Crystallography*, eds. R. K. Harris, R. E. Wasylishen and M. J. Duer, Wiley-Blackwell, Oxford, 2009, pp. 275–287.
25. P. E. Blöchl, *Phys. Rev. B: Condens. Matter Mater. Phys.*, 1994, **50**, 17953–17979.
26. J. R. Yates, C. J. Pickard and F. Mauri, *Phys. Rev. B: Condens. Matter Mater. Phys.*, 2007, **76**, 024401.
27. C. Bonhomme, C. Gervais, F. Babonneau, C. Coelho, F. Pourpoint, T. Azaïs, S. E. Ashbrook, J. M. Griffin, J. R. Yates, F. Mauri and C. J. Pickard, *Chem. Rev.*, 2012, **112**, 5733–5779.
28. S. E. Ashbrook and D. McKay, *Chem. Commun.*, 2016, **52**, 7186–7204.
29. M. Kibalchenko, D. Lee, L. Shao, M. C. Payne, J. J. Titman and J. R. Yates, *Chem. Phys. Lett.*, 2010, **498**, 270–276.
30. D. V. Dudenko, J. R. Yates, K. D. M. Harris and S. P. Brown, *CrystEngComm*, 2013, **15**, 8797–8807.
31. M. U. Schmidt, J. Brüning, J. Glinnemann, M. W. Hützler, P. Mörschel, S. N. Ivashevskaya, J. van de Streek, D. Braga, L. Maini, M. R. Chierotti and R. Gobetto, *Angew. Chem., Int. Ed.*, 2011, **50**, 7924–7926.
32. A. Bacchi, G. Cantoni, M. R. Chierotti, A. Girlando, R. Gobetto, G. Lapadula, P. Pelagatti, A. Sironi and M. Zecchini, *CrystEngComm*, 2011, **13**, 4365–4375.
33. L. Maini, D. Braga, F. Grepioni, G. I. Lampronti, K. Gaglioti, R. Gobetto and M. R. Chierotti, *CrystEngComm*, 2016, **18**, 4651–4657.
34. M. Baias, J.-N. Dumez, P. H. Svensson, S. Schantz, G. M. Day and L. Emsley, *J. Am. Chem. Soc.*, 2013, **135**, 17501–17507.
35. R. K. Harris, R. E. Wasylishen and M. J. Duer, *NMR Crystallography*, Wiley, 2012.
36. R. K. Harris, P. Hodgkinson, C. J. Pickard, J. R. Yates and V. Zorin, *Magn. Reson. Chem.*, 2007, **45**(suppl. 1), S174–S186.
37. M. R. Chierotti and R. Gobetto, *CrystEngComm*, 2013, **15**, 8599–8612.
38. M. H. Levitt, *J. Chem. Phys.*, 2008, **128**, 124503.
39. T. Gullion and J. Schaefer, *J. Magn. Reson.*, 1989, **81**, 196–200.
40. M. Hong and K. Schmidt-Rohr, *Acc. Chem. Res.*, 2013, **46**, 2154–2163.
41. M. H. Levitt, in *Encyclopedia of Nuclear Magnetic Resonance. Advances in NMR*, eds. D. M. Grant and R. K. Harris, Wiley, 2002, vol. 9, pp. 165–196.
42. R. Tycko and G. Dabbagh, *Chem. Phys. Lett.*, 1990, **173**, 461–465.
43. A. W. Hing, S. Vega and J. Schaefer, *J. Magn. Reson.*, 1992, **96**, 205–209.
44. A. E. Bennett, R. G. Griffin, J. H. Ok and S. Vega, *J. Chem. Phys.*, 1992, **96**, 8624–8627.
45. D. K. Sodickson, M. H. Levitt, S. Vega and R. G. Griffin, *J. Chem. Phys.*, 1993, **98**, 6742–6748.

46. C. P. Grey, W. S. Veeman and A. J. Vega, *J. Chem. Phys.*, 1993, **98**, 7711–7724.
47. E. Hughes, T. Gullion, A. Goldbourt, S. Vega and A. J. Vega, *J. Magn. Reson.*, 2002, **156**, 230–241.
48. P. Thureau, G. Mollica, F. Ziarelli and S. Viel, *Annu. Rep. NMR Spectrosc.*, 2014, **82**, 217–249.
49. D. Braga, G. Palladino, M. Polito, K. Rubini, F. Grepioni, M. R. Chierotti and R. Gobetto, *Chem.–Eur. J.*, 2008, **14**, 10149–10159.
50. E. Vinogradov, P. K. Madhu and S. Vega, *Chem. Phys. Lett.*, 1999, **314**, 443–450.
51. A. Lesage, D. Sakellariou, S. Hediger, B. Eléna, P. Charmont, S. Steuernagel and L. Emsley, *J. Magn. Reson.*, 2003, **163**, 105–113.
52. D. Sakellariou, A. Lesage, P. Hodgkinson and L. Emsley, *Chem. Phys. Lett.*, 2000, **319**, 253–260.
53. R. E. Wasylishen, S. E. Ashbrook and S. Wimperis, *NMR of Quadrupolar Nuclei in Solid Materials*, John Wiley & Sons, 2012.
54. A. Samoson, E. Lippmaa and A. Pines, *Mol. Phys.*, 1988, **65**, 1013–1018.
55. K. T. Mueller, B. Q. Sun, G. C. Chingas, J. W. Zwanziger, T. Terao and A. Pines, *J. Magn. Reson.*, 2011, **213**, 298–315.
56. S. E. Ashbrook and S. Wimperis, *Prog. Nucl. Magn. Reson. Spectrosc.*, 2004, **45**, 53–108.
57. L. Frydman and J. S. Harwood, *J. Am. Chem. Soc.*, 1995, **117**, 5367–5368.
58. F. H. Larsen, H. J. Jakobsen, P. D. Ellis and N. C. Nielsen, *J. Magn. Reson.*, 1998, **131**, 144–147.
59. A. E. Aliev, S. E. Mann, D. Iuga, C. E. Hughes and K. D. M. Harris, *J. Phys. Chem. A*, 2011, **115**, 5568–5578.
60. M. Chan-Huot, S. Wimperis, C. Gervais, G. Bodenhausen and L. Duma, *ChemPhysChem*, 2015, **16**, 204–215.
61. M. Dračínský and P. Hodgkinson, *RSC Adv.*, 2015, **5**, 12300–12310.
62. G. Soda and T. Chiba, *J. Chem. Phys.*, 1969, **50**, 439–455.
63. A. E. Aliev and K. D. M. Harris, in *Supramolecular Assembly via Hydrogen Bonds I*, ed. D. M. P. Mingos, Springer-Verlag Berlin, Berlin, 2004, vol. 108, pp. 1–53.
64. A. Bērziņš and P. Hodgkinson, *Solid State Nucl. Magn. Reson.*, 2015, **65**, 12–20.
65. M. Dračínský, M. Šála and P. Hodgkinson, *CrystEngComm*, 2014, **16**, 6756–6764.
66. G. R. Desiraju, *Angew. Chem., Int. Ed. Engl.*, 1995, **34**, 2311–2327.
67. G. R. Desiraju, *Cryst. Growth Des.*, 2011, **11**, 896–898.
68. D. Braga, F. Grepioni and L. Maini, *Chem. Commun.*, 2010, **46**, 6232–6242.
69. G. A. Jeffrey, *An Introduction to Hydrogen Bonding*, Oxford University Press, Oxford, 1997.
70. T. Steiner, *Angew. Chem., Int. Ed.*, 2002, **41**, 48–76.
71. M. R. Chierotti and R. Gobetto, *Chem. Commun.*, 2008, 1621–1634.
72. S. P. Brown, *Solid State Nucl. Magn. Reson.*, 2012, **41**, 1–27.
73. P. A. Frey, *Magn. Reson. Chem.*, 2001, **39**, S190–S198.

74. U. Sternberg and E. Brunner, *J. Magn. Reson., Ser. A*, 1994, **108**, 142–150.
75. D. Braga, L. Maini, G. de Sanctis, K. Rubini, F. Grepioni, M. R. Chierotti and R. Gobetto, *Chem.–Eur. J.*, 2003, **9**, 5538–5548.
76. M. R. Chierotti, R. Gobetto, C. Nervi, A. Bacchi, P. Pelagatti, V. Colombo and A. Sironi, *Inorg. Chem.*, 2014, **53**, 139–146.
77. G. C. Levy and R. L. Lichter, *Nitrogen-15 Nuclear Magnetic Resonance Spectroscopy*, John Wiley & Sons, New York, 1979.
78. A. M. Orendt and J. C. Facelli, in *Annual Reports on NMR Spectroscopy*, ed. G. A. Webb, Academic Press, 2007, vol. 62, pp. 115–178.
79. H. K. Miah, D. A. Bennett, D. Iuga and J. J. Titman, *J. Magn. Reson.*, 2013, **235**, 1–5.
80. D. H. Brouwer and J. A. Ripmeester, *J. Magn. Reson.*, 2007, **185**, 173–178.
81. M. K. Pandey, M. Malon, A. Ramamoorthy and Y. Nishiyama, *J. Magn. Reson.*, 2015, **250**, 45–54.
82. L. Duma, D. Abergel, P. Tekely and G. Bodenhausen, *Chem. Commun.*, 2008, 2361–2363.
83. C. M. Rohlfing, L. C. Allen and R. Ditchfield, *J. Chem. Phys.*, 1983, **79**, 4958–4966.
84. G. Wu, C. J. Freure and E. Verdurand, *J. Am. Chem. Soc.*, 1998, **120**, 13187–13193.
85. G. Hou, S. Paramasivam, S. Yan, T. Polenova and A. J. Vega, *J. Am. Chem. Soc.*, 2013, **135**, 1358–1368.
86. K. Loth, P. Pelupessy and G. Bodenhausen, *J. Am. Chem. Soc.*, 2005, **127**, 6062–6068.
87. L. Yao, A. Grishaev, G. Cornilescu and A. Bax, *J. Am. Chem. Soc.*, 2010, **132**, 10866–10875.
88. R. E. Stark, L. W. Jelinski, D. J. Ruben, D. A. Torchia and R. G. Griffin, *J. Magn. Reson.*, 1983, **55**, 266–273.
89. N. Asakawa, S. Kuroki, H. Kurosu, I. Ando, A. Shoji and T. Ozaki, *J. Am. Chem. Soc.*, 1992, **114**, 3261–3265.
90. T. Kameda, N. Takeda, S. Kuroki, H. Kurosu, S. Ando, I. Ando, A. Shoji and T. Ozaki, *J. Mol. Struct.*, 1996, **384**, 17–23.
91. N. Takeda, S. Kuroki, H. Kurosu and I. Ando, *Biopolymers*, 1999, **50**, 61–69.
92. N. Tjandra and A. Bax, *J. Am. Chem. Soc.*, 1997, **119**, 9576–9577.
93. T. Kameda and I. Ando, *J. Mol. Struct.*, 1997, **412**, 197–203.
94. Y. Wei, D.-K. Lee and A. Ramamoorthy, *J. Am. Chem. Soc.*, 2001, **123**, 6118–6126.
95. B. J. Wylie and C. M. Rienstra, *J. Chem. Phys.*, 2008, **128**, 052207.
96. Z. Gu and A. McDermott, *J. Am. Chem. Soc.*, 1993, **115**, 4282–4285.
97. M. Ilczyszyn, D. Godzisz, M. M. Ilczyszyn and K. Mierzwicki, *Chem. Phys.*, 2006, **323**, 231–242.
98. R. Gobetto, C. Nervi, E. Valfre, M. R. Chierotti, D. Braga, L. Maini, F. Grepioni, R. K. Harris and P. Y. Ghi, *Chem. Mater.*, 2005, **17**, 1457–1466.
99. R. Gobetto, C. Nervi, M. R. Chierotti, D. Braga, L. Maini, F. Grepioni, R. K. Harris and P. Hodgkinson, *Chem.–Eur. J.*, 2005, **11**, 7461–7471.

100. M. Singh and J. Schaefer, *J. Am. Chem. Soc.*, 2011, **133**, 2626–2631.
101. K. Schmidt-Rohr, W. Hu and N. Zumbulyadis, *Science*, 1998, **280**, 714–717.
102. T. Asakura, T. Konakazawa, M. Demura, T. Ito and Y. Maruhashi, *Polymer*, 1996, **37**, 1965–1973.
103. B. Fortier-McGill, V. Toader and L. Reven, *Macromolecules*, 2012, **45**, 6015–6026.
104. K. Schmidt-Rohr and H. W. Spiess, *Multidimensional Solid-state NMR and Polymers*, Elsevier, New York, 1994.
105. J. D. van Beek, L. Beaulieu, H. Schäfer, M. Demura, T. Asakura and B. H. Meier, *Nature*, 2000, **405**, 1077–1079.
106. C. M. Gowda, E. R. H. van Eck, A. M. van Buul, E. Schwartz, G. W. P. van Pruissen, J. J. L. M. Cornelissen, A. E. Rowan, R. J. M. Nolte and A. P. M. Kentgens, *Macromolecules*, 2012, **45**, 2209–2218.
107. W. L. Jarrett, G. C. Johnson and L. J. Mathias, *J. Magn. Reson., Ser. A*, 1995, **116**, 156–160.
108. K. L. Anderson-Altmann, C. G. Phung, S. Mavromoustakos, Z. Zheng, J. C. Facelli, C. D. Poulter and D. M. Grant, *J. Phys. Chem.*, 1995, **99**, 10454–10458.
109. J. Z. Hu, J. C. Facelli, D. W. Alderman, R. J. Pugmire and D. M. Grant, *J. Am. Chem. Soc.*, 1998, **120**, 9863–9869.
110. Y. Wei, A. C. de Dios and A. E. McDermott, *J. Am. Chem. Soc.*, 1999, **121**, 10389–10394.
111. I. Schnell and H. W. Spiess, *J. Magn. Reson.*, 2001, **151**, 153–227.
112. S. P. Brown, *Prog. Nucl. Magn. Reson. Spectrosc.*, 2007, **50**, 199–251.
113. S. Dusold and A. Sebald, *Annu. Rep. NMR Spectrosc.*, 2000, **41**, 185–264.
114. G. Morris, in *Encyclopedia of Nuclear Magnetic Resonance. Advances in NMR*, ed. D. M. Grant and R. K. Harris, Wiley, 2002, vol. 9, pp. 35–44.
115. W. Sommer, J. Gottwald, D. E. Demco and H. W. Spiess, *J. Magn. Reson., Ser. A*, 1995, **113**, 131–134.
116. I. Schnell, S. P. Brown, H. Y. Low, H. Ishida and H. W. Spiess, *J. Am. Chem. Soc.*, 1998, **120**, 11784–11795.
117. K. Schmidt-Rohr, *Macromolecules*, 1996, **29**, 3975–3981.
118. Ü. Akbey, S. Granados-Focil, E. B. Coughlin, R. Graf and H. W. Spiess, *J. Phys. Chem. B*, 2009, **113**, 9151–9160.
119. Y. J. Lee, B. Bingöl, T. Murakhtina, D. Sebastiani, W. H. Meyer, G. Wegner and H. W. Spiess, *J. Phys. Chem. B*, 2007, **111**, 9711–9721.
120. S. R. Benhabbour, R. P. Chapman, G. Scharfenberger, W. H. Meyer and G. R. Goward, *Chem. Mater.*, 2005, **17**, 1605–1612.
121. M. R. Chierotti, L. Ferrero, N. Garino, R. Gobetto, L. Pellegrino, D. Braga, F. Grepioni and L. Maini, *Chem.–Eur. J.*, 2010, **16**, 4347–4358.
122. C. Ochsenfeld, F. Koziol, S. P. Brown, T. Schaller, U. P. Seelbach and F.-G. Klärner, *Solid State Nucl. Magn. Reson.*, 2002, **22**, 128–153.
123. T. Schaller, U. P. Büchele, F.-G. Klärner, D. Bläser, R. Boese, S. P. Brown, H. W. Spiess, F. Koziol, J. Kussmann and C. Ochsenfeld, *J. Am. Chem. Soc.*, 2007, **129**, 1293–1303.

124. S. P. Brown, I. Schnell, J. D. Brand, K. Müllen and H. W. Spiess, *J. Mol. Struct.*, 2000, **521**, 179–195.
125. P. Avenier, A. Lesage, M. Taoufik, A. Baudouin, A. De Mallmann, S. Fiddy, M. Vautier, L. Veyre, J.-M. Basset, L. Emsley and E. A. Quadrelli, *J. Am. Chem. Soc.*, 2007, **129**, 176–186.
126. J. M. Griffin, D. R. Martin and S. P. Brown, *Angew. Chem., Int. Ed.*, 2007, **46**, 8036–8038.
127. S. P. Brown, X. X. Zhu, K. Saalwächter and H. W. Spiess, *J. Am. Chem. Soc.*, 2001, **123**, 4275–4285.
128. S. D. Gumbert, M. Koerbitzer, E. Alig, M. U. Schmidt, M. R. Chierotti, R. Gobetto, X. Li and J. van de Streek, *Dyes Pigm.*, 2016, **131**, 364–372.
129. S. Cavadini, S. Antonijevic, A. Lupulescu and G. Bodenhausen, *J. Magn. Reson.*, 2006, **182**, 168–172.
130. J. R. Patel, R. A. Carlton, T. E. Needham, C. O. Chichester and F. G. Vogt, *Int. J. Pharm.*, 2012, **436**, 685–706.
131. M. Hong, K. J. Fritzsching and J. K. Williams, *J. Am. Chem. Soc.*, 2012, **134**, 14753–14755.
132. V. Agarwal, R. Linser, U. Fink, K. Faelber and B. Reif, *J. Am. Chem. Soc.*, 2010, **132**, 3187–3195.
133. A. K. Chattah, R. Zhang, K. H. Mroue, L. Y. Pfund, M. R. Longhi, A. Ramamoorthy and C. Garnero, *Mol. Pharm.*, 2015, **12**, 731–741.
134. A. Wong, K. J. Pike, R. Jenkins, G. J. Clarkson, T. Anupõld, A. P. Howes, D. H. G. Crout, A. Samoson, R. Dupree and M. E. Smith, *J. Phys. Chem. A*, 2006, **110**, 1824–1835.
135. F. G. Vogt, H. Yin, R. G. Forcino and L. Wu, *Mol. Pharm.*, 2013, **10**, 3433–3446.
136. A. W. Tang, X. Kong, V. Terskikh and G. Wu, *J. Phys. Chem. B*, 2016, **120**, 11142–11150.
137. G. R. Desiraju, P. S. Ho, L. Kloo, A. C. Legon, R. Marquardt, P. Metrangolo, P. Politzer, G. Resnati and K. Rissanen, *Pure Appl. Chem.*, 2013, **85**, 1711–1713.
138. P. Politzer, J. S. Murray and T. Clark, *Phys. Chem. Chem. Phys.*, 2010, **12**, 7748–7757.
139. L. C. Gilday, S. W. Robinson, T. A. Barendt, M. J. Langton, B. R. Mullaney and P. D. Beer, *Chem. Rev.*, 2015, **115**, 7118–7195.
140. P. C. Vioglio, M. R. Chierotti and R. Gobetto, *CrystEngComm*, 2016, **18**, 9173–9184.
141. C. M. Widdifield, G. Cavallo, G. A. Facey, T. Pilati, J. Lin, P. Metrangolo, G. Resnati and D. L. Bryce, *Chem.–Eur. J.*, 2013, **19**, 11949–11962.
142. P. C. Vioglio, L. Catalano, V. Vasylyeva, C. Nervi, M. R. Chierotti, G. Resnati, R. Gobetto and P. Metrangolo, *Chem.–Eur. J.*, 2016, **22**, 16817–16826.
143. J. Viger-Gravel, S. Leclerc, I. Korobkov and D. L. Bryce, *CrystEngComm*, 2013, **15**, 3168–3177.

144. J. Viger-Gravel, S. Leclerc, I. Korobkov and D. L. Bryce, *J. Am. Chem. Soc.*, 2014, **136**, 6929–6942.
145. R. Glaser, N. Chen, H. Wu, N. Knotts and M. Kaupp, *J. Am. Chem. Soc.*, 2004, **126**, 4412–4419.
146. M. Kaupp, O. L. Malkina, V. G. Malkin and P. Pyykkö, *Chem.–Eur. J.*, 1998, **4**, 118–126.
147. M. Baldrighi, G. Cavallo, M. R. Chierotti, R. Gobetto, P. Metrangolo, T. Pilati, G. Resnati and G. Terraneo, *Mol. Pharm.*, 2013, **10**, 1760–1772.
148. M. Baldrighi, D. Bartesaghi, G. Cavallo, M. R. Chierotti, R. Gobetto, P. Metrangolo, T. Pilati, G. Resnati and G. Terraneo, *CrystEngComm*, 2014, **16**, 5897–5904.
149. F. Marchetti, J. Palmucci, C. Pettinari, R. Pettinari, F. Condello, S. Ferraro, M. Marangoni, A. Crispini, S. Scuri, I. Grappasonni, M. Cocchioni, M. Nabissi, M. R. Chierotti and R. Gobetto, *Chem.–Eur. J.*, 2015, **21**, 836–850.
150. R. Pettinari, C. Pettinari, F. Marchetti, R. Gobetto, C. Nervi, M. R. Chierotti, E. J. Chan, B. W. Skelton and A. H. White, *Inorg. Chem.*, 2010, **49**, 11205–11215.
151. R. Pettinari, F. Marchetti, C. Pettinari, F. Condello, B. W. Skelton, A. H. White, M. R. Chierotti and R. Gobetto, *Dalton Trans.*, 2016, **45**, 3974–3982.
152. F. Marchetti, C. Pettinari, R. Pettinari, A. Cingolani, R. Gobetto, M. R. Chierotti, A. Drozdov and S. I. Troyanov, *Inorg. Chem.*, 2006, **45**, 3074–3085.
153. M. Weingarth, N. Raouafi, B. Jouvelet, L. Duma, G. Bodenhausen, K. Boujlel, B. Schöllhorn and P. Tekely, *Chem. Commun.*, 2008, 5981–5983.
154. P. M. J. Szell and D. L. Bryce, *Annu. Rep. NMR Spectrosc.*, 2015, **84**, 115–162.
155. P. M. J. Szell and D. L. Bryce, *J. Phys. Chem. C*, 2016, **120**, 11121–11130.
156. R. J. Attrell, C. M. Widdifield, I. Korobkov and D. L. Bryce, *Cryst. Growth Des.*, 2012, **12**, 1641–1653.
157. R. Ramachandran and E. Oldfield, *J. Chem. Phys.*, 1984, **80**, 674–677.
158. G. S. Harbison, A. Slokenbergs and T. M. Barbara, *J. Chem. Phys.*, 1989, **90**, 5292–5298.
159. G. K. Semin, T. A. Babushkina, S. P. Khrlakyan, E. Y. Pervova, V. V. Shokina and I. L. Knunyants, *Theor. Exp. Chem.*, 1971, **4**, 179–181.
160. G. A. Bowmaker and S. Hacobian, *Aust. J. Chem.*, 1968, **21**, 551–564.
161. G. A. Bowmaker and S. Hacobian, *Aust. J. Chem.*, 1969, **22**, 2047–2059.
162. G. A. Bowmaker, *J. Chem. Soc., Faraday Trans. 2*, 1976, **72**, 1964–1969.
163. S. K. Burley and G. A. Petsko, *Science*, 1985, **229**, 23–28.
164. K. M. Guckian, B. A. Schweitzer, R. X.-F. Ren, C. J. Sheils, P. L. Paris, D. C. Tahmassebi and E. T. Kool, *J. Am. Chem. Soc.*, 1996, **118**, 8182–8183.
165. E. J. Corey and M. C. Noe, *J. Am. Chem. Soc.*, 1996, **118**, 319–329.

166. S. P. Brown, T. Schaller, U. P. Seelbach, F. Koziol, C. Ochsenfeld, F.-G. Klärner and H. W. Spiess, *Angew. Chem., Int. Ed.*, 2001, **40**, 717–720.
167. S. P. Brown, I. Schnell, J. D. Brand, K. Müllen and H. W. Spiess, *J. Am. Chem. Soc.*, 1999, **121**, 6712–6718.
168. C. Ochsenfeld, *Phys. Chem. Chem. Phys.*, 2000, **2**, 2153–2159.
169. D. Wasserfallen, I. Fischbach, N. Chebotareva, M. Kastler, W. Pisula, F. Jäckel, M. D. Watson, I. Schnell, J. P. Rabe, H. W. Spiess and K. Müllen, *Adv. Funct. Mater.*, 2005, **15**, 1585–1594.
170. W. Pisula, M. Kastler, D. Wasserfallen, M. Mondeshki, J. Piris, I. Schnell and K. Müllen, *Chem. Mater.*, 2006, **18**, 3634–3640.

CHAPTER 9

Quantitative Analysis of Weak Non-covalent σ-Hole and π-Hole Interactions

SAIKAT KUMAR SETH*[a], ANTONIO BAUZÁ[b] AND ANTONIO FRONTERA*[b]

[a]Department of Physics, Jadavpur University, Kolkata, 700032, India;
[b]Department of Chemistry, Universitat de les Illes Balears, Crta. De Valldemossa km 7.5, 07122 Palma de Mallorca, Baleares, Spain
*E-mail: saikatim@yahoo.co.in, skseth@phys.jdvu.ac.in, toni.frontera@uib.es

9.1 Introduction and Historical Perspective

Supramolecular chemistry can be regarded as one of the most multidisciplinary fields of research. The rapid development of this discipline has been related to the increasing efficiency and success in the synthesis of structures of different sizes, shapes and functionalities. Supramolecular chemists rely on the comprehension of the non-covalent forces that are the basis of highly specific recognition, transport and regulation mechanisms. Many biological and chemical processes are often controlled by an intricate combination of non-covalent interactions.[1] This is exemplified by the foundation of the life process itself, which is considered as the ultimate expression of function. In particular, in host-guest chemistry, interactions between targeted guests and rationally designed receptors govern the formation of assemblies of different

Monographs in Supramolecular Chemistry No. 26
Understanding Intermolecular Interactions in the Solid State: Approaches and Techniques
Edited by Deepak Chopra

Published by the Royal Society of Chemistry, www.rsc.org

shapes, sizes and affinities.[2] Therefore, a proper description of the relevant interactions between molecules is necessary for the understanding of the various supramolecular processes that relies on strong, directional interactions such as hydrogen bonding and halogen bonding, and less directional forces like ion pairing.

The first halogen-bonded complex was synthesized by Colin ($I_2 \cdots NH_3$) approximately two centuries ago. In 1814, Colin reported the formation of a blue-black color upon combination of iodine with amilose[3] and of a liquid upon mixing dry iodine with gaseous ammonia.[4] The molecular composition of this liquid was established fifty years later by Guthrie.[5] The formation of such a complex was not understood until the discovery of charge–transfer interactions by. Mulliken[6] and Hassel.[7] After Colin's discovery, in 1819 Pelletier and Caventou[8] reported experimental evidence showing the ability of I_2 to interact in an attractive fashion with I^- to form the I_3^- anion. Moreover, in 1883, Rhoussopoulos[9] reported the formation of an adduct between quinoline and iodoform, evidencing that CHI_3, exhibits similar behavior to I_2 as regards forming adducts with Lewis bases. Later on, Remsen and Norris, who described the formation of different 1:1 dimers by Br_2 and Cl_2 with various amines, also reported similar behavior for bromine and chlorine.[10]

In the 20th century, the research activity in this field increased exponentially. There has been a phenomenal increase in the number of experimental studies and phenomena that depict the role of halogen bonding interactions. It was of extreme relevance to explain the behavior of I_2 when it was dissolved in organic solvents, since solutions of different colors were formed.[11] In 1948, Benesi and Hildebrand explained these phenomena as cases of intermolecular donor–acceptor interactions.[12] Some years later Mulliken[13] rationalized them as a subclass of the family of electron donor–acceptor molecular complexes. X-ray analyses reported by Hassel in the 1950s were important to identify the structural features of the intermolecular interaction occurring in several complexes formed by dihalogens or halocarbons with electron donor molecules.[14] In the next decades, the directionality of the interaction was confirmed,[15] and Kochi showed that π-donating moieties are also able to form complexes with halocarbons.[16] Bent gave a comprehensive discussion of the X-ray structures known at that time in 1968, giving the geometric features of the interactions (short interatomic distances and high directionality).[17] These geometric features were confirmed two decades later through statistical analysis of the structures in the Cambridge Structural Database (CSD) by Parthasarathy and Murray-Rust.[18] Legon and co-workers[19] also undertook a systematic analysis of halogen-bonded adducts formed in the gas phase *via* microwave spectroscopy and their results revealed close similarity with the results in condensed phases. Moreover, in the late 1980s, a number of crystallographic works were devoted towards the analysis of non-spherical effective shapes of atoms in crystals, termed as "polar flattening".[20] By the examination of the nearest non-bonded interatomic distances, it was shown that where the two atoms are each covalently bound to only one other atom they exhibit non-spherical effective shapes. Such atoms behave as if flattened at their poles.

An important contribution towards the understanding of the halogen bond (XB) was obtained by computational studies on the distribution of the electron density in halogen atoms in the early 1990s. The studies by Politzer and Murray were very significant as they demonstrated the anisotropic charge distribution on halogen atoms forming one covalent bond[21] and paved the way to the definition of the "σ-hole": a region of depleted and often positive electrostatic potential on the surface of the halogen atom. After that, "halogen bonding" was used to unify the concept of donor–acceptor interactions described so far, between a variety of donors (lone pair, anion, π-system) and halogens,[22] and boosted the interest of the scientific community in this particular topic and the number of papers on XB increased rapidly.[23]

The σ-hole concept was introduced by Clark *et al.*[24] in the context of halogen bonding. More recently, in general σ-hole interactions, including covalently bonded atoms of groups IV–VI, have been also considered as an important addition to the cluster of well-known directional non-covalent interactions.[25,26] As a natural consequence of the success of halogen bonding, the usefulness of the σ-hole bonding interactions involving these atoms in crystal engineering and supramolecular chemistry has been recently investigated, starting from a series of computational studies.[27] It was shown that covalently bonded atoms of groups IV–VI can possess regions (from two to four) of positive electrostatic potential on the extension of their covalent bonds attributed to the anisotropy of the distribution of the electronic density.[28] More recently, it has been shown that covalently bonded atoms of group VIII can also participate in σ-hole bonding.[29] The size and magnitude of these σ-hole potentials are governed by the same factors as for the halogens, as discussed in this chapter.

An important and relatively unexplored group of non-covalent interactions can be classified as π-hole bonding. A π-hole is a region of positive electrostatic potential involving π-orbitals that are perpendicular to a portion of a molecular framework. In the case that this portion is an aromatic π-system substituted with electron withdrawing groups (characterized by a positive electrostatic surface potential region perpendicular to the aromatic molecular plane), the interaction is known as either lone pair-π[30] or anion-π[31] depending on the nature of the electron donor. As a matter of fact, the anion-π interaction[32] has been investigated by a plethora of theoretical studies[33] in addition to an increasing number of experimental investigations.[34] Anion-π interactions have become important in fields as diverse as medicine, environmental chemistry and biochemical processes.[31a,35] Furthermore, their application in the design of selective anion receptors and transport channels confirms their significance in the field of supramolecular chemistry.[36]

The identification of π-hole interactions in crystal structures was carried out in 1973 by Bürgi and Dunitz in a series of studies wherein they analyzed the trajectory along which a nucleophile attacks the π-hole present in a

C=O group.[37] In a closely related study, the ability of guanosine to act as a π-hole was pointed out by Egli and co-workers in 1995, when they studied the crystal structure of Z-DNA.[38] Recently, the study of π-hole interactions has been extended to acyl carbons,[39] sulphur containing moieties,[40] RNO_2 molecules,[41] XCN, XZO_2 (X = halogen, Z = pnicogen).[42] Finally, it is also noteworthy to mention the studies carried out by Grabowski and coworkers on the physical nature of π-hole interactions involving triel atoms.[43] The groups of Liu[44] and Yourdkhani[45] have analyzed the interplay between this novel interaction and pnicogen and tetrel bonding interactions.

9.2 Nature of σ-Hole and Π-Hole Interactions

9.2.1 σ-Hole Interactions

A σ-hole can be viewed as the region of positive electrostatic potential located along the vector of a covalent bond.[46] This concept of a σ-hole was originally used to explain the halogen bonding wherein the electropositive potential is found on the halogen atom at the end of the X–Hlg vector (X can be any atom, although it is typically carbon).[25a,47] Such electronic anisotropy is not unique to atoms of the halogen family due to the discovery of σ-holes on covalently bonded atoms of the pnicogen (Pn) and chalcogen (Ch) families.[27,48]

In analogy to hydrogen bonding, the atom on which the electropositive potential is located is commonly referred to as the donor atom.[49,50] Hence, σ-hole interactions can be written as X–D···A (where X can be any atom, D = donor atom and A = acceptor moiety; *i.e.* similar to hydrogen bonding).[49a] The magnitude of a σ-hole depends on two main factors: it becomes more positive when X–D is more polarizable[51] and when X is more electron withdrawing.[25b,28b,52,53]

Any enhancement of the polarizability increases the magnitude of the σ-hole and strengthens the resulting σ-hole complex. The molecular electrostatic potential surface plots of several YF_2 molecules (Y = O, S and Se) and F_2C=Se are shown in Figure 9.1. Two σ-holes are present for the YF_2 molecule on the extension of each covalent Y–F bond. The molecular electrostatic potential (MEP) value at the σ-hole of the SeF_2 molecule is the highest on account of the greater polarizability of the C–Se bond. Interestingly, in F_2C=Se

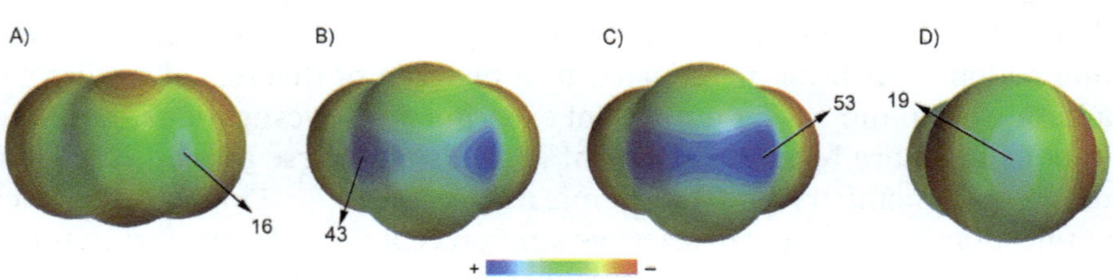

Figure 9.1 Molecular electrostatic potential of representative chalogen bonding donor molecules at the B3LYP/6-31+G* level of theory. Energies at the σ-holes are given in kcal mol^{-1}. (A) O_2F, (B) SF_2, (C) SeF_2, (D) F_2C=Se.

a single σ-hole is observed along the C=Se bond (in addition to a π-hole above and below the C atom). Chalcogen bonding interactions involving a single σ-hole have been studied in C=S/Se and P=S/Se derivatives.[54]

The ability of a σ-hole to interact with an electron-rich atom depends not only on the magnitude of the electropositive potential, but also on its steric environment. Indeed, while the value of the MEP at the σ-hole is similar in representative compounds of the same row (see Figure 9.2, BrF, SeF_2, AsF_3 and GeF_4), the σ-holes are sterically more accessible with lower valence, *i.e.* BrF > SeF_2 > AsF_3 > GeF_4. This steric hindrance is particularly relevant in the case of carbon, which is the smallest tetrel atom. Thus, in a sp^3 hybridized electron-deficient C atom, such as C_2F_6 (see Figure 9.3, left) there is only a very limited space available for the electron-rich guest molecule to nest itself. In addition, the negative electrostatic potential at the van der Waals surface of the interacting guest would be very close to the possible negative

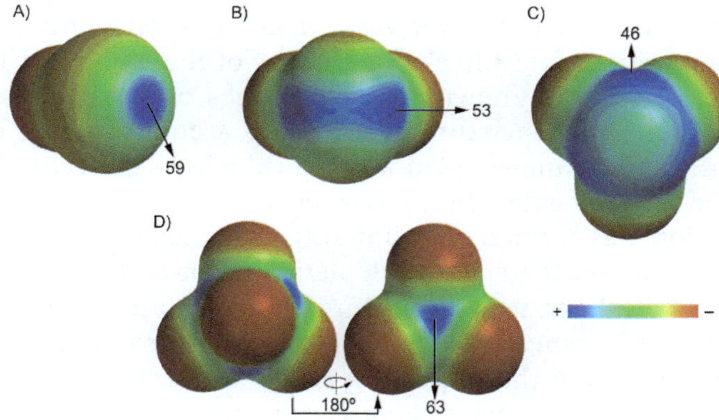

Figure 9.2 Molecular electrostatic potential surfaces of representative halogen. (A) BrF, chalcogen, (B) SeF_2, pnicogen, (C) AsF_3 and tetrel, (D) GeF_4 bond donors at the B3LYP/6-31+G* level of theory. Energies at the σ-holes are given in kcal mol^{-1}.

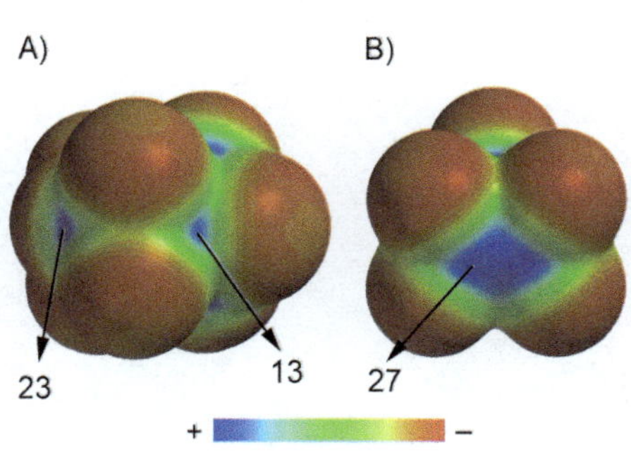

Figure 9.3 Molecular electrostatic potential surface of (A) hexafluoroethane and (B) hexafluorocyclopropane at the B3LYP/6-31+G* level of theory. Energies at the σ-holes are given in kcal mol^{-1}.

belt of the electrostatic potential of (in this example) F. In small rings, such as hexafluorocyclopropane (C_3F_6, see Figure 9.3, right), the σ-hole on C is more exposed due to the small FCCF dihedral angle. In this context, experimental evidence of σ-hole interactions in CF_3 groups was reported by Hathwar *et al.*[55] Moreover, Esterhuysen *et al.* have shown that the trifluoromethyl groups can act as both electrophiles and nucleophiles to form non-covalent interactions.[56]

It is worth pointing out that in some cases the σ-bond donor can also bear lone pair electrons, meaning that it can act concurrently as the σ-bond donor and acceptor. Such interactions have been pointed out for atoms of groups V–VII (in particular sulfur and phosphorus),[28c] but could in principle also occur with tetrel atoms (*i.e.* a carbene).

9.2.2 π-Hole Interactions

A π-hole is a region of positive electrostatic potential typically located perpendicular to a molecular framework. Depending on the nature of the electron-rich partner and the characteristics of the molecular fragment, π-hole interactions are known by other names or sub-classes. Politzer *et al.*[25a] used the nomenclature of π-hole to account for the depletion of charge density on unoccupied π-type orbitals on the central atom in SO_2 and similar molecules. In the particular case of group III elements, the interaction was termed triel bonding by Grabowski.[43b] In this type of π-hole interaction the electron depletion is located on a single atom (similar to σ-holes) and usually appears in the literature using the name of the group, for instance pnicogen/chalcogen π-hole interaction. Several studies have appeared that report on π-hole interactions involving SO_3, XNO_2, $XONO_2$, *etc.*[40–42] Shown in Figures 9.4 and 9.5 are the MEP surfaces of four representative planar molecules where the π-hole is located over the central atom belonging to the second (F_2CO and FNO_2 in Figure 9.4) or third (F_2**Si**O and FPO_2 in Figure 9.5) row of the periodic table. The MEP at the π-hole is significantly more positive as the bond donor atom becomes

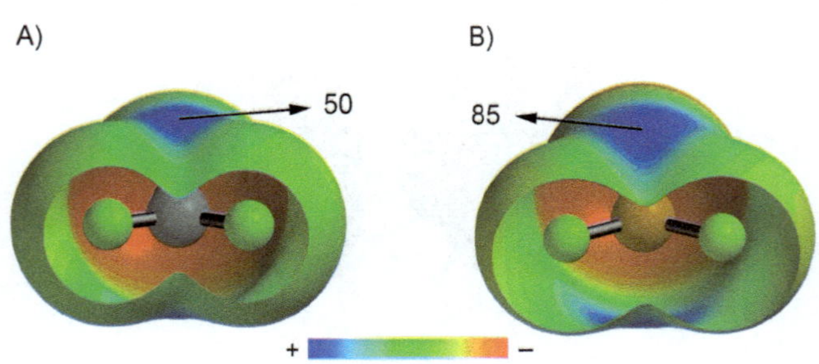

Figure 9.4 Molecular electrostatic potential of representative tetrel (A) F_2CO and (B) F_2SiO bond donors at the B3LYP/6-31+G* level of theory. Energies at the π-holes are given in kcal mol^{-1}.

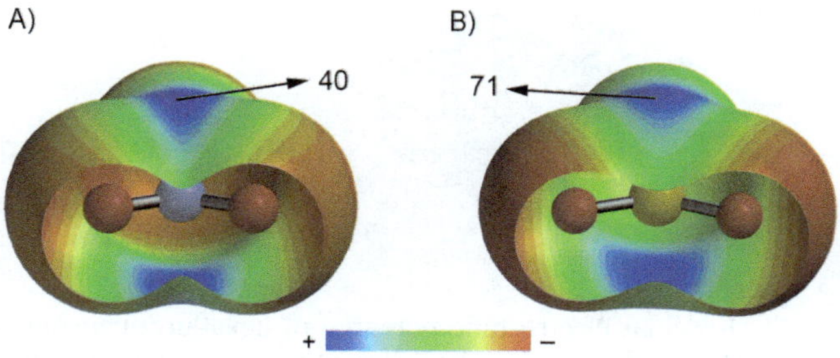

Figure 9.5 Molecular electrostatic potential of representative pnicogen (A) FNO_2 and (B) FPO_2 bond donors at the B3LYP/6-31+G* level of theory. Energies at the π-holes are given in kcal mol^{-1}.

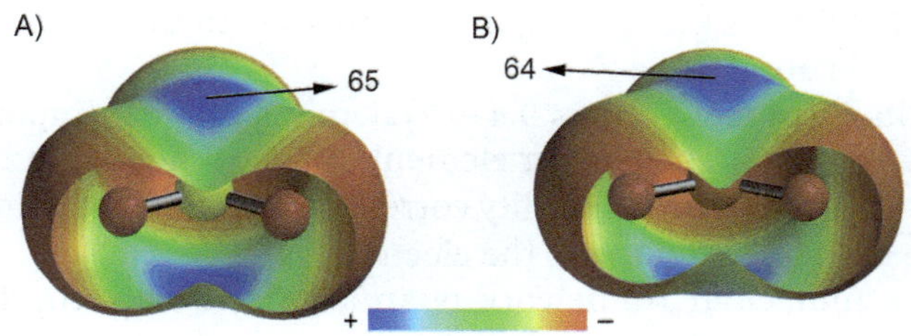

Figure 9.6 Molecular electrostatic potential of representative chalcogen (A) SO_3 and (B) SeO_3 bond donors at the B3LYP/6-31+G* level of theory. Energies at the π-holes are given in kcal mol^{-1}.

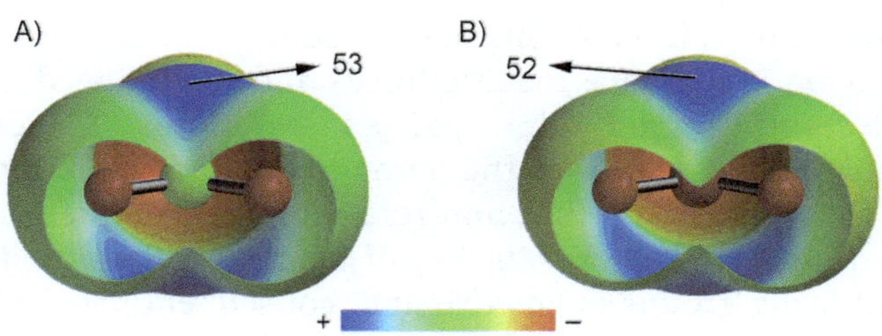

Figure 9.7 Molecular electrostatic potential of representative halogen (A) $FClO_2$ and (B) $FBrO_2$ bond donors at the B3LYP/6-31+G* level of theory. Energies at the π-holes are given in kcal mol^{-1}.

increasingly polarizable (heavier atoms). This is especially important on going from elements of row 2 to elements of rows 3–5. In addition, the π-hole is more accessible (more extended) in heavier (typically also larger) atoms (see Figures 9.6 and 9.7).

Since the second row atoms of groups VI and VII do not participate in π-hole bonding, the MEP surfaces of four representative π-hole donor molecules of rows 3 and 4 are represented in Figures 9.6 and 9.7. It can be observed that all compounds exhibit similar electrostatic potential values, indicating that the ability of S/Cl and Se/Br to participate in π-hole bonding interactions

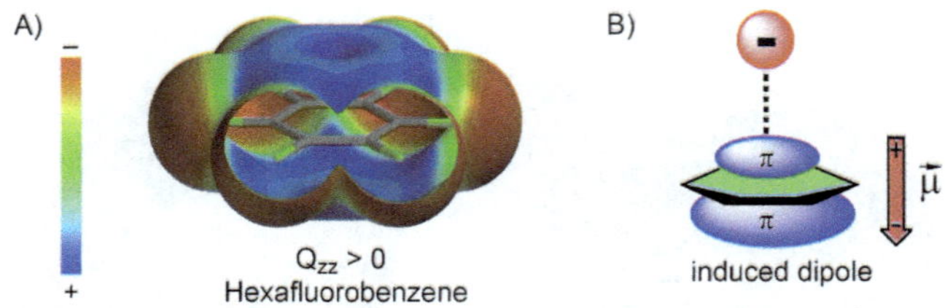

Figure 9.8 (A) Molecular electrostatic potential of hexafluorobenzene at the B3LYP/6-31+G* level of theory. (B) Schematic representation of the ion-induced dipole.

is similar. Interestingly, the results derived from the MEP values shown in Figures 9.4–9.7 clearly indicate that the π-hole potentials decrease on going from Si to Cl (85 to 53 kcal mol^{-1}), in agreement with the atomic polarizability of the π-hole donor.

The results obtained (Figures 9.4–9.7) demonstrate the important role of the electronegativity of the donor element, having a π-hole, influencing the MEP value. In fact, the polarizability correlates with the ionization potential, and thus to a certain extent with the electronegativity.

When the molecular framework is an aromatic ring, the interactions are commonly referred to as anion-π or lone pair-π interactions (depending on whether the electron-rich guest is anionic or neutral). Anion/lone pair-π interactions have attracted much attention in recent years and several reviews have been published describing their physical nature.[30–32,57] In addition, several energy partitioning schemes have been used to decompose the total interaction energy into individual terms.[58] The general conclusion is that electrostatic forces and ion-induced polarization dictate the stability of anion/lone pair-π interactions.[59] The electrostatic term is explained by means of the permanent quadrupole moment (Q_{zz}) of the arene, which is the first non-zero multipole moment in symmetric arenes (see Figure 9.8A). The Q_{zz} of benzene substituted with electron withdrawing groups is positive and, consequently, the electrostatic charge–quadrupole interaction between an anion and an aromatic ring is attractive.

On the other hand, the ion-induced polarization of the π-system by the anion is also important, inducing a dipole in it (see Figure 9.8B). Therefore, the polarization contribution to the total interaction energy derives from the interaction of the anion with the induced dipole.[58,59] The electrostatic term is governed by the magnitude of the Q_{zz} and the polarization term by the magnitude of the molecular polarizability parallel to the main symmetry axis (denoted as $\alpha_{||}$) in symmetric arenes (or perpendicular to the ring plane in asymmetric arenes, denoted as α_{zz}), which are inherent properties of the π-system. In contrast to the intuitive view where the substituent induces changes in the aryl π-system, Wheeler and Houk proposed a model where substituent effects in these systems can be attributed mainly to direct interactions between the anion and local C–Y dipoles.[60] A similar explanation has been provided for π-acidic azines.[61]

The comprehension that electrostatic and induced polarization governs anion/lone pair-π interactions is essential towards the understanding of the dual binding mode exhibited by arenes with small quadrupole moments.[62] For instance, 1,3,5-trifluorobenzene (Q_{zz} = 0.57 B) and *s*-triazine (Q_{zz} = 0.90 B) are able to interact with both anions and cations because the electrostatic term is negligible and the interaction is thus dominated by the polarization term, which is always attractive. Similarly, the anion-π interaction energy of benzene with chloride is very small (2.4 kcal mol^{-1}), due to the compensating effect between the electrostatic (unfavorable) and ion-induced polarization (favorable) forces that approximately cancel each other out.[63]

9.3 Hirshfeld Surface Technique

9.3.1 Crystal Engineering and Models to Describe Crystal Packing

The term "crystal engineering" was defined by Desiraju as "the understanding of intermolecular interactions in the context of crystal packing and the utilization of such understanding in the design of new solids with desirable physical and chemical properties".[64] A systematic and thorough analysis of different factors that direct the packing geometry of solid crystalline materials has been investigated in detail.[65] The features influencing the crystal packing are described by strong, classic intermolecular interactions such as hydrogen bonding. Such analyses do not represent a complete picture, however; subtler effects, such as weaker H-bonding interactions or van der Waals interactions, may be present and overlooked. To predict a crystal structure, an accurate understanding and absolute control over the interplay of intermolecular interactions is obligatory.[66] In this context, several approaches have been taken into account for the understanding and prediction of the final crystal structure of small molecules. These approaches are: (a) the model proposed by Gavezzotti for crystal structure prediction and to calculate the lattice energies;[67] (b) Etter proposed the model based on strong hydrogen bonding interactions to form the molecular crystal;[68] (c) the model based on supramolecular synthons has gained fabulous interest in which the degree of occurrence of a particular kind of interaction were taken into account;[69] (d) the model proposed by Hunter based on molecular electrostatic potentials through the estimation of relative hydrogen bonding parameters[70] and (e) Hirshfeld surface[71] analysis is a novel tool based on the complete quantification of all intermolecular interactions in a crystal structure in a visual manner. This technique was proposed and developed by Spackman *et al.*[72]

9.3.2 Theoretical Background for Hirshfeld Surface Calculation

Molecular crystal structures are often represented through various models such as ball-and-stick, space filling models, solvent accessible surfaces and electron density iso-surfaces. These representations often have some drawbacks so that they do not precisely depict the space filled by molecules in crystals, or the deformation of molecular space within the crystalline environment of the molecules of solid-state structures. Hirshfeld surfaces basically measure the space occupied by a molecule in crystals, and represent vital information regarding intermolecular interactions and various contacts within the structure. The method of Hirshfeld surface calculation was developed recently for studying crystal structures, which includes the calculations of the molecular surface depending on Hirshfeld's stockholder partitioning scheme, in which the crystal space is partitioned into smooth, non-overlapping and interlocking volumes.[73] In 1997, Spackman and Byrom first introduced this stockholder partitioning scheme and applied it to small molecules such as urea, formamide and ice.[74] They observed that Hirshfeld surface partitioning gave more accurate values for the dipole moments of the aforementioned molecules in comparison with generalized Wigner-Seitz (WS) partitioning and the quantum theory of atoms in molecules. In this particular scheme, the molecular surfaces are constructed depending on the electron distribution and are calculated as the sum of spherical atom electron densities.[74,75] The Hirshfeld surface enclose the volume in which the electron distribution due to sum of spherical atoms for the molecule (pro-molecule) dominates the corresponding sum over the crystal (pro-crystal).[74]

The Hirshfeld surfaces signify the partition into molecular fragments of the total electron density of the crystalline material. The other partition schemes, such as QTAIM[76] or WS[77] are also well known. Spackman *et al.*[78] described that the QTAIM[76] partition gives rise to numerical integration problems caused by the sudden discontinuities of the zero-flux boundary surface. The Hirshfeld surface divides the electron density of a molecule into continuous atomic fragments. The WS[77] method also displays irregularities of the molecule surface and also depends on the atomic radii. The accuracy of the molecular surface also depends on two factors: (i) the partition may be applied pair wise with atoms in the molecule; (ii) the surface generation test may be generated over aggregates of atoms.[74] So, the Hirshfeld surface is basically an extension of the Hirshfeld concept in which the electron density of a molecule is divided into continuous atomic fragments. From the analogy of this particular concept, a molecule within a crystal is defined by a weight function:[74]

$$w(\mathbf{r}) = \sum_{i \in \text{molecule}} \rho_i(r) / \sum_{i \in \text{crystal}} \rho_i(r) = \rho_{\text{promolecule}}(r) / \rho_{\text{procrystal}}(r) \qquad (9.1)$$

where $\rho(r)$ is a spherically averaged Hartree–Fock atomic electron density function centered at the ith nucleus. The weight function $w(\mathbf{r})$ is cut off at

0.5 Å *i.e.* the ratio of pro-molecule to pro-crystal electron densities equals 0.5. This clearly signifies the maximum proximity of neighboring molecular volumes and prevents overlap.[79] Based on this consideration, Hirshfeld surfaces differ from other molecular volumes and surfaces such as fused-sphere van der Waals volumes and solvent-accessible surfaces. The molecular surfaces will tend to touch but never overlap. In generally, they tend to fill approximately 95% of the volume of organic structures.[80] Thus, the surfaces are packed tightly but a gap will be evident between the molecular surfaces where no molecule dominates the electron density. For a given crystal structure and set of spherical atomic electron densities, the Hirshfeld surface is unique, which indicates the prospects of gaining detailed insight into intermolecular interactions present in the crystal structures.

9.3.3 Various Surfaces and Associated Fingerprint Plots

The general information about all intermolecular interactions involved within a crystal structure is easily available through a Hirshfeld surface analysis in a novel visual manner. The main advantage of this particular technique is the aesthetically pleasing and accessible manner through which the information about all type of contacts is easily available. Different properties can be encoded on the Hirshfeld surface since it is unique for a given crystal structure and a set of spherical atomic electron densities. In the first attempt, the gray-scale representation of the Hirshfeld surface was evidenced. However, keeping in mind the aim to visualize the presence of different intermolecular interactions, Spackman and coworkers[80] explored successfully five different functions (d_e, d_i, d_{norm}, shape-index and curvedness) mapped in color on the Hirshfeld surface of the molecules.

As explored by Spackman and coworkers,[80] for each point on the surface two colored distance properties are defined: (i) the distance from the point to the nearest nucleus external to the surface is named as d_e, and (ii) the distance to the nearest nucleus internal to the surface is d_i. The mapping of the function d_e on the surface represents an interpretable picture, which includes all close intermolecular contacts involved with the structure of the investigating molecule. These close intermolecular contacts appear as bright red areas on the surface. d_i is also mapped in a similar way to d_e but the application of this particular property is quite different. d_i plays a vital role in determining fingerprint plots and also in the study of atomic and ionic size.[80] Based on these two contact distances (d_e and d_i), a normalized contact distance d_{norm} is defined, which is given by eqn (9.2) and enables identifying the particular region on the total surface that is involved in intermolecular interactions.[71] The value of d_{norm} may be either negative or positive. When intermolecular contacts are shorter than vdw separations, the value of d_{norm} is negative and it shows positive value for longer contacts compare to vdw separations. The main advantage of this particular method is that owing to the symmetry of d_{norm} in d_e and d_i, the short contact has been evidenced on the Hirshfeld surface by two identical red spots, although not necessarily on the same molecule. Due to the symmetry information involved within d_e and d_i in the expression of d_{norm} (eqn 9.2), where two Hirshfeld surfaces touch, both display a red colored spot on the surface. This d_{norm} surface is represented by a red-white-blue color scheme as follows: red regions with negative values of d_{norm} represent the intermolecular contacts shorter than the sum of the van der Waals radii; blue regions with positive values of d_{norm} represent the intermolecular contacts longer than the sum of the van der Waals radii; white regions denote the distance of contacts exactly corresponding to the van der Waals separation with d_{norm} values of zero. So, basically, to identify the regions involve in vital intermolecular interactions and to compare these intermolecular interactions in the crystal lattice, the normalized contact distance (d_{norm}) was mapped onto the Hirshfeld surface.

$$d_{\text{norm}} = \frac{d_i - r_i^{\text{vdw}}}{r_i^{\text{vdw}}} + \frac{d_e - r_e^{\text{vdw}}}{r_e^{\text{vdw}}}. \tag{9.2}$$

Apart from the above-mentioned surfaces, two further colored properties based on the local curvature of the Hirshfeld surface can be specified.[81] These colored properties are named as 'shape-index' and 'curvedness'. The shape-index surface is extremely useful for identifying the self-complementary nature of interactions involved within a supramolecular structure. Due to two shapes of the shape-index that differ by sign are represented by complementary "*stamp*" and "*mould*" pairs that represent the blue bumps and red hollows on the Hirshfeld surface. In this context, the shape index is known as a dimensionless measure of "*which*" shape. Moreover, it can be sensitive to slight changes in surface shape, particularly in areas where the total curvature is very low. The formal range of shape-index varying from (−1) to (+1) and this range is particularly very suitable for visual display of the surface. The π–π interactions are very clearly evidenced by the red and blue triangles on the same region of the surface of the molecule. This also gives an outline of how the molecules overlap to interconnect with each other in a face-to-face orientation. Furthermore, the red and blue triangles on both sides of the shape-index surface indicate how the partner molecules within the structure are related by translation. These blue triangles are represented by the convex regions due to carbon atoms of the rings within the surface and the red triangle represents the concave regions due to the ring carbon atoms of the partner molecule, which is juxtaposed through π-stacking interaction above the parent molecule.

Curvedness can also be mapped on the Hirshfeld surface and known as a measure of "*how much shape*". For flat regions of a surface, the value of the curvedness is low and high curvedness is evidenced for the areas with sharp curvature. The curvedness surface enables the extraction of knowledge regarding the number of nearest neighbors or coordination sphere of the molecule. The curvedness is measured in the range −4.0–0.4 and this range is most suitable in building molecular surfaces. The curvedness surfaces are represented by hefty green areas is separated by dark blue edges.

Sometimes, the yellow or red colours are also observed on the flat region of the surface, which correspond the hydrogen bonding interactions. Moreover, the existence of π–π interactions is also clearly evidenced by the curvedness surface by the flat region toward the bottom of the both sides of the investigating molecule.

The two-dimensional graphical plots associated with the Hirshfeld surface named 'fingerprint plot' also represent the summary of the information of intermolecular interactions gained in the Hirshfeld surface calculation with d_{norm} or other functions.[71] The size and shape of the fingerprint plot is very unique for each investigating molecule. In general, one Hirshfeld surface includes tens of thousands of individual points, which contain the information relating to intermolecular interactions within the structure. As established by Spackman and coworkers,[80] to generate a fingerprint plot, d_e and d_i are calculated for each point and then the data is binned into certain intervals. The 2D plots were created by binning (d_e, d_i) pairs with a width of 0.01 Å. The coloring of each bin (essentially a pixel) of the 2D plot is a function of the fraction of surface points in that bin, which varies from blue (few points) through green (moderate number of points) to red (many points) and white for no points.[71,80] Complementary regions are visible in the fingerprint plots where one molecule acts as a donor ($d_e > d_i$) and the other as an acceptor ($d_e < d_i$). The fingerprint plots can be decomposed to highlight and to quantify the contacts between particular atom pairs.[74] The fingerprint plot is pseudo-symmetrical about the diagonal $d_i = d_e$ for the structure that contains only one molecule in the asymmetric unit. This is only due to the close packing of the molecules on the Hirshfeld surface, which ensures that the points (d_e, d_i) and (d_i, d_e) appear on the plot. By binning these (d_i, d_e) pairs, fingerprint plots allow the determination of various intermolecular interactions easily by observing the shape of the plot.

All the Hirshfeld surfaces and corresponding 2D fingerprint plots presented in this chapter were generated using *CrystalExplorer 3.1*. In *CrystalExplorer*,[78] the internal consistency is important when comparing one structure with another; for the generation of the Hirshfeld surface all bond lengths to hydrogen (or deuterium) atoms are set to typical neutron values (C–H = 1.083 Å, O–H = 0.983 Å, N–H = 1.009 Å).[82] *CrystalExplorer* accepts a structure input file in CIF format. The fingerprint plots do not scale with molecular size. The size of the plot is constant and does not depend on how many atoms are within the molecule. Thus, this method is extremely suitable to compare molecular structures that have pronounced variations in size.

Hirshfeld surface calculations[71–75,80] can be applied to various types of molecules, including small, large, organic and inorganic molecules. The most significant application of this analysis is the "screening of polymorphs". Polymorphism mainly attracts greater attention in chemistry, pharmaceutical and materials science,[83] mainly due to the relevance in the associated physiochemical properties, and it can often be protected by patent and therefore is of crucial economic importance.[84] Due to the special features and uniqueness of the Hirshfeld surface, one can easily obtain a better understanding of the intermolecular interactions involved in polymorphic structures. In general, all types of well-known non-covalent interactions are easily traceable by this unique Hirshfeld surface analysis.[85] This analysis is also very sensitive to structural abnormalities, and incorrect crystal structures can be easily identified. In this chapter, we have successfully utilized the Hirshfeld surface analysis to explore the contributions of unconventional weak interactions involving π-hole and σ-hole (*vide infra*) by representing the contributions involved within the investigating structures.

9.4 Computational Methods

The energies of all complexes included in this chapter were computed at the RI-MP2/def2-TZVP level of theory. We have used the crystallographic coordinates for the theoretical analysis of the non-covalent interactions present in the solid state. The calculations were performed by using the program TURBOMOLE [version 7.0].[86] The interaction energies were calculated with correction for the basis set superposition error (BSSE) by using the Boys–Bernardi counterpoise technique.[87] Bader's "atoms in molecules" theory has been used to study the interactions discussed herein by means of the AIMALL package.[88] The calculations on the MEPs were performed using the SPARTAN software.[89]

9.5 Exploration of σ-Hole Interactions

9.5.1 Group VII Interactions (Halogen Bonding)

Significant attention is paid to halogen bonding due to its applications in crystal engineering, design of functional materials and biological systems.[90,91] A halide R–X (X = halogen) will frequently form a weakly bound complex R–X···B with a negative site B and this is termed as "halogen bonding".[92] The extensive crystallographic surveys on halide crystal structures contained in the CSD by Murray-Rust *et al.* were particularly significant.[93] Their studies revealed that covalently bonded halogen atoms can interact attractively and in a highly directional manner with both nucleophiles and electrophiles. In this chapter, we describe the contribution of halogen bonding interaction of specific crystal structures from CSD[94] by using Hirshfeld surface and theoretical calculations. Hirshfeld surface analysis has been recently used to characterize halogen-bonding interactions in diiodomethane.[95] Moreover, it has been also used to study halogen bonding in 2-trichloromethyl-3H-4-quinazoline polymorphs.[96]

A wide number of studies[97] devoted to the analysis of solid-state architectures have shown the predominance of halogen bonding interactions as self-assembly driving forces. Clear evidence to illustrate the dominant role of halogen bonding interactions in solid state chemistry is related to the bis(*N*-heterocycle)diacetylene unit. In this regard, two examples are worth mentioning, involving ditopic iodo donors[98] and bifunctional molecules (both of them containing halogen bond donor and acceptor moieties),[99] forming a chain-like structure.[100] The formation of chain-like structures is useful in the preparation of materials with conducting and magnetic properties as well as non-centrosymmetric crystals capable of second harmonic generation in optical devices. In this regard, Metrangolo and Resnati *et al.*[101] and van der Boom *et al.*[102] demonstrated the formation of a 1D chain in a helical disposition using dibromopyridinium helices and stilbazole-based helices, respectively (see Figure 9.9).

As noted, both crystal structures ADOLIP[102] and BEYQIG[101] (Figure 9.9) contain dual halide atoms and we are interested in exploring the contribution of halogen bonding interactions by Hirshfeld surface calculation. The d_{norm}, d_i and d_e surfaces have been mapped over (−0.385 Å–1.903 Å, 1.000 Å–2.926 Å, 1.000 Å–2.889 Å) and (−0.199 Å–1.343 Å, 0.910 Å–2.792 Å, 1.000 Å–2.898 Å) for ADOLIP and BEYQIG respectively (Figures 9.10 and 9.11). The surfaces are shown as transparent to allow visualization of the moieties.

The information on halogen bonding interactions is clearly evidenced by the circular depressions on the d_{norm} surfaces of both the structures. The circular depression (deep red) that is evidenced on the d_{norm} surface of ADOLIP corresponding to the I atom is larger than the depression corresponding to the F atom which may arise due to the fact that the I···I contact is more distant than the I···F or F···F interactions within the structure (Figure 9.9). The large depression on the d_{norm} surface corresponding to the Br atom of BEYQIG is a characteristic of Br···Br and Br···I contacts (Figure 9.11). The properties of d_e and d_i also support the halogen bonding interactions of the structures (Figures 9.10 and 9.11). The halogen bonding interactions are also evident by the two-dimensional fingerprint plots (Figure 9.10 and Figure 9.11). The translated view of the total contribution of all interactions and the decomposed plot of halide–halide interactions (I···I, I···F/F···I and F···F in ADOLIP and Br···Br, Br···I/I···Br and I···I in BEYQIG) are depicted in Figures 9.10 and 9.11. In the ADOLIP structure the I···I contacts comprises of 3.9% of the total Hirshfeld surface area of the molecule and that is evidenced from the breakdown of the fingerprint plot and the decomposition of the d_e, d_i and d_{norm} surfaces (Figure 9.10). The I···I interaction is reflected by the distribution of scattered points in the fingerprint plot that spreads only up to $d_i = d_e = 1.852$ Å. The proportion of I···F contribution is higher (4.8%) than the F···I contribution (3.0%) in ADOLIP (Figure 9.10). The I···F interaction is represented by the spike in the bottom left area ($d_i = 2.068$ Å and $d_e = 1.416$ Å) whereas the F···I contribution is represented by the bottom right region of the fingerprint plot ($d_i = 1.411$ Å and $d_e = 2.073$ Å). The proportion of I···F/F···I interaction comprises 7.8% of the total Hirshfeld surface of each molecule of ADOLIP. The decomposition of the Hirshfeld surfaces due to I···F/F···I interaction as well as the decomposed fingerprint plot are depicted in Figure 9.10. The F···F interaction comprises only 3.2% of the total Hirshfeld surface area of the molecule, and that is clearly evidenced by the fingerprint plot in the scattered plot of $d_i = d_e = 2.146$ Å (Figure 9.10).

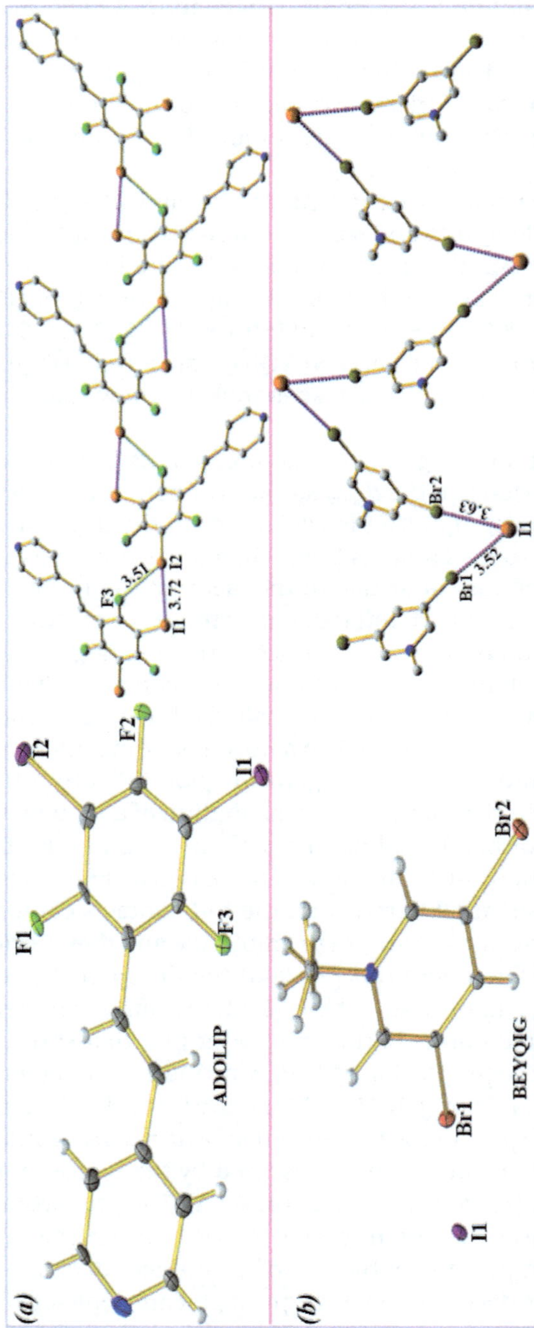

Figure 9.9 Molecular views and halogen bonded chains of (a) ADOLIP and (b) BEYQIG. The CSD reference codes are indicated and distance is in Å.

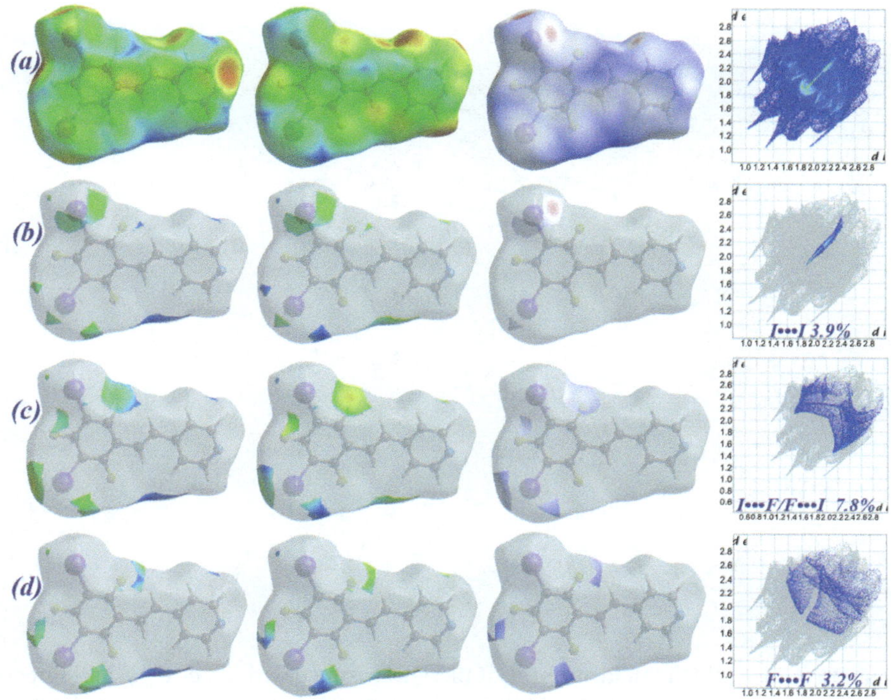

Figure 9.10 (a) Hirshfeld surfaces of ADOLIP mapped with d_e, d_i, d_{norm} and fingerprint plot. Contribution to the Hirshfeld surfaces and breakdown of the fingerprint plot due to (b) I⋯I interaction, (c) I⋯F/F⋯I contacts and (d) F⋯F contact.

In BEYQIG the ring substituted Br atom interacts with the I atom in the molecule at $(1 - x, 1/2 + y, 1/2 - z)$ and the I atom again acts as an acceptor to another Br atom at $(1/2 - x, 1 - y, -1/2 + z)$, thus forming a 1D chain along (0 1 0) direction through halogen bonding interactions. By using Hirshfeld surface analysis, we have examined the existence of Br⋯Br along with I⋯Br and Br⋯I contributions to the molecule. The Br⋯Br contribution comprises 4.7% and is depicted by the decomposed fingerprint plot by scattered point in the region $d_i = d_e = 1.988$ Å. The I⋯Br contribution to the Hirshfeld surface is 0.0% (no contribution) whereas the Br⋯I contribution comprises 4.6% of the total Hirshfeld surface area of each molecule. The Br⋯I interaction is represented by the spike (d_i = 1.672 Å; d_e = 1.822 Å) in the donor area of the decomposed fingerprint plot (Figure 9.11). These X-bonded interactions in BEYQIG are shown by the decomposed d_e, d_i, d_{norm} surfaces along with scattered fingerprint plots (Figure 9.11). Moreover, we have also analyzed the Hirshfeld surface calculation for the free I atom of the asymmetric unit, which shows that there is I⋯Br contribution to the Hirshfeld surface that comprises 16.3%. Thus, all the interactions involving the halogen atoms are critically analyzed and shown by different surfaces, which provide rapid quantitative insights into the halogen⋯halogen bond.

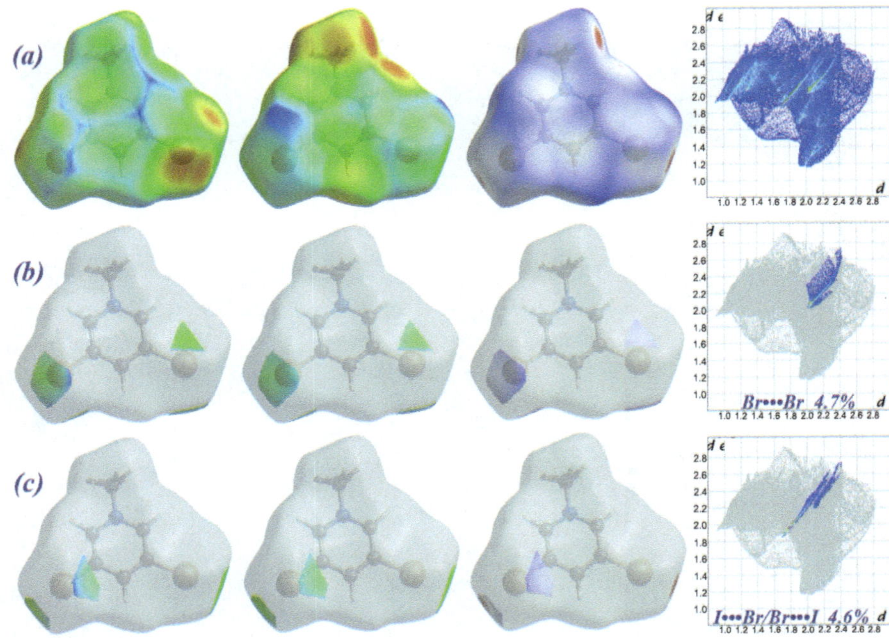

Figure 9.11 (a) Hirshfeld surfaces of BEYQIG mapped with d_e, d_i, d_{norm} and fingerprint plot. Contribution to the Hirshfeld surfaces and breakdown of the fingerprint plot due to (b) Br···Br interaction and (c) I···Br/Br···I contacts.

We have computed the interaction energy of halogen-bonded dimers retrieved from the X-ray structures of both compounds. Moreover, we have also computed the MEP surface of the monomeric compounds plotted onto the van der Waals surface (0.002 a.u.) to examine the MEP value at the σ-hole of the halogen atom. As noted, the binding energies indicate that the halogen bonding interaction is considerably stronger in the BEYQIG complex (−61.9 kcal mol^{-1}) than in ADOLIP homodimer (−3.5 kcal mol^{-1}) (Figure 9.12). This is due to the favorable electrostatic contribution of BEYQIG (ion-pair nature of the interaction between iodide and pyridinium). These results are in agreement with the MEP surfaces that evidence a more positive σ-hole MEP value at the Br atom for BEYQIG than at the I atom in ADOLIP, thus expecting a stronger interaction from an electrostatic point of view. We also performed Bader's "atoms in molecules" analysis for both structures. The presence of a bond critical point (CP) and bond path connecting two atoms is unambiguous evidence of the interaction.[103] The distribution of bond CPs and bond paths in both structures reveals the presence of a bond CP (red sphere) and bond path (dashed line) that interconnect both halogen atoms, thus characterizing the halogen bonding interactions.

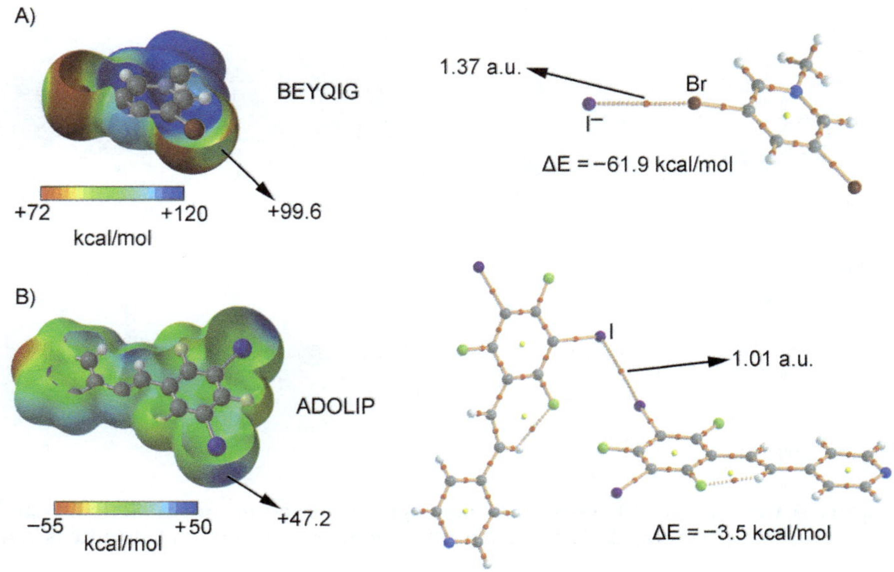

Figure 9.12 Interaction energies, MEP surfaces (0.002 a.u. isosurface) and AIM analyses (bond CP in red and ring CP in yellow) showing halogen bonding interactions of (A) BEYQIG and (B) ADOLIP structures. The value of the density at the bond CP as well as both CSD codes are also indicated.

9.5.2 Group VI Interactions (Chalcogen Bonding)

The interaction between the S or Se atom with an electronegative atom like N or O has attracted tremendous attention in exploring the σ-hole interaction.[104] The σ-hole interaction among divalent sulfur and oxygen atoms plays a crucial role in structural and biological applications.[105] It has also been explored that folded protein structures can be stabilized through S···O interactions.[106] Initially Rosenfield *et al.*[107] showed the structural features of both intra- and intermolecular interactions between S and O atoms for organic and inorganic structures. They showed that the nucleophilic O atom tend to interact with the divalent S atom (Y–S–Z) from the backside of S–Y and S–Z bonds, which shows the direction of the σ-hole interaction. Kucsman and Kapovits[108] describe the details of the relative directional preference of S···O interactions to the O atom. In the case of S···O=C intramolecular interactions, it was shown that the S atom tends to lie in the direction of the lone pairs of O atom and can be described as $n \rightarrow \sigma^*$ orbital interaction. We have focused our analysis on two interesting structures that illustrate the ability of S atoms to interact by means of the σ-hole with an aromatic moiety and a discrete lone pair. The first example corresponds to the work from Hobza and coworkers[104f] (see ellipsoid view in Figure 9.13a), where S···π type chalcogen bonds are present in the crystal structure of phenyl substituted thiaborane.

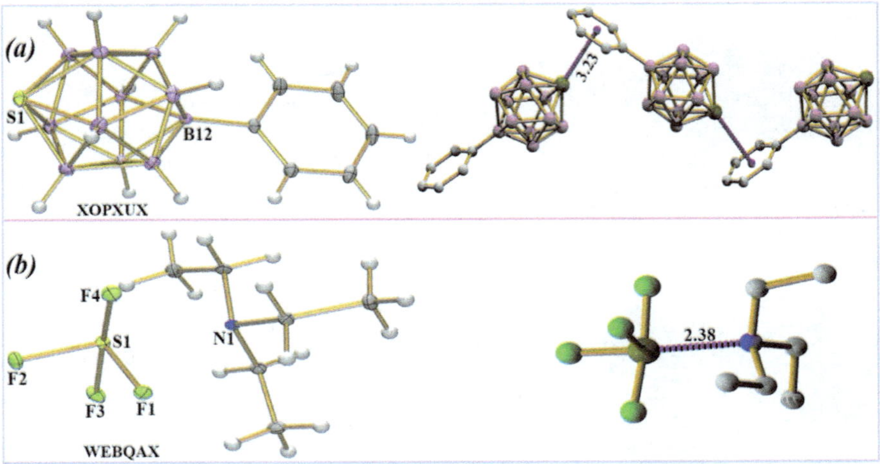

Figure 9.13 An ORTEP view and chalcogen interaction in (a) XOPXUX and (b) WEBQAX. The CSD reference codes are indicated and distance is in Å.

Theoretical calculations revealed that the most stable binding motif in the crystal structure of 12-Ph-*closo*-1-$SB_{11}H_{10}$ corresponds to a very strong B–S···π chalcogen bond exceeding −8 kcal mol^{-1}. On the other hand, Nziko and Scheiner[109] studied the chalcogen N···S interaction between tetravalent SF_4 and a series of amines (alkyl and arylamines) by theoretical calculations. It was also proposed that the N···S bonding interaction in the dimethylamine···SF_4 complex should be designated as a weak covalent bond owing to the shortness of the bond and the strength. As experimental evidence, it had been shown that sulphur tetrafluoride and triethylamine react at low temperatures to form a 1 : 1 adduct (N···S interaction), which was characterized by X-ray crystallography (see Figure 9.13b).[110] The S···N bond has a length of 2.384(2) Å and it is an archetypical example of a dative N → S^{IV} bond. Therefore, in this section we have considered the structures XOPXUX[104f] and WEBQAX[110] retrieved from the CSD to explore both S···π and N···S interactions observed in their solid-state assembly.

We have explored the contribution of S···C_π/C_π···S and N···S type chalcogen bonding in the solid-state structure of each molecule of XOPXUX and WEBQAX respectively by using Hirshfeld surface analysis. The d_e, d_i and d_{norm} surfaces are mapped over the normal range and are depicted in Figure 9.14. A detailed inspection of the d_e and d_i surfaces of XOPXUX shows that there is a clear greenish yellow flat region over the aryl ring and the region over the S atom indicates that there is some interaction in between the S atom and the aryl ring. The small depressions on the d_{norm} surface are due to short contacts but one can easily visualize that the region over the S atom and the aryl ring is more distant than the region over the molecule as a whole. The interaction of the S atom and the centroid of the aryl ring is designated as the S···C_π/C_π···S chalcogen interaction. The splitting of the Hirshfeld surface clearly identifies the chalcogen interaction and the contribution to the total

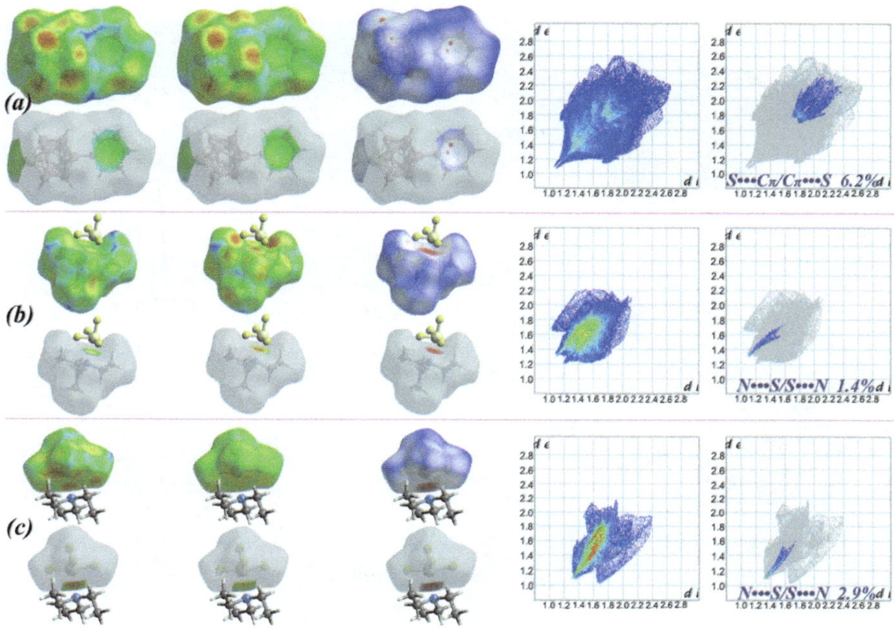

Figure 9.14 (a) Hirshfeld surfaces are mapped with d_e, d_i, d_{norm} and full fingerprint plot, and resolved into S···C_π/C_π···S contacts for the complex XOPXUX; (b) Hirshfeld surfaces (d_e, d_i, d_{norm}), full fingerprint plot and contribution to N···S/S···N interaction for moiety A (triethylamine) of WEBQAX; (c) Hirshfeld surfaces (d_e, d_i, d_{norm}), full fingerprint plot and contribution to N···S/S···N interaction for moiety B (SF_4) of WEBQAX.

Hirshfeld surface of the molecule. Each surface (d_e, d_i and d_{norm}) shows colored patches only over those atoms that take part in the S···C_π/C_π···S chalcogen interaction. This interaction is also evident from the 2D fingerprint plot wherein both the S···C_π interaction (d_i = 1.752 Å; d_e = 1.687 Å) and C_π···S counterpart (d_i = 1.677 Å; d_e = 1.767 Å) contributes 3.1% each to the total Hirshfeld surface. The breakdown of the fingerprint plot in Figure 9.14 shows that the S···C_π/C_π···S contact contributes 6.2% of the total Hirshfeld surface area of the molecule. This interaction is represented by the scattered points in the regions of (d_i = d_e = 1.737) (Figure 9.14a).

Next, we analyzed the Hirshfeld surfaces for WEBQAX to explore N···S/S···N chalcogen bonding. Since the structure has two individual moieties of SF_4 and triethylamine, each moiety was considered individually to analyze their surfaces (Figure 9.14b and c). In both the moieties, the N···S/S···N interaction is distinctly evident by the yellow-green spot, red-yellow spot and red spot on the d_e, d_i and d_{norm} surfaces respectively. These spots are clearly visible on the breakdown plots of the surfaces. The triethyl amine surface does not show a S···N contribution and shows a N···S contribution only, which contributes 1.4% of the total Hirshfeld surface area. The sharp spike in the region (d_i = 1.081 Å, d_e = 1.277 Å) designates the N···S chalcogen bonding

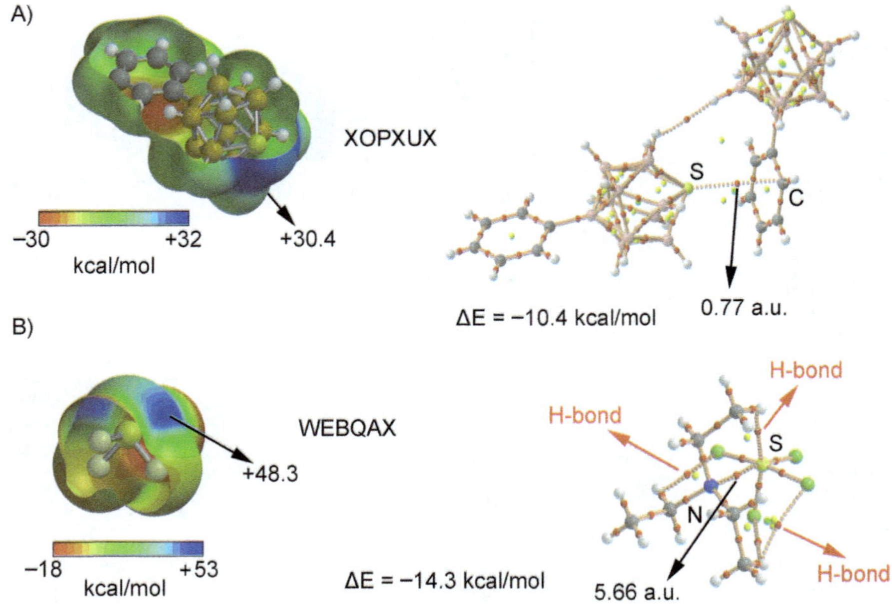

Figure 9.15 Interaction energies, MEP surfaces (0.002 a.u. isosurface) and AIM analyses (bond CP in red, ring CP in yellow and cage CP in green) showing chalcogen bonding interactions of (A) XOPXUX and (B) WEBQAX structures. The value of the density at the bond CP as well as both CSD codes are also indicated.

in the triethylamine moiety (Figure 9.14b). Similarly, we have analyzed the surface for the SF_4 moiety and it shows that there is no N···S contribution. The total contribution of the N···S/S···N interaction comes from the S···N contribution, which contributes 2.9% of the total surface area (Figure 9.14c). The sharp spike in the region (d_i = 1.282 Å; d_e = 1.076 Å) designates the N···S/S···N chalcogen interaction.

We have also studied theoretically both compounds by computing the MEP surfaces, interaction energies and AIM analyses (Figure 9.15). The complex involving WEBQAX (Et_3N···SF_4) exhibits a more favorable interaction energy value, mainly due to the more positive σ-hole over the S atom observed for this moiety (see MEP surfaces). In addition, the AIM analysis of WEBQAX reveals that the chalcogen bond is characterized by the presence of a bond CP and a bond path connecting the S atom to the N atom. Moreover, the AIM distribution also shows the existence of several H-bonding interactions between the H atoms of the ethyl groups and the F atoms of the SF_4 moiety that further contribute to the stabilization of the system. In particular, three bond CPs and bond paths connect three H atoms of the Et_3N Lewis base to three F atoms of SF_4. On the other hand, the AIM analysis of XOPXUX reveals that the S···π interaction is characterized by the presence of a bond CP and a bond path that connect the S atom of the thiaborane moiety to one carbon atom of the phenyl ring.

9.5.3 Group V Interactions (Pnictogen Bonding)

The pnictogen interactions in different complexes have been reported earlier, which exhibit both inter[54,111] and intramolecular[112] interactions. The pnictogen interaction involving nitrogen, phosphorus and arsenic atoms have also been reported.[113,114] Experimentally, in several crystal structures the existence of pnictogen bonds such as N···N,[115] P···P[116] and As···As[117] varieties have been explored. Recently Grabowski *et al.*[118] explored the pnictogen interaction for the charged complexes of ZH_4^+, ZFH_3^+ and ZF_4^+ (Z = N, P and As). Moreover, pnictogen interactions involving radical cations have also been studied recently.[119] In this regard, the solid state 1,4-dimethylpiperazine[120] forms a ditopic complex with PBr_3 exhibiting a short intermolecular P···N distance of 2.80 Å (see Figure 9.16a), close to the theoretical value.[121] On the other hand, the interaction between π-electrons of arenes and the heavier pnicogen atoms (As, Sb and Bi) has been known for a long time, but its study in the context of supramolecular chemistry is more recent.[122] An investigation by Johnson *et al.* has highlighted the potential use of this weak interaction to regulate self-assembly processes to form macrocycles.[123] Mainly using X-ray analysis[124] it has been demonstrated that post-transition metals with stereochemically active lone pairs, such as heavier pnicogens in low oxidation states, yield stable complexes with arenes (see Figure 9.16b). In this section we explore pnictogen bonding by using Hirshfeld surface analysis and theoretical calculations for both structures (MEWYES[120] and LAHJAF[124a]) from the CSD.

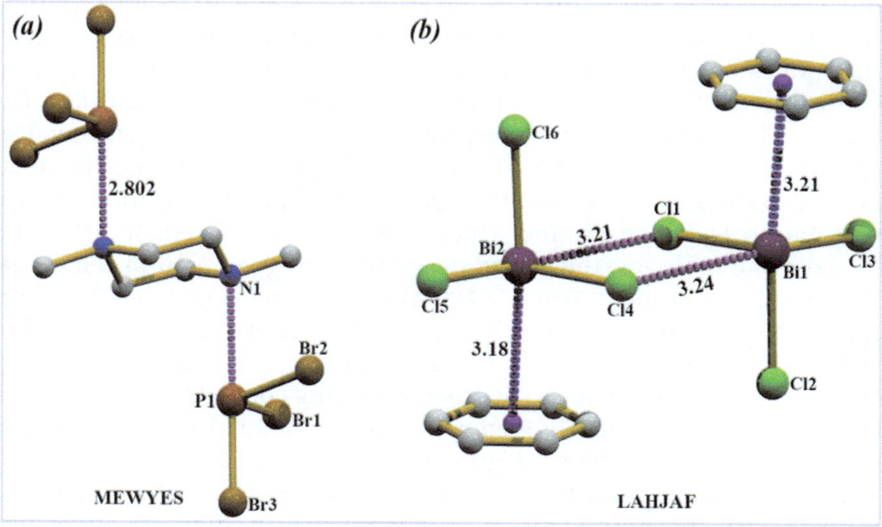

Figure 9.16 Pnictogen interactions P···N and Bi···π for the complexes MEWYES (a) and LAHJAF (b) respectively. The CSD reference codes are indicated and distance is in Å.

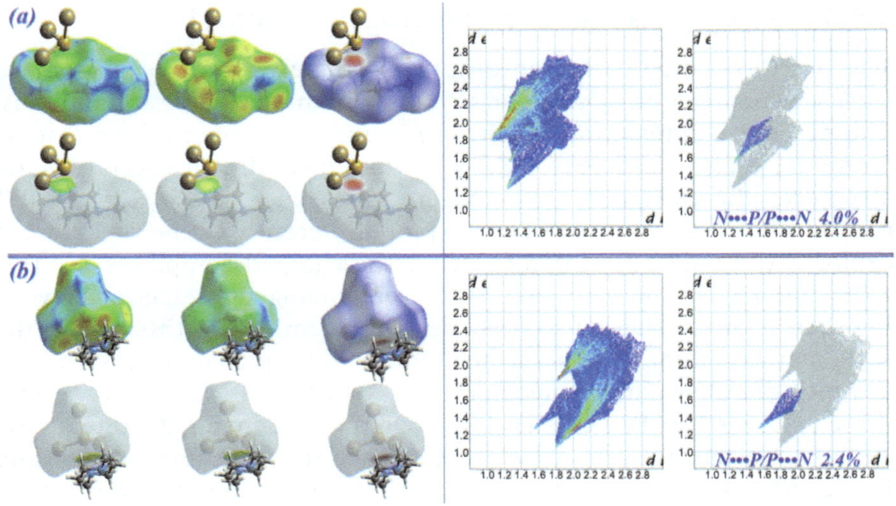

Figure 9.17 Hirshfeld surfaces are mapped with d_e, d_i and d_{norm} and full fingerprint plot, and resolved into Si⋯I/I⋯Si contacts [(a) moiety A and (b) moiety B] for the complex MEWYES.

In the MEWYES structure, we studied the P⋯N contacts whereas in LAHJAF the As⋯π contact was explored by using Hirshfeld surface analyses. The molecular views and the pnictogen bonding contacts are shown in Figure 9.16. For both the structures, the d_{norm} surfaces were generated in the range (−0.245 Å–1.579 Å) (Figures 9.17 and 9.18). The individual moieties of each structure were taken into account to generate and to calculate the contribution of pnictogen contacts. For the MEWYES structure, we calculated the d_e, d_i and d_{norm} surfaces whereas we analyzed the contribution of pnictogen contacts in LAHJAF by generating the d_{norm} and shape-index surfaces only. Since, four individual moieties are involved within the structure of LAHJAF, the d_e and d_i surfaces are not included here. The surfaces are shown transparent for clear visualization and the orientation is shown around which they were calculated. In MEWYES, for both the moieties, the d_e and d_i surfaces are generated in the ranges (1.000 Å–2.719 Å) and (1.000 Å–2.799 Å) respectively (Figure 9.17). In both moieties, the P⋯N/N⋯P contribution is evident by the greenish yellow, yellow-green and red spots in the d_e, d_i and d_{norm} surfaces respectively. Other visible spots are due to hydrogen bonding contacts and H⋯H short contacts. Only the P⋯N/N⋯P contribution to the total Hirshfeld surface of d_e, d_i and d_{norm} is shown in Figure 9.17 to locate the contacts very easily. The contributions from the P⋯N/N⋯P contact for both the moieties in MEWYES are shown in Figure 9.17. The fingerprint plots do not exhibit the distinct spikes in both the donor and acceptor regions of the plot since we have analyzed each moiety individually. For moiety-A, the N⋯P interaction comprises of 4.0%, which is clearly evident in the breakdown of the

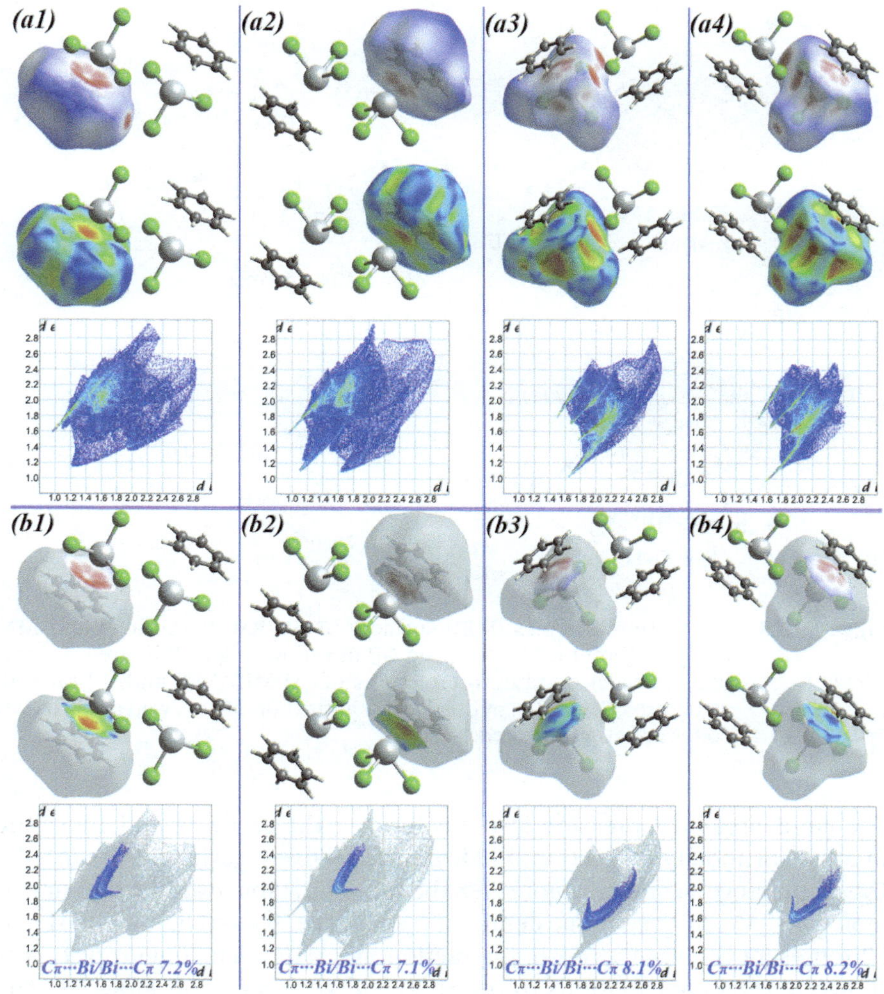

Figure 9.18 Hirshfeld surfaces of LAHJAF are mapped with d_{norm}, shape-index and full fingerprint plots for (a1) moiety A1; (a2) moiety A2; (a3) moiety B1 and (a4) moiety B2. The Hirshfeld surfaces and associated fingerprint plots are resolved into $C_\pi\cdots Bi/Bi\cdots C_\pi$ contacts [(b1) moiety A1; (b2) moiety A2; (b3) moiety B1 and (b4) moiety B2].

fingerprint plot in the region ($d_i = 1.252$ Å; $d_e = 1.533$ Å). In the case of PBr_3 *i.e.* moiety B, the P···N contact contribution comprises 2.4% of the total Hirshfeld surface area. The spike in the region ($d_i = 1.527$ Å; $d_e = 1.252$ Å) designates the P···N/N···P contact, which is clearly evidenced by the breakdown of the fingerprint plot (Figure 9.17).

The asymmetric unit of LAHJAF comprises two aryl rings and two $BiCl_3$ moieties, which are designated as (A1, A2) and (B1, B2) respectively. All the surfaces were generated in the normal range and are depicted in Figure 9.18. $Bi\cdots C_\pi/C_\pi\cdots Bi$ contacts are evidenced by the d_{norm} and shape-index surfaces.

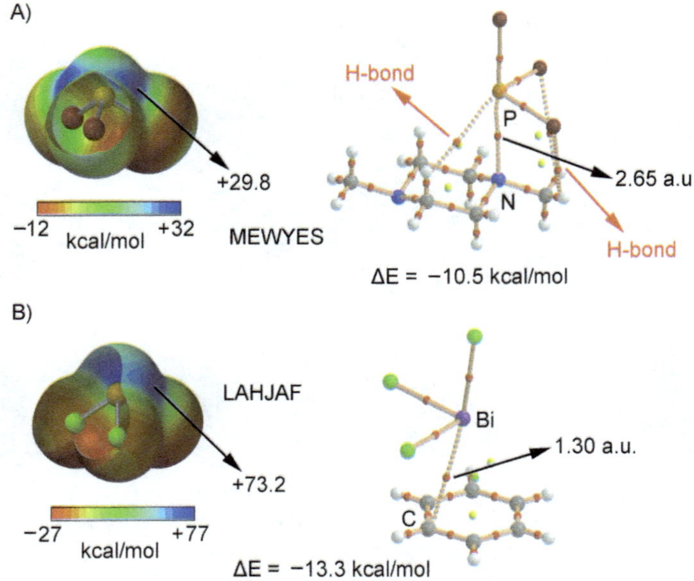

Figure 9.19 Interaction energies, MEP surfaces (0.002 a.u. isosurface) and AIM analyses (bond CP in red, ring CP in yellow, cage CP in green) showing pnicogen bonding interactions of (A) MEWYES and (B) LAHJAF structures. The value of the density at the bond CP as well as both CSD codes are also indicated.

The large circular depression in red color over the aryl ring and above the Bi atom for moiety (A1, A2) and (B1, B2) respectively designate the Bi···C_π/C_π···Bi pnictogen contacts. Moreover, the red-yellow spot over the aryl ring in moiety A1, A2 and green-blue depression above the Bi atom in moiety B1 and B2 represent the Bi···C_π/C_π···Bi interaction in LAHJAF. The fingerprint plots show the total contributions to the Hirshfeld surfaces of all moieties. The Hirshfeld surfaces and fingerprint plots for moieties A1, A2, B1 and B2 are depicted in Figure 9.18(a1)–(a4) respectively. The aryl rings *i.e.* moieties A1 and A2 shows only C_π···Bi contacts, which comprises 7.2% and 7.1% respectively of the total surface area. In the case of moieties B1 and B2, the contributions to the total Hirshfeld surface of moieties B1 and B2 due to Bi···C_π contact are 8.1% and 8.2% respectively. The total C_π···Bi/Bi···C_π contacts are represented by the breakdown of d_{norm}, shape-index surfaces and fingerprint plots in Figure 9.18(b1)–(b4) for moieties A1–B2 respectively.

For both structures, we computed the interaction energies, MEP surfaces and AIM analyses. The results gathered in Figure 9.19 show a larger interaction energy value for the complex involving the LAHJAF structure, as anticipated by the MEP surface. The value over Bi atom is far more positive than the one obtained for the P atom in the MEWYES structure. However, the AIM analysis indicated that several ancillary hydrogen bonding interactions are also established between the PBr_3 molecule and the hydrogen atoms of

1,4-dimethylpiperazine (see dashed bond paths in Figure 9.19A), thus compensating to some extent the lower MEP value in MEWYES and yielding quite similar interaction energy values. The pnicogen bonding interaction in LAHJAF is characterized by the presence of a single bond CP that connects the Bi atom to one carbon atom of the aromatic ring.

9.5.4 Group IV Interactions (Tetrel Bonding)

The term "tetrel bonding" was coined by our group in 2013[125,126] for σ-hole interactions where atoms of group IV act as Lewis acids, although this interaction had been studied before by other groups.[127–129] The interactions involving Si, Ge and Sn have been explored earlier, while "non-covalent carbon bonding" is an issue in which much more attention needs to be paid to explore in detail. In this regard, Alkorta *et al.*[127] explored molecular complexes between silicon derivatives (SiX_4, X = halogen) and electron-rich groups (NH_3, H_2O, *etc.*) as well as the impact of this interaction in aminopropylsilanes.[128] This non-covalent force has also been explored in complexes involving S_N2 processes for the elements of group IV.[125,126,130] In addition, the stability of complexes involving methanol as a tetrel bond donor with Lewis bases Y = H_2O, H_2S, HF, HCl, HBr, ClF, LiF, LiCl, LiBr, NH_3 and PH_3 has been theoretically studied by Mani and Arunan.[131] The existence of tetrel bonding has been investigated experimentally and the importance of tetrel bonding has been reported.[132–134] It has been shown in experimental work that tetrel bonding interactions induce planarity in five- and six-membered silicon rings. Due to the concurrent tetrel bonds, the "inverse sandwich" complexes in perhalogenated cyclohexasilanes or perchlorocyclopentasilane interact with the anion or lone-pair donor molecules.[135,136] The evidence for the existence of tetrel bonding involving carbon as well as silicon has been growing extensively[125,137] and quantum calculations have already confirmed the presence of carbon tetrel bonds between electron deficient carbon atoms and electron rich tetrel bond acceptors.[138] Very recently, Thomas *et al.* identified 716 compounds that may exhibit C···O tetrel bonding in the compounds in the CSD.[132] We have considered two crystal structures from the CSD to explore Si···I and C···N tetrel bonding for AZEBAJ[139] and TCYCPR[140] respectively. The molecular views and packing through tetrel bonding are depicted in Figure 9.20.

In the AZEBAJ structure, the interaction in between Si and I prompted us to explore the contribution by Hirshfeld surface analysis. We analyzed the surfaces by considering three moieties individually. Moiety A contains the Si ring and moieties B and C includes two I atoms above and below the Si ring respectively. The d_{norm}, shape-index and curvedness surfaces are depicted in Figure 9.21, which are drawn in the range (−0.395–1.652 Å), (−1.0–1.0 Å) and (−4.0–0.4 Å) respectively. The interaction in between Si and I atoms as shown in Figure 9.20a is also clearly evident from the Hirshfeld surfaces of each moiety (Figure 9.21a–c). The large reddish depression in the d_{norm} surface, the red-yellow depression in the shape-index and the green-blue depression in

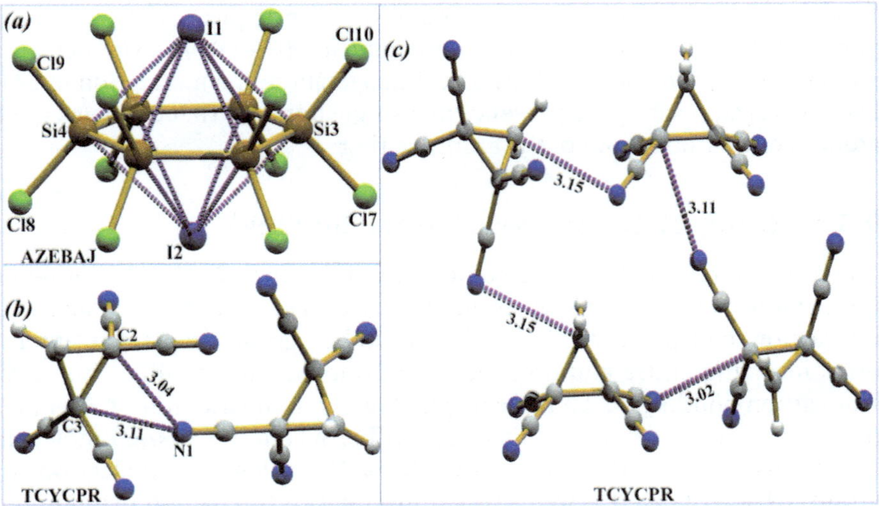

Figure 9.20 (a) X-ray structure of AZEBAJ showing Si···I/I···Si tetrel bonding; (b) X-ray structure of the dimer of TCYCPR; (c) carbon-bonding network in TCYCPR. The CSD reference codes are indicated and distance is in Å.

the curvedness surface of moiety A represent Si···I tetrel bonding. In the case of moiety B and moiety C, the reddish, bluish and greenish depressions in d_{norm}, shape-index and curvedness surfaces respectively represent the tetrel bonding contacts. It was observed that moiety A does not participate in the I···Si contribution, whereas the moieties B and C do not exhibit any contribution for the Si···I contact. The proportion of Si···I/I···Si contacts comprise 7.0%, 25.9% and 26.0% of the Hirshfeld surface of the moieties A, B and C respectively. The breakdown of the surfaces along with fingerprint plots for all moieties is depicted in Figure 9.21. The Si···I interaction is represented by the spike (d_i = 1.552 Å, d_e = 1.657 Å) in moiety A and I···Si interactions are evident in the region (d_i = 1.667, d_e = 1.547 Å) and (d_i = 1.677, d_e = 1.577 Å) for moieties B and C respectively (Figure 9.21).

The Hirshfeld surface of TCYCPR is illustrated in Figure 9.22, showing the surface that was mapped over d_{norm} only. The tetrel bonding interactions depicted in Figure 9.20 for TCYCPR are summarized effectively in the spots, with the circular red colored depressions on the d_{norm} surface and other visible spots due to H···H contacts. The d_{norm} surface was mapped in the range (−0.131–1.034 Å). In this structure, the C···N interaction has a larger contribution than its N···C counterparts. The breakdown of the fingerprint plot shows that the C···N interaction contributes 21.3% of the total Hirshfeld surface area whereas the N···C contact comprises only 19.7% of the total surface. Thus, the C···N/N···C tetrel bonding contacts comprise 41.0% of the total Hirshfeld surface for each molecule of TCYCPR. The points in the (d_i, d_e) regions of (1.442 Å, 1.598 Å) in the fingerprint plot is due to N···C contacts whereas the region (d_i = 1.593 Å, d_e = 1.442 Å) indicates the C···N contact.

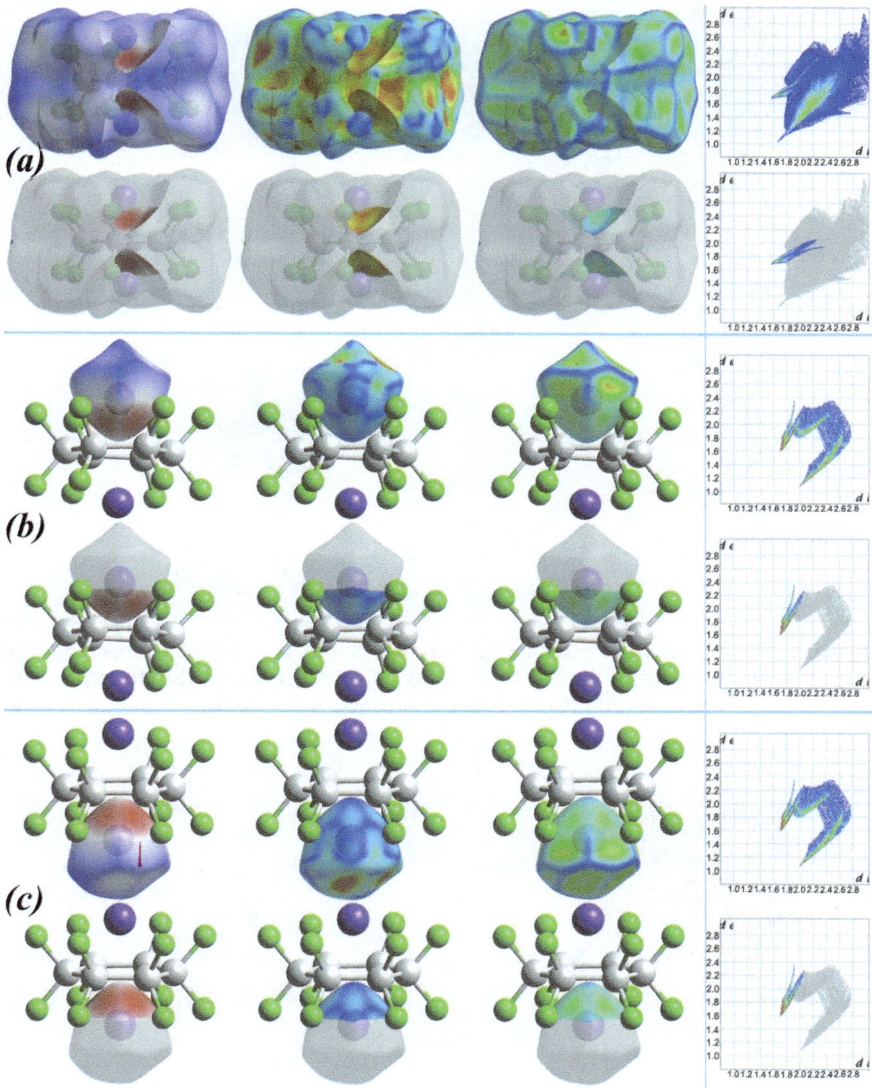

Figure 9.21 Hirshfeld surfaces are mapped with d_{norm}, shape-index, curvedness and full fingerprint plot, and resolved into Si⋯I/I⋯Si contacts [(a) moiety A; (b) moiety B and (c) moiety C] for the complex AZEBAJ.

The total C⋯N/N⋯C tetrel bonding is reflected in the distribution of the scattered points of the fingerprint plot that spreads around $d_i = d_e = 1.596$ Å (Figure 9.22).

For both structures, we computed the interaction energies, MEP surfaces and performed the AIM analyses. The results gathered in Figure 9.23 show that the MEP resulting from the combination of six Si σ-holes in the AZEBAJ structure is very large and positive, more than twice the value obtained along the C–C bond in the TCYCPR structure. In addition, the iodide presents a

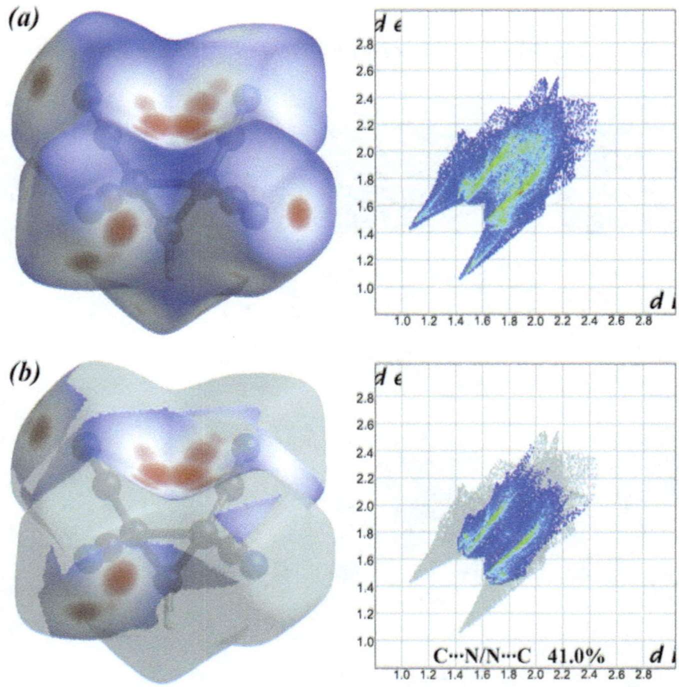

Figure 9.22 (a) Hirshfeld surfaces are mapped with d_{norm} and the corresponding fingerprint plot of TCYCPR; (b) d_{norm} surface and the fingerprint plot are resolved into C···N/N···C contacts in TCYCPR.

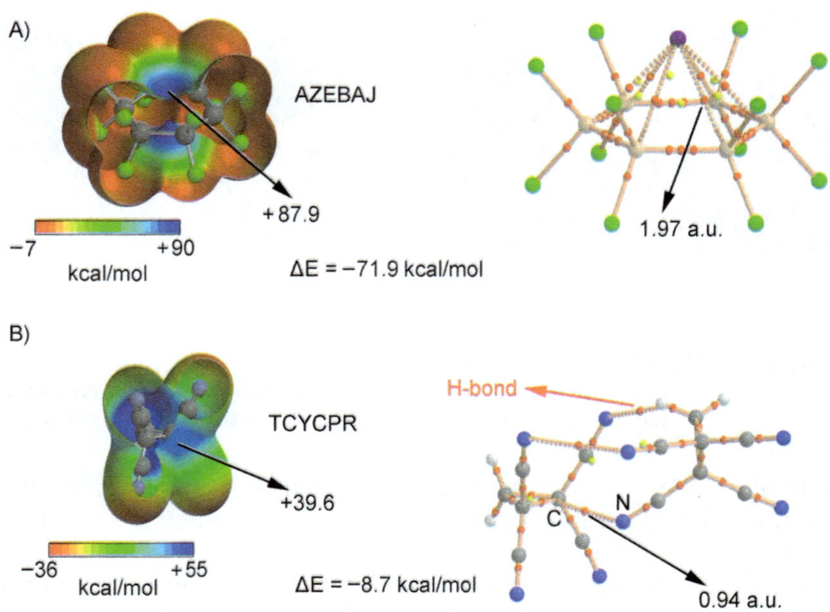

Figure 9.23 Interaction energies, MEP surfaces (0.002 a.u. isosurface) and AIM analyses (bond CP in red, ring CP in yellow and cage CP in green) showing tetrel bonding interactions of (A) AZEBAJ and (B) TCYCPR structures. The value of the density at the bond CP as well as both CSD codes is also indicated.

higher basicity than the neutral N atom in TCYCPR, thus leading to a more favorable interaction energy value for the complex involving the AZEBAJ structure, which exhibits six concurrent σ-hole interactions. The AIM analysis shows the existence of six bond CPs and bond paths connecting the I atom to the six silicon atoms. Moreover, the interaction is further characterized by the presence of six ring CPs (yellow spheres) as a consequence of the formation of six supramolecular rings. In TCYCPR the N···C interaction is characterized by the presence of a single bond CP and bond path connecting the N atom to one carbon atom of the CC bond.

9.6 Exploration of π-Hole Interactions

9.6.1 Group III Interactions (Triel Bonding)

The triel centers (group III of elements) in trihydrides and trihalides have been characterized as efficient π-hole bond donors[141] since they possess a region of positive electrostatic potential perpendicular to the molecular plane.[142] The triel atom in the species mentioned above is an electron deficient center since it presents six electrons in the outer shell, which are the electrons of three σ-bonds between the triel atom and hydrogen or halogen atoms. The electron deficient region is related to the outer vacant p orbital, perpendicularly located to the plane of the molecule. In general, the deficiency of valence electrons is commonly known as the hypovalency,[143] and the octet rule is not obeyed here. As noted in Figure 9.24, in NOBQAX[144] the solid-state architecture of $ClSiNMeBCl_2$ is governed by the triel bonding interactions established between two dihaloborane units, thus creating an infinite 1D column, and has been identified as a potential low-cost single-source precursor for quaternary high-temperature Si/C/B/N ceramics. On the other hand, in the GANMUD[145] structure, synthesized by the reactions of alkylhalogenoboranes $R^nBX_3 - n$ ($n = 0$–2) with 1,3-dimethyl-1,3,2-diazaborolidines, a triel bonding

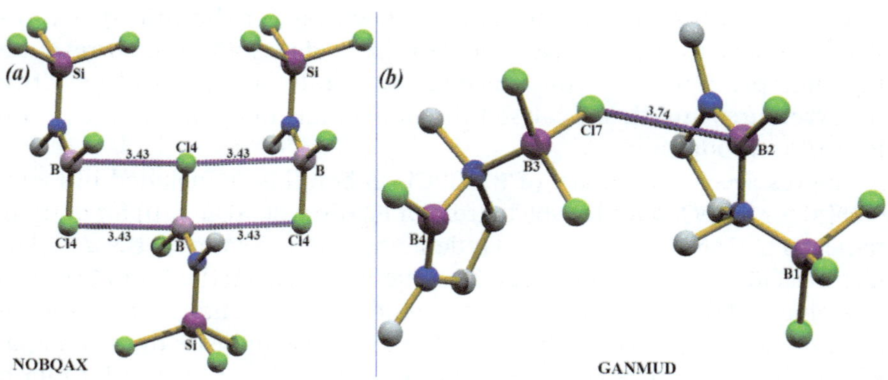

Figure 9.24 (a) Triel bonding network of NOBQAX showing B···Cl/Cl···B contacts; (b) X-ray structure of the dimer of GANMUD shows B···Cl/Cl···B interaction. The CSD reference codes are indicated and distances are in Å.

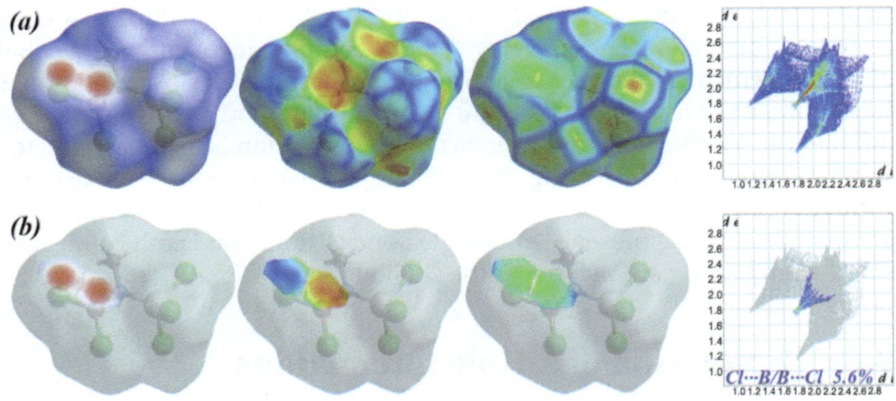

Figure 9.25 (a) Hirshfeld surfaces are mapped with d_{norm}, shape-index, curvedness and corresponding fingerprint plot of NOBQAX; (b) Hirshfeld surfaces and fingerprint plot are resolved into Cl···B/B···Cl contacts in NOBQAX.

interaction involving the chlorine and boron atoms of two neighboring moieties is established. In this section, we explore the triel bonding by using Hirshfeld surface analysis and theoretical calculations of both structures (NOBQAX and GANMUD), retrieving the crystallographic information from the CSD.

The X-ray structure of GANMUD contains two individual moieties in the asymmetric unit. We have calculated the Hirshfeld surface for both moieties and renamed them as moiety A and moiety B. For both the structures, Hirshfeld surfaces are mapped with d_{norm}, shape-index and curvedness and are depicted in Figures 9.25 and 9.26. The d_{norm} surface of NOBQAX has been drawn in the range (−0.160–0.934 Å) whereas for GANMUD the range was (0.002–1.228 Å) for moiety A and (0.002–1.239 Å) for moiety B. The shape-index and curvedness surfaces are mapped in the range (−1.0–1.0 Å) and (−4.0–0.4 Å) respectively for the title structures. For both structures, we have included the surfaces and the corresponding fingerprint plot that contains all types of interactions involved within the structures and then analysed the individual contribution due to triel B···Cl/Cl···B bonding.

The resolved counterpart of B···Cl/Cl···B bonding is included in Figure 9.25(b) for NOBQAX and at the bottom of Figure 9.26(a) and (b) for both the moieties of GANMUD. The circular depression in red color on the d_{norm} surface of NOBQAX clearly indicates the contact in between the B and Cl atoms. The blue and orange spot on the shape-index surface and greenish spot on the curvedness surface on the B and Cl atoms also indicate the interaction within the atoms. The corresponding two-dimensional fingerprint plot of NOBQAX contains two distinct shorter crossed spikes in between the longer spikes in the donor and acceptor region of the plot (Figure 9.25a). These shorter spikes indicate the B···Cl and Cl···B interactions of each molecule of NOBQAX and are evident in the region of (d_i = 1.682 Å, d_e = 1.722 Å) and

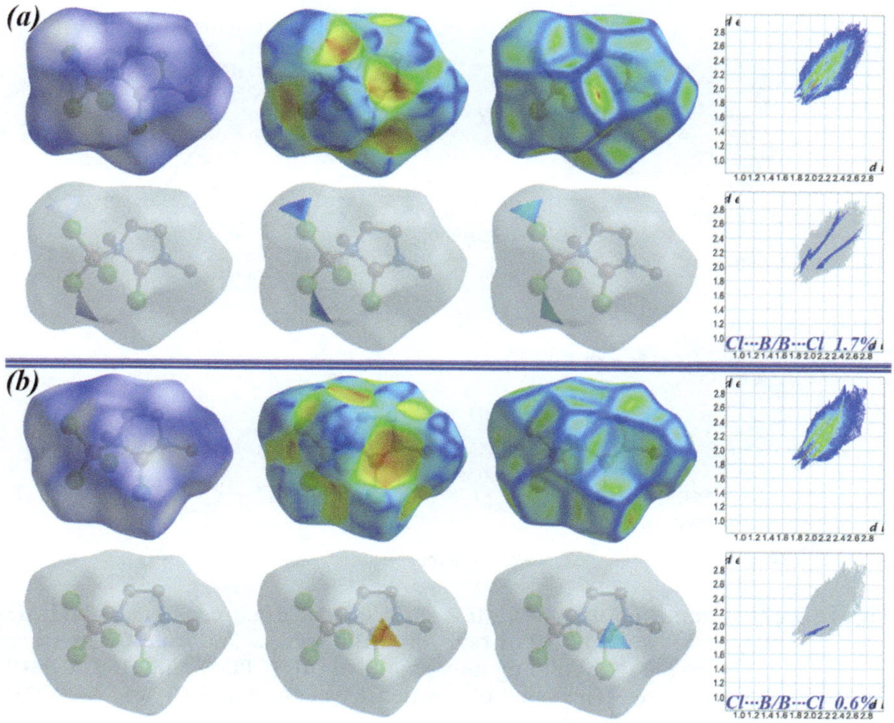

Figure 9.26 Hirshfeld surfaces are mapped with d_{norm}, shape-index, curvedness and full fingerprint plot, and are resolved into Cl···B/B···Cl contacts for the complex GANMUD (a) moiety A; (b) moiety B.

(d_i = 1.722 Å, d_e = 1.682 Å) respectively. Interestingly, the Cl···B and B···Cl contacts contribute 2.8% to the Hirshfeld surface and thus B···Cl/Cl···B triel bonding comprises 5.6% of the total Hirshfeld surface area of NOBQAX (Figure 9.25).

For the structure GANMUD, the interaction in between the Cl and B atoms is not clearly evident by the Hirshfeld surface of both moieties (Figure 9.26) due to the much lower contribution. No such red circular depressions are evident on the d_{norm} surfaces but the existence of Cl···B/B···Cl interaction is evident by the surfaces whenever we split the surfaces corresponding to Cl···B/B···Cl interaction. The resolved surfaces (d_{norm}, shape-index and curvedness) and associated fingerprint plots for both the moieties clearly indicate the existence of triel bonding within the structure of GANMUD (Figure 9.26). In moiety A, the Cl···B interaction (1.2%) dominates the counterpart B···Cl interaction (0.5%) and is evident in the regions (d_i = 1.832 Å, d_e = 1.882 Å) and (d_i = 2.058 Å, d_e = 1.953 Å) respectively. The total contribution from the Cl···B/B···Cl interaction comprises only 1.7% to the total Hirshfeld surface area of moiety A. For moiety B, no contribution is evidenced from the Cl···B contact. Moreover, the B···Cl contact also has a much lower contribution (0.6%) to the Hirshfeld surface and is evident in the bottom end region of the fingerprint plot (d_i = 1.882 Å, d_e = 1.837 Å) (Figure 9.26b).

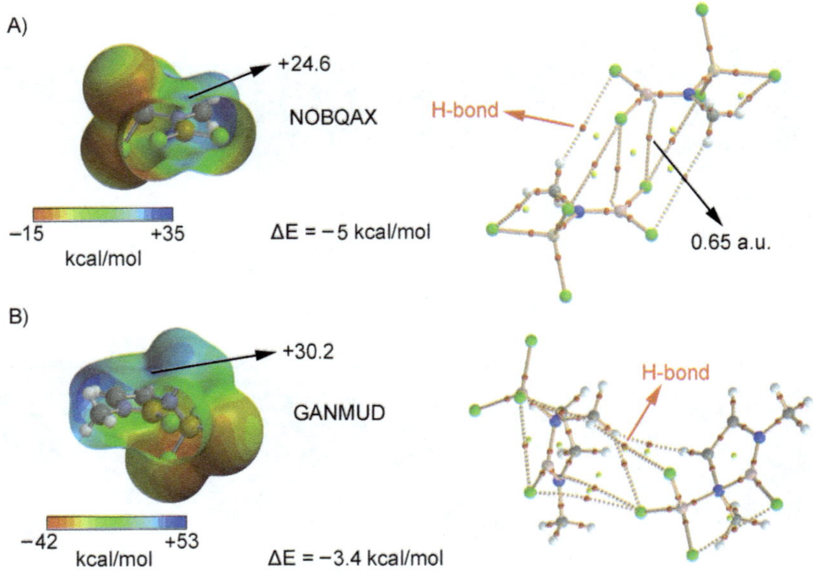

Figure 9.27 Interaction energies, MEP surfaces (0.002 a.u. isosurface) and AIM analyses (bond CP in red, ring CP in yellow and cage CP in green) showing triel bonding interactions of (A) NOBQAX and (B) GANMUD structures. The value of the density at the bond CP and both CSD codes are also indicated.

For both the structures, we have computed the interaction energies, MEP surfaces and performed AIM analyses of the homodimers. The results gathered in Figure 9.27 show that the MEP values over the boron atoms are similar in both compounds, thus explaining the similar interaction energy values obtained for both complexes. In addition, the AIM analysis reveals that for both structures several ancillary hydrogen bonding interactions can be observed (highlighted in red in the AIM analyses) in addition to the triel bonding interactions, characterized by the bond CPs and bond paths connecting the boron and chlorine atoms. Therefore, the contribution of the triel bonding to the total interaction energy is modest; in agreement with the Hirsfeld analysis commented on above that reveals a contribution lower the 2% of the triel bonding.

9.6.2 Group V Interactions (Pnicogen Bonding)

There a several examples in the literature devoted to the role of nitro-moieties to act as efficient and highly directional pnicogen bond donors, due to their planar geometry and the presence of two electron-withdrawing O atoms, which ensure a positive and easily accessible electrostatic potential region over the N atom. The analysis of the directionality and the strength of the pnicogen bonding interaction involving nitrobenzene and nitromethane derivatives, using the CSD as a source for experimental data, has also

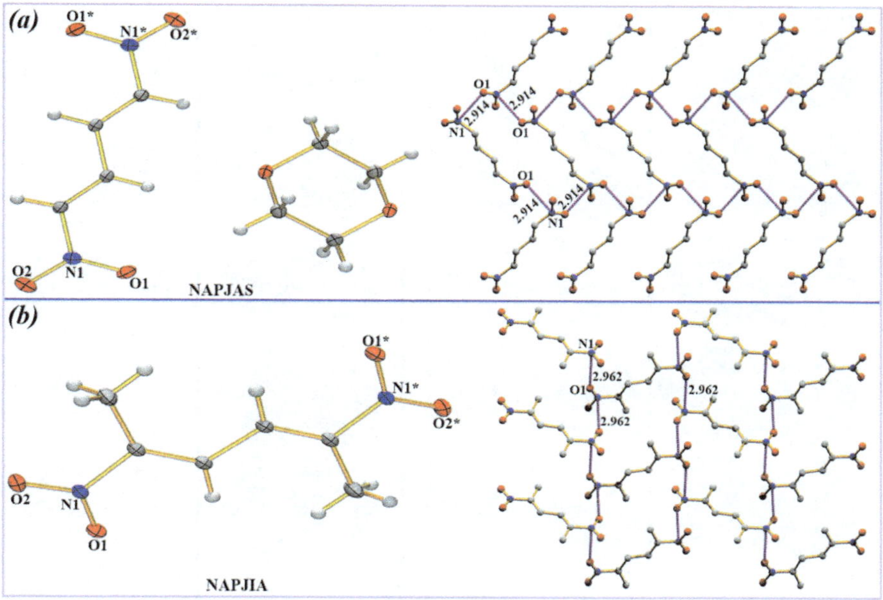

Figure 9.28 ORTEP view and network of N···O/O···N interaction in (a) NAPJAS; (b) NAPJIA. The CSD reference codes are indicated and distance is in Å.

been performed.[146] In addition, the directional character of this type of interaction involving NO_3^-[147] and nitrate esters[148] was also been unveiled by the same authors. Finally, the work of Bauzá and coworkers[149] is a nice example of the predominant character that pnicogen π-hole interactions possess in solid-state chemistry. Furthermore, several theoretical studies have been devoted to the analysis of the physical nature of the interaction,[41a,150] as well as the interplay between π-hole pnicogen bonds and other σ- and π-hole interactions.[111b]

In this section, we explored the pnicogen bonding by using Hirshfeld surface analysis and theoretical calculations of two structures (NAPJAS[149] and NAPJIA[149]) from the CSD. The molecular views and pnicogen-bonded network of the two structures are depicted in Figure 9.28.

The patterns of intermolecular interactions are similar in the title structures (NAPJAS and NAPJIA), which prompted us to explore the contributions of the weak pnicogen bonding forces in the crystal packing, as well as the importance of pnicogen bonding in establishing the organization of the extended structure. In this study, the pnicogen contacts (N···O/O···N), which are responsible in the crystal packing, were evaluated with respect to their contribution to the overall stability of the crystal structures. In this context, the Hirshfeld surfaces of NAPJAS and NAPJIA are illustrated in Figure 9.29 showing surfaces that have been mapped over d_e, d_i, d_{norm}, shape-index and curvedness. The d_e, d_i and d_{norm} surfaces of NAPJAS were drawn in the range (1.000–2.276 Å), (1.000–2.351 Å) and (−0.146–1.027 Å) respectively whereas these ranges were (1.000–2.457 Å), (1.000–2.436 Å) and (−0.071–0.926 Å) for

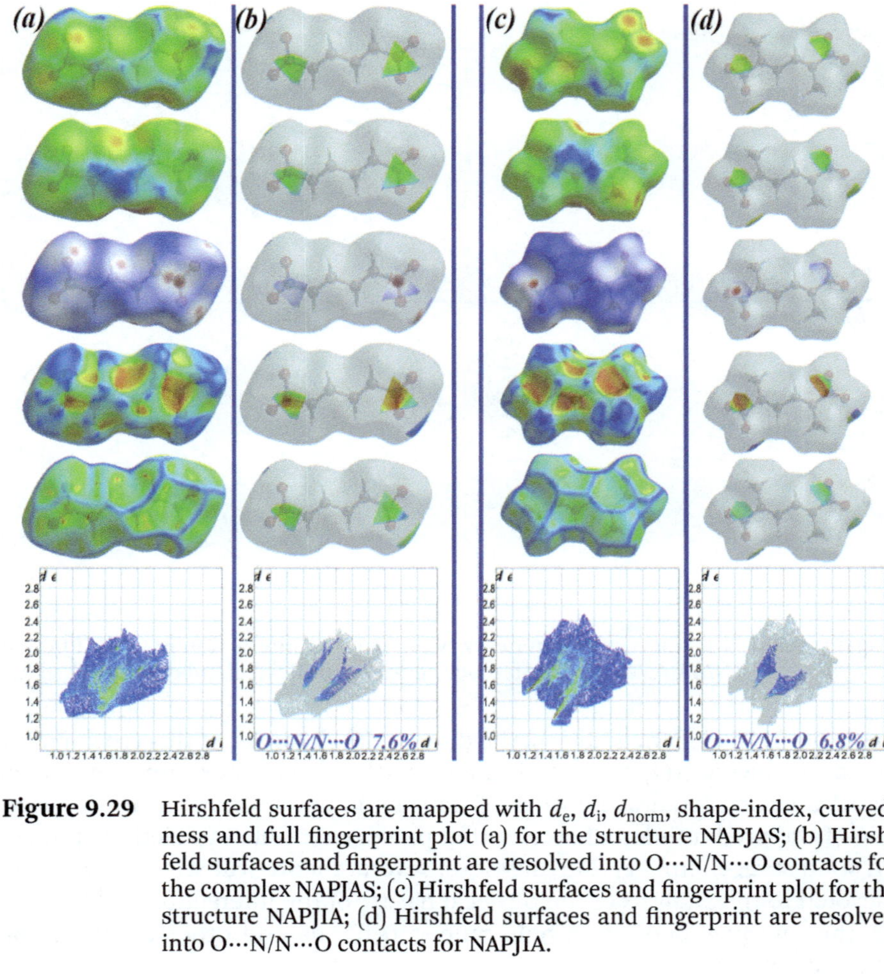

Figure 9.29 Hirshfeld surfaces are mapped with d_e, d_i, d_{norm}, shape-index, curvedness and full fingerprint plot (a) for the structure NAPJAS; (b) Hirshfeld surfaces and fingerprint are resolved into O···N/N···O contacts for the complex NAPJAS; (c) Hirshfeld surfaces and fingerprint plot for the structure NAPJIA; (d) Hirshfeld surfaces and fingerprint are resolved into O···N/N···O contacts for NAPJIA.

NAPJIA. For both structures, the shape-index and curvedness were drawn in the normal range *i.e.* (−1.0–1.0 Å) and (−4.0–0.4 Å). The red circular depressions on the d_{norm} surface of both structures indicate the interaction in between the O atom and N atom of the nitro group.

The colored patches (above the O and N atoms) on the d_e, d_i, shape-index and curvedness surfaces indicate the interaction between O and N atoms. The total interactions involved within the structures are also visible on the 2D fingerprint plots. We have decomposed the fingerprint plots to quantify the N···O/O···N contribution related to each structure of NAPJAS and NAPJIA. In both structures, the N···O interaction has a larger contribution that its O···N counterpart. The N···O/O···N interactions appear as two small spikes within the fingerprint plots (Figure 9.29) of both structures where the proportion of N···O/O···N interactions comprise 7.6% and 6.8% of the total Hirshfeld surfaces of NAPJAS and NAPJIA respectively. The points in the (d_i, d_e) regions of (1.512 Å, 1.386 Å) and (d_i, d_e) regions of (1.381 Å, 1.507 Å) in the fingerprint

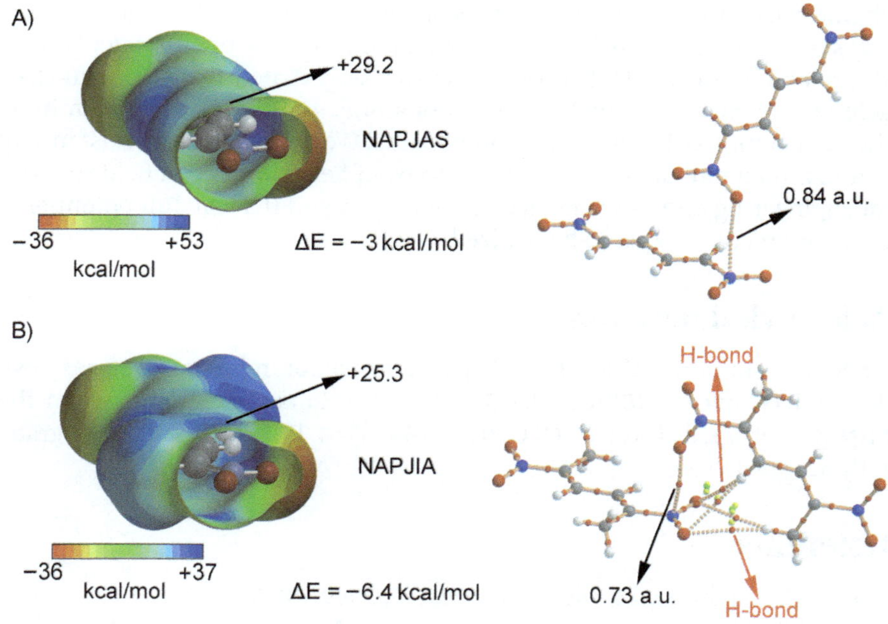

Figure 9.30 Interaction energies, MEP surfaces (0.002 a.u. isosurface) and AIM analyses (bond CP in red, ring CP in yellow and cage CP in green) showing π-hole pnicogen bonding interactions of (A) NAPJAS and (B) NAPJIA structures. The values of the density at the bond CP as well as both CSD codes are also indicated.

plot of NAPJAS are due to N···O and O···N interactions (Figure 9.29) whereas the region ($d_i = 1.532$ Å, $d_e = 1.411$ Å and $d_i = 1.411$ Å, $d_e = 1.532$ Å) indicate the N···O and O···N pnicogen bonding in NAPJIA (Figure 9.29d).

For both nitro-derivatives, we have computed the interaction energies, MEP surfaces and performed AIM analyses of the homodimers. As noted in Figure 9.30, the MEP value at the N atom in both compounds is very similar, thus slight differences in their dimerization energies are expected. However, the interaction energy in the NAPJIA structure is larger (more than double) in absolute value than in NAPJAS. The AIM analysis reveals that this is due to several ancillary hydrogen bonding interactions involving the methyl and aliphatic C–H groups and the nitro-moiety (see the AIM analysis in Figure 9.30B). The σ-hole interaction is characterized in both complexes by the presence of a bond CP and bond path that connect the O donor atom to the N atom of the nitro group.

9.7 Conclusions

In this chapter, we have analyzed several X-ray structures from the CSD exhibiting σ-hole and π-hole interactions that have a prominent role in their crystal packing. For the first time in the literature, they have been described and characterized using Hirshfeld surface analysis. Moreover, the energetic features of the interactions have been evaluated by means of DFT calculations. The σ- and π-hole contacts observed in the X-ray structures have been characterized by means of "Bader's theory of atoms in molecules" by examining the distribution of bond CPs and bond paths. We believe that the existence of σ- and π-hole interactions is expected to be of key relevance in fields of crystal engineering and supramolecular chemistry, and that the full potential of these interactions is yet to be realized.

Acknowledgements

S K Seth is grateful to the SERB-DST (Government of India) for Overseas Postdoctoral Fellowship (SB/OS/PDF-524/2015-16). Financial support from the MINECO of Spain (project CTQ2014-57393-C2-1-P, FEDER funds) is gratefully acknowledged.

References

1. H. J. Schneider, *Angew. Chem., Int. Ed.*, 2009, **48**, 3924–3977.
2. (a) J. M. Lehn, *Supramolecular Chemistry Concepts and Perspectives*, Wiley-VCH, Weinheim, 1995; (b) A. S. Mahadevi and G. N. Sastry, *Chem. Rev.*, 2016, **116**, 2775–2825.
3. J. J. Colin and H. G. de Claubry, *Ann. Chim.*, 1814, **90**, 87–100.
4. J. J. Colin, *Ann. Chim.*, 1814, **91**, 252–272.
5. F. Guthrie, *J. Chem. Soc.*, 1863, **16**, 239–244.
6. R. S. Mulliken, *J. Am. Chem. Soc.*, 1950, **72**, 600–608.
7. O. Hassel, *Structural Aspects of Interatomic Charge-transfer Bonding*, http://www.nobelprize.org/nobel_prizes/chemistry/laureates/1969/hassel-lecture.pdf.
8. P. Pelletier and J. Caventou, *Ann. Chim. Phys.*, 1819, **10**, 142–177.
9. O. Rhoussopoulos, *Ber. Dtsch. Chem. Ges.*, 1883, **16**, 202–203.
10. I. Remsen and J. F. Norris, *Am. Chem. J.*, 1896, **18**, 90–95.
11. J. Kleinberg and A. W. Davidson, *Chem. Rev.*, 1948, **42**, 601–609.
12. (a) H. A. Benesi and J. H. Hildebrand, *J. Am. Chem. Soc.*, 1948, **70**, 2832–2833; (b) H. A. Benesi and J. H. Hildebrand, *J. Am. Chem. Soc.*, 1949, **71**, 2703–2707.
13. (a) R. S. Mulliken, *J. Am. Chem. Soc.*, 1952, **74**, 811–824; (b) R. S. Mulliken, *J. Phys. Chem.*, 1952, **56**, 801–822.
14. (a) O. Hassel and J. Hvoslef, *Acta Chem. Scand.*, 1954, **8**, 873; (b) O. Hassel and C. Romming, *Q. Rev., Chem. Soc.*, 1962, **16**, 1–18; (c) O. Hassel and K. O. Stromme, *Acta Chem. Scand.*, 1958, **12**, 1146; (d) O. Hassel and K. O. Stromme, *Acta Chem. Scand.*, 1959, **13**, 1781–1786.
15. (a) R. S. Brown, *Acc. Chem. Res.*, 1997, **30**, 131–137; (b) D. Lenoir and C. Chiappe, *Chem.–Eur. J.*, 2003, **9**, 1036–1044.

16. S. Rosokha and J. Kochi, X-Ray Structures and Electronic Spectra of the π-Halogen Complexes between Halogen Donors and Acceptors with π-Receptors, in *Halogen Bonding. Fundamentals and Applications*, ed. P. Metrangolo and G. Resnati, Springer-Verlag, Berlin, Heidelberg, 2008, vol. 126, pp. 137–160.
17. H. Bent, *Chem. Rev.*, 1968, **68**, 587–648.
18. N. Ramasubbu, R. Parthasarathy and P. Murray-Rust, *J. Am. Chem. Soc.*, 1986, **108**, 4308–4314.
19. (a) A. C. Legon, *Chem.–Eur. J.*, 1998, **4**, 1890–1897; (b) A. C. Legon, *Angew. Chem., Int. Ed.*, 1999, **38**, 2686–2714.
20. (a) S. C. Nyburg and C. H. Faerman, *Acta Crystallogr., Sect. B: Struct. Sci.*, 1985, **41**, 274–279; (b) S. C. Nyburg, C. H. Faerman and L. Prasad, *Acta Crystallogr., Sect. B: Struct. Sci.*, 1987, **43**, 106–110; (c) M. Ishikawa, S. Ikuta, M. Katada and H. Sano, *Acta Crystallogr., Sect. B: Struct. Sci.*, 1990, **46**, 592–598.
21. (a) T. Brinck, J. S. Murray and P. Politzer, *Int. J. Quantum Chem.*, 1992, **44**, 57–64; (b) T. Brinck, J. S. Murray and P. Politzer, *Int. J. Quantum Chem.*, 1993, **48**, 73–88; (c) J. S. Murray, K. Paulsen and P. Politzer, *Proc.–Indian Acad. Sci., Chem. Sci.*, 1994, **106**, 267–275.
22. (a) P. Metrangolo and G. Resnati, *Chem.–Eur. J.*, 2001, **7**, 2511–2519; (b) P. Metrangolo, H. Neukirch, T. Pilati and G. Resnati, *Acc. Chem. Res.*, 2005, **38**, 386–395; (c) E. Parisini, P. Metrangolo, T. Pilati, G. Resnati and G. Terraneo, *Chem. Soc. Rev.*, 2011, **40**, 2267–2278.
23. M. Erdelyi, *Nat. Chem.*, 2014, **6**, 762–764.
24. (a) T. Clark, M. Hennemann, J. S. Murray and P. A. Politzer, *J. Mol. Model.*, 2007, **13**, 291–296; (b) J. S. Murray, P. Lane, T. Clark and P. Politzer, *J. Mol. Model.*, 2007, **13**, 1033–1038.
25. (a) P. Politzer, J. S. Murray and T. Clark, *Phys. Chem. Chem. Phys.*, 2013, **15**, 11178–11189; (b) L. P. Wolters, P. Schyman, M. J. Pavan, W. L. Jorgensen, F. M. Bickelhaupt and S. Kozuch, *Wiley Interdiscip. Rev.: Comput. Mol. Sci.*, 2014, **4**, 523–540; (c) K. El Hage, V. Pandyarajan, N. B. Phillips, B. J. Smith, J. G. Menting, J. Whittaker, M. C. Lawrence, M. Meuwly and M. A. Weiss, *J. Biol. Chem.*, 2016, **291**, 27023–27041; (d) P. Politzer, J. S. Murray and T. Clark, *Top. Curr. Chem.*, 2015, **358**, 19–42.
26. (a) P. Politzer, J. S. Murray and T. Clark, *Phys. Chem. Chem. Phys.*, 2010, **12**, 7748–7757; (b) K. E. Riley, J. S. Murray, J. Franfrlík, J. Rezáč, R. J. Solá, M. C. Concha, F. M. Ramos and P. Politzer, *J. Mol. Model.*, 2011, **17**, 3309–3318; (c) B. V. Pandiyan, P. Deepa and P. Kolandaivel, *Mol. Phys.*, 2016, **114**, 3629–3642.
27. (a) J. S. Murray, P. Lane, T. Clark and P. Politzer, *J. Mol. Model.*, 2007, **13**, 1033–1038; (b) J. S. Murray, P. Lane and P. Politzer, *Int. J. Quantum Chem.*, 2007, **107**, 2286–2292; (c) J. S. Murray, P. Lane and P. Politzer, *J. Mol. Model.*, 2009, **15**, 723–729.

28. (a) S. C. Nyburg and C. H. Faerman, *Acta Crystallogr., Sect. B: Struct. Sci.*, 1985, **41**, 274–279; (b) P. Politzer, K. E. Riley, F. A. Bulat and J. S. Murray, *Comput. Theor. Chem.*, 2012, **998**, 2–8; (c) T. N. G. Row and R. Parthasarathy, *J. Am. Chem. Soc.*, 1981, **103**, 477–479; (d) N. Ramasubbu and R. Parthasarathy, *Phosphorus Sulfur Relat. Elem.*, 1987, **31**, 221–229.
29. (a) A. Frontera and A. Bauzá, *Phys. Chem. Chem. Phys.*, 2017, **19**, 30063–30068; (b) A. Bauzá and A. Frontera, *Angew. Chem., Int. Ed.*, 2015, **54**, 7340–7343; (c) A. Bauzá and A. Frontera, *Phys. Chem. Chem. Phys.*, 2015, **17**, 24748–24753; (d) A. Bauzá and A. Frontera, *ChemPhysChem*, 2015, **16**, 3625–3630.
30. (a) M. Egli and S. Sarkhel, *Acc. Chem. Res.*, 2007, **40**, 197–205; (b) T. J. Mooibroek, P. Gamez and J. Reedijk, *CrystEngComm*, 2008, **10**, 1501–1515.
31. (a) A. Frontera, P. Gamez, M. Mascal, T. J. Mooibroek and J. Reedijk, *Angew. Chem., Int. Ed.*, 2011, **50**, 9564–9583; (b) B. L. Schottel, H. T. Chifotides and K. R. Dunbar, *Chem. Soc. Rev.*, 2008, **37**, 68–83; (c) C. Caltagirone and P. A. Gale, *Chem. Soc. Rev.*, 2009, **38**, 520–563.
32. (a) P. Gamez, T. J. Mooibroek, S. J. Teat and J. Reedijk, *Acc. Chem. Res.*, 2007, **40**, 435–444; (b) B. P. Hay and V. S. Bryantsev, *Chem. Commun.*, 2008, **44**, 2417–2428.
33. (a) D. Quiñonero, C. Garau, C. Rotger, A. Frontera, P. Ballester, A. Costa and P. M. Deyà, *Angew. Chem., Int. Ed.*, 2002, **41**, 3389–3392; (b) M. Mascal, A. Armstrong and M. D. Bartberger, *J. Am. Chem. Soc.*, 2002, **124**, 6274–6276; (c) I. Alkorta, I. Rozas and J. Elguero, *J. Am. Chem. Soc.*, 2002, **124**, 8593–8598; (d) D. Y. Kim, N. J. Singh and K. S. Kim, *J. Chem. Theory Comput.*, 2008, **4**, 1401–1407.
34. (a) B. L. Schottel, J. Bacsa and K. R. Dunbar, *Chem. Commun.*, 2005, **41**, 46–47; (b) B. Han, J. Lu and J. K. Kochi, *Cryst. Growth Des.*, 2008, **8**, 1327–1334; (c) M. Mascal, I. Yakovlev, E. B. Nikitin and J. C. Fettinger, *Angew. Chem., Int. Ed.*, 2007, **46**, 8782–8784; (d) R. J. Götz, A. Robertazzi, I. Mutikainen, U. Turpeinen, P. Gamez and J. Reedijk, *Chem. Commun.*, 2008, **44**, 3384–3386; (e) M. Albrecht, M. Müller, O. Mergel, K. Rissanen and A. Valkonen, *Chem.–Eur. J.*, 2010, **16**, 5062–5069.
35. (a) C. Estarellas, A. Frontera, D. Quiñonero and P. M. Deyà, *Angew. Chem., Int. Ed.*, 2011, **50**, 415–418; (b) S. Chakravarty, Z. Z. Sheng, B. Iverson and B. Moore, *FEBS Lett.*, 2012, **586**, 4180–4185; (c) D. D. Jenkins, J. B. Harris, E. E. Howell, R. J. Hinde and J. Baudry, *J. Comput. Chem.*, 2013, **34**, 518–522.
36. (a) R. E. Dawson, A. Hennig, D. P. Weimann, D. Emery, V. Ravikumar, J. Montenegro, T. Takeuchi, S. Gabutti, M. Mayor, J. Mareda, C. A. Schalley and S. Matile, *Nat. Chem.*, 2010, **2**, 533–538; (b) N. Sakai, J. Mareda, E. Vauthey and S. Matile, *Chem. Commun.*, 2010, **46**, 4225–4237.
37. (a) H. B. Burgi, *Inorg. Chem.*, 1973, **12**, 2321–2325; (b) H. B. Burgi, J. D. Dunitz and E. Shefter, *J. Am. Chem. Soc.*, 1973, **95**, 5065–5067; (c) H. B. Burgi, J. D. Dunitz, J. M. Lehn and G. Wipff, *Tetrahedron*, 1974, **30**, 1563–1572.

38. M. Egli and R. V. Gessner, *Proc. Natl. Acad. Sci. U. S. A.*, 1995, **92**, 180–184.
39. P. Sjoberg and P. Politzer, *J. Phys. Chem.*, 1990, **94**, 3959–3961.
40. (a) L. M. Azofra, I. Alkorta and S. Scheiner, *Theor. Chem. Acc.*, 2014, **133**, 1586, 8p; (b) M. D. Esrafili and R. Nurazar, *Mol. Phys.*, 2016, **114**, 276–282; (c) X. Guo, Y. W. Liu, Q. Z. Li, W. Z. Li and J. B. Cheng, *Chem. Phys. Lett.*, 2015, **620**, 7–12; (d) D. Quiñonero, *Molecules*, 2015, **20**, 11632–11659.
41. (a) A. Bauzá, R. Ramis and A. Frontera, *J. Phys. Chem. A*, 2014, **118**, 2827–2834; (b) D. Quiñonero, A. Bauzá, G. Sánchez-Sanz, C. Trujillo, I. Alkorta and J. Elguero, *New J. Chem.*, 2016, **40**, 9060–9072.
42. (a) G. Sánchez-Sanz, C. Trujillo, M. Solimannejad, I. Alkorta and J. Elguero, *Phys. Chem. Chem. Phys.*, 2013, **15**, 14310–14318; (b) J. E. Del Bene, I. Alkorta and J. Elguero, *J. Phys. Chem. A*, 2013, **117**, 6893–6903.
43. (a) S. L. Grabowski, *Molecules*, 2015, **20**, 11297–11316; (b) S. L. Grabowski, *ChemPhysChem*, 2015, **16**, 1470–1479; (c) S. L. Grabowski, *ChemPhysChem*, 2014, **15**, 2985–2993.
44. M. X. Liu, H. Y. Zhuo, Q. Z. Li, C. Wen-Zuo and J. B. Cheng, *J. Mol. Model.*, 2016, **22**, 1–7.
45. S. Yourdkhani, T. Korona and N. L. Hadipour, *J. Comput. Chem.*, 2015, **36**, 2412–2428.
46. T. Clark, *Wiley Interdiscip. Rev.: Comput. Mol. Sci.*, 2013, **3**, 13–20.
47. (a) T. Clark, M. Hennemann, J. S. Murray and P. Politzer, *J. Mol. Model.*, 2007, **13**, 291–296; (b) T. Brinck, J. S. Murray and P. Politzer, *Int. J. Quantum Chem.*, 1992, **44**, 55–64; (c) P. Auffinger, F. A. Hays, E. Westhof and P. S. Ho, *Proc. Natl. Acad. Sci. U. S. A.*, 2004, **101**, 16789–16794; (d) P. Politzer, P. Lane, M. C. Concha, Y. Ma and J. S. Murray, *J. Mol. Model.*, 2007, **13**, 305–311.
48. (a) P. Politzer and J. S. Murray, *Practical Aspects of Computational Chemistry*, Springer, Heidelberg, 2009; (b) J. S. Murray, K. E. Riley, P. Politzer and T. Clark, *Aust. J. Chem.*, 2010, **63**, 1598–1607.
49. (a) E. Arunan, G. R. Desiraju, R. A. Klein, J. Sadlej, S. Scheiner, I. Alkorta, D. C. Clary, R. H. Crabtree, J. J. Dannenberg, P. Hobza, H. G. Kjaergaard, A. C. Legon, B. Mennucci and D. J. Nesbitt, *Pure Appl. Chem.*, 2011, **83**, 1637–1641; (b) G. R. Desiraju, P. S. Ho, L. Kloo, A. C. Legon, R. Marquardt, P. Metrangolo, P. Politzer, G. Resnati and K. Rissanen, *Pure Appl. Chem.*, 2013, **85**, 1711–1713.
50. T. Clark, *Faraday Discuss.*, 2017, **203**, 9–27, and references cited therein.
51. T. Clark, P. Politzer and J. A. Murray, *Wiley Interdiscip. Rev.: Comput. Mol. Sci.*, 2015, **5**, 169–177.
52. J. S. Murray, P. Lane, T. Clark, K. E. Riley and P. Politzer, *J. Mol. Model.*, 2012, **18**, 541–548.
53. A. J. Stone, *J. Am. Chem. Soc.*, 2013, **135**, 7005–7009.
54. A. Bauzá, I. Alkorta, A. Frontera and J. Elguero, *J. Chem. Theory Comput.*, 2013, **9**, 5201–5210.
55. V. R. Hathwar, D. Chopra, P. Panini and T. N. Guru Row, *Cryst. Growth Des.*, 2014, **14**, 5366–5369.

56. C. Esterhuysen, A. Heßelmann and T. Clark, *ChemPhysChem*, 2017, **18**, 772.
57. (a) A. Frontera, D. Quiñonero and P. M. Deyà, *Wiley Interdiscip. Rev.: Comput. Mol. Sci.*, 2011, **1**, 440–459; (b) A. Frontera, *Coord. Chem. Rev.*, 2013, **257**, 1716–1727; (c) J. J. Fiol, M. Barceló-Oliver, A. Tasada, A. Frontera, À. Terrón and Á. García-Raso, *Coord. Chem. Rev.*, 2013, **257**, 2705–2715.
58. (a) D. Kim, P. Tarakeshwar and K. S. Kim, *J. Phys. Chem. A*, 2004, **108**, 1250–1258; (b) D. Y. Kim, N. J. Singh, J. W. Lee and K. S. Kim, *J. Chem. Theory Comput.*, 2008, **4**, 1162–1169.
59. (a) C. Garau, A. Frontera, D. Quiñonero, P. Ballester, A. Costa and P. M. Deyà, *ChemPhysChem*, 2003, **4**, 1344–1348; (b) D. Quiñonero, C. Garau, A. Frontera, P. Ballester, A. Costa and P. M. Deyà, *Chem. Phys. Lett.*, 2002, **359**, 486–492; (c) A. Bauzá, P. M. Deyà, A. Frontera and D. Quiñonero, *Phys. Chem. Chem. Phys.*, 2014, **16**, 1322–1326.
60. S. E. Wheeler and K. N. Houk, *J. Phys. Chem. A*, 2010, **114**, 8658–8664.
61. S. E. Wheeler and J. W. G. Bloom, *Chem. Commun.*, 2014, **50**, 11118–11121.
62. (a) C. Garau, D. Quiñonero, A. Frontera, P. Ballester, A. Costa and P. M. Deyà, *Org. Lett.*, 2003, **5**, 2227–2229; (b) C. Garau, A. Frontera, D. Quiñonero, P. Ballester, A. Costa and P. M. Deyà, *J. Phys. Chem. A*, 2004, **108**, 9423–9427.
63. (a) C. Garau, D. Quiñonero, A. Frontera, P. Ballester, A. Costa and P. M. Deyà, *New J. Chem.*, 2003, **27**, 211–214; (b) D. Quiñonero, A. Frontera, C. Garau, P. Ballester, A. Costa and P. M. Deyà, *ChemPhysChem*, 2006, **7**, 2487–2491.
64. G. R. Desiraju, *Crystal Engineering: The Design of Organic Solids*, Elsevier, Amsterdam, 1989.
65. G. R. Desiraju, *Chem. Commun.*, 1997, 1475–1482.
66. (a) J. Y. Lee, B. H. Hong, W. Y. Kim, S. K. Min, Y. Kim, M. V. Jouravlev, R. Bose, K. S. Kim, I. C. Hwang, L. J. Kaufman, C. W. Wong, P. Kim and K. S. Kim, *Nature*, 2009, **460**, 498–501; (b) T. S. Thakur and G. R. Desiraju, *Cryst. Growth Des.*, 2008, **8**, 4031–4044; (c) S. L. Price, *Acc. Chem. Res.*, 2009, **42**, 117–126.
67. J. D. Dunitz and A. Gavezzotti, *Chem. Soc. Rev.*, 2009, **38**, 2622–2633.
68. M. C. Etter, *Acc. Chem. Res.*, 1990, **23**, 120–126.
69. G. R. Desiraju, *Angew. Chem., Int. Ed.*, 1995, **34**, 2311–2327.
70. C. A. Hunter, *Angew. Chem., Int. Ed.*, 2004, **43**, 5310–5324.
71. M. A. Spackman and J. J. McKinnon, *CrystEngComm*, 2002, **4**, 378–392.
72. (a) J. J. McKinnon, D. Jayatilaka and M. A. Spackman, *Chem. Commun.*, 2007, 3814–3816; (b) M. A. Spackman, J. J. McKinnon and D. Jayatilaka, *CrystEngComm*, 2008, **10**, 377–388; (c) M. A. Spackman and D. Jayatilaka, *CrystEngComm*, 2009, **11**, 19–32.
73. (a) J. J. McKinnon, F. P. A. Fabbiani and M. A. Spackman, *Cryst. Growth Des.*, 2007, **7**, 755–769; (b) F. L. Hirshfeld, *Theor. Chim. Acta*, 1977, **44**, 129–138.
74. M. A. Spackman and P. G. Byrom, *Chem. Phys. Lett.*, 1997, **267**, 215–220.

75. J. J. McKinnon, A. S. Mitchell and M. A. Spackman, *Chem.-Eur. J.*, 1998, **4**, 2136–2141.
76. R. F. W. Bader, *Atoms in Molecules–a Quantum Theory*, Oxford University Press, Oxford, 1990.
77. E. Wigner and F. Seitz, *Phys. Rev.*, 1933, **43**, 804–810.
78. S. K. Wolff, D. J. Grimwood, J. J. McKinnon, M. J. Turner, D. Jayatilaka and M. A. Spackman, *Crystal Explorer (Version 3.0)*, University of Western Australia, 2010.
79. A. Altomare, G. Cascarano, C. Giacovazzo and A. Guagliardi, *J. Appl. Crystallogr.*, 1994, **27**, 435–436.
80. J. J. McKinnon, M. A. Spackman and A. S. Mitchell, *Acta Crystallogr., Sect. B: Struct. Sci.*, 2004, **60**, 627–668.
81. J. J. Koenderink and A. J. van Doorn, *Image Vis. Comput.*, 1992, **10**, 557–564.
82. F. H. Allen, O. Kennard, D. G. Watson, L. Brammer, A. G. Orpen and R. Taylor, *J. Chem. Soc., Perkin Trans. 1*, 1987, **2**, S1–S19.
83. (a) J. Bernstein, *Cryst. Growth Des.*, 2011, **11**, 632–650; (b) R. J. Davey, *Chem. Commun.*, 2003, 1463–1467.
84. J. O. Henck, U. J. Griesser and A. Burger, *Pharm. Ind.*, 1997, **59**, 165–169.
85. (a) S. K. Seth, I. Saha, C. Estarellas, A. Frontera, T. Kar and S. Mukhopadhyay, *Cryst. Growth Des.*, 2011, **11**, 3250–3265; (b) S. K. Seth, D. Sarkar, A. D. Jana and T. Kar, *Cryst. Growth Des.*, 2011, **11**, 4837–4849; (c) S. K. Seth, D. Sarkar and T. Kar, *CrystEngComm*, 2011, **13**, 4528–4535; (d) S. K. Seth, D. Sarkar, A. Roy and T. Kar, *CrystEngComm*, 2011, **13**, 6728–6741; (e) S. K. Seth, P. C. Mandal, T. Kar and S. Mukhopadhyay, *J. Mol. Struct.*, 2011, **994**, 109–116; (f) S. K. Seth, G. C. Maity and T. Kar, *J. Mol. Struct.*, 2011, **1000**, 120–126; (g) S. K. Seth, N. C. Saha, S. Ghosh and T. Kar, *Chem. Phys. Lett.*, 2011, **506**, 309–314; (h) P. Manna, S. K. Seth, A. Das, J. Hemming, R. Prendergast, M. Helliwell, S. R. Choudhury, A. Frontera and S. Mukhopadhyay, *Inorg. Chem.*, 2012, **51**, 3557–3571; (i) S. K. Seth, G. C. Maity and T. Kar, *J. Mol. Struct.*, 2012, **1021**, 89–94; (j) M. Mitra, P. Manna, S. K. Seth, A. Das, J. Meredith, M. Helliwell, A. Bauzá, S. R. Choudhury, A. Frontera and S. Mukhopadhyay, *CrystEngComm*, 2013, **15**, 686–696; (k) S. K. Seth, *CrystEngComm*, 2013, **15**, 1772–1781; (l) S. K. Seth, N. K. Das, K. Aich, D. Sen, H. K. Fun and S. Goswami, *J. Mol. Struct.*, 2013, **1048**, 157–165; (m) M. Mitra, S. K. Seth, S. R. Chudhury, P. Manna, A. Das, M. Helliwell, A. Bauzá, A. Frontera and S. Mukhopadhyay, *Eur. J. Inorg. Chem.*, 2013, 4679–4685; (n) P. Manna, S. K. Seth, M. Mitra, A. Das, N. J. Singh, S. R. Choudhury, T. Kar and S. Mukhopadhyay, *CrystEngComm*, 2013, **15**, 7879–7886; (o) S. K. Seth, *Inorg. Chem. Commun.*, 2014, **43**, 60–63; (p) S. K. Seth, *J. Mol. Struct.*, 2014, **1064**, 70–75; (q) T. Samanta, L. Dey, J. Dinda, S. K. Chattopadhyay and S. K. Seth, *J. Mol. Struct.*, 2014, **1068**, 58–70; (r) S. K. Seth, *J. Mol. Struct.*, 2014, **1070**, 65–74; (s) P. Manna, S. R. Choudhury, M. Mitra, S. K. Seth, M. Helliwell, A. Bauzá, A. Frontera and S. Mukhopadhyay, *J. Solid State Chem.*, 2014,

220, 149–156; (t) M. Mitra, P. Manna, A. Bauzá, P. Ballester, S. K. Seth, S. R. Choudhury, A. Frontera and S. Mukhopadhyay, *J. Phys. Chem. B*, 2014, **118**, 14713–14726; (u) S. K. Seth, V. S. Lee, J. Yana, S. M. Zain, A. C. Cunha, V. F. Ferreira, A. K. Jordão, M. C. B. V. de Souza, S. M. S. V. Wardell, J. L. Wardell and E. R. T. Tiekink, *CrystEngComm*, 2015, **17**, 2255–2266; (v) M. Servati-Gargari, S. K. Seth, R. L. LaDuca, O. Z. Yesilel, A. Pochodylo, A. Bauzá, B. C. Jana, T. Arslan, A. Frontera and G. Mahmoudi, *Inorg. Chim. Acta*, 2015, **438**, 220–231; (w) S. Adhikari, T. Kar and S. K. Seth, *RSC Adv.*, 2016, **6**, 99139–99148; (x) G. Mahmoudi, L. Dey, H. Chowdhury, A. Bauzá, B. K. Ghosh, A. M. Kirillov, S. K. Seth, A. V. Gurbanov and A. Frontera, *Inorg. Chim. Acta*, 2017, **461**, 192–205; (y) K. Jana, T. Maity, S. C. Debnath, B. C. Samanta and S. K. Seth, *J. Mol. Struct.*, 2017, **1130**, 844–854.

86. R. Ahlrichs, M. Bär, M. Hacer, H. Horn and C. Kömel, *Chem. Phys. Lett.*, 1989, **162**, 165–169.
87. S. B. Boys and F. Bernardi, *Mol. Phys.*, 1970, **19**, 553–566.
88. T. A. Keith, *AIMAll (Version 13.05.06)*, TK Gristmill Software, Overland Park KS, USA, 2013.
89. Y. Shao, L. F. Molnar, Y. Jung, J. Kussmann, C. Ochsenfeld, S. T. Brown, A. T. B. Gilbert, L. V. Slipchenko, S. V. Levchenko, D. P. O'Neill, R. A. DiStasio Jr, R. C. Lochan, T. Wang, G. J. O. Beran, N. A. Besley, J. M. Herbert, C. Y. Lin, T. Van Voorhis, S. H. Chien, A. Sodt, R. P. Steele, V. A. Rassolov, P. E. Maslen, P. P. Korambath, R. D. Adamson, B. Austin, J. Baker, E. F. C. Byrd, H. Dachsel, R. J. Doerksen, A. Dreuw, B. D. Dunietz, A. D. Dutoi, T. R. Furlani, S. R. Gwaltney, A. Heyden, S. Hirata, C. P. Hsu, G. Kedziora, R. Z. Khalliulin, P. Klunzinger, A. M. Lee, M. S. Lee, W. Z. Liang, I. Lotan, N. Nair, B. Peters, E. I. Proynov, P. A. Pieniazek, Y. M. Rhee, J. Ritchie, E. Rosta, C. D. Sherrill, A. C. Simmonett, J. E. Subotnik, H. L. Woodcock III, W. Zhang, A. T. Bell, A. K. Chakraborty, D. M. Chipman, F. J. Keil, A. Warshel, W. J. Hehre, H. F. Schaefer, J. Kong, A. I. Krylov, P. M. W. Gill and M. Head-Gordon, *Phys. Chem. Chem. Phys.*, 2006, **8**, 3172.
90. (a) L. C. Allen, *J. Am. Chem. Soc.*, 1989, **111**, 9003–9014; (b) P. Politzer, Z. P. I. Shields, F. A. Bulat and J. S. Murray, *J. Chem. Theory Comput.*, 2011, **7**, 377–384.
91. L. P. Wolters, P. Schyman, M. J. Pavan, W. L. Jorgensen, F. M. Bickelhaupt and S. Kozuch, *Wiley Interdiscip. Rev.: Comput. Mol. Sci.*, 2014, **4**, 523–540.
92. P. Politzer and J. S. Murray, *ChemPhysChem*, 2013, **14**, 278–294.
93. (a) P. Murray-Rust and W. D. S. Motherwell, *J. Am. Chem. Soc.*, 1979, **101**, 4374–4376; (b) P. Murray-Rust, W. C. Stallings, C. T. Monti, R. K. Preston and J. P. Glusker, *J. Am. Chem. Soc.*, 1983, **105**, 3206–3214.
94. F. H. Allen, *Acta Crystallogr., Sect. B: Struct. Sci.*, 2002, **58**, 380–388.
95. A. S. Novikov, D. M. Ivanov, M. S. Avdontceva and V. Yu. Kukushkin, *CrystEngComm*, 2017, **19**, 2517–2525.

96. A. J. Rybarczyk-Pirek, L. Chęcińska, M. Małecka and S. Wojtulewski, *Cryst. Growth Des.*, 2013, **13**, 3913–3924.
97. (a) O. Dumele, B. Schreib, U. Warzok, N. Trapp, C. A. Schalley and F. Diederich, *Angew. Chem., Int. Ed.*, 2017, **56**, 1152–1157; (b) F. Pan, N. K. Beyeh, R. H. A. Ras and K. Rissanen, *Cryst. Growth Des.*, 2016, **16**, 6729–6733; (c) S. T. Nguyen, A. L. Rheingold, G. S. Tschumper and D. L. Watkins, *Cryst. Growth Des.*, 2016, **16**, 6648–6653; (d) F. Fernández-Palacio, M. Poutanen, M. Saccone, A. Siiskonen, G. Terraneo, G. Resnati, O. Ikkala, P. Metrangolo and A. Priimagi, *Chem. Mater.*, 2016, **28**, 8314–8321; (e) G. Kawaguchi, M. Maesato, T. Komatsu, T. Imakubo, A. Kiswandhi, D. Graf and H. Kitagawa, *Chem. Mater.*, 2016, **28**, 7276–7286; (f) P. S. Salini, K. Shinaj and M. Hariharan, *Cryst. Growth Des.*, 2016, **16**, 5822–5830.
98. A. Crihfield, J. Hartwell, D. Phelps, R. B. Walsh, J. L. Harris, J. F. Payne, W. T. Pennington and T. W. Hanks, *Cryst. Growth Des.*, 2003, **3**, 313–320.
99. T. Shirman, J. F. Lamere, L. J. W. Shimon, T. Gupta, J. M. L. Martin and M. E. Van Der Boom, *Cryst. Growth Des.*, 2008, **8**, 3066–3072.
100. (a) R. Gajda and A. Katrusiak, *Acta Crystallogr., Sect. B: Struct. Sci.*, 2007, **63**, 896–902; (b) C. B. Aakeroy, J. Desper, B. A. Helfrich, P. Metrangolo, T. Pilati, G. Resnati and A. Stevenazzi, *Chem. Commun.*, 2007, **43**, 4236–4238; (c) A. K. Przybył and M. J. Kubicki, *J. Chem. Crystallogr.*, 2012, **42**, 685–690.
101. T. A. Logothetis, F. Meyer, P. Metrangolo, T. Pilati and G. Resnati, *New J. Chem.*, 2004, **28**, 760–763.
102. A. C. B. Lucassen, T. Zubkov, L. J. W. Shimon and M. E. Van Der Boom, *CrystEngComm*, 2007, **9**, 538–540.
103. R. F. W. Bader, *Chem. Rev.*, 1991, **91**, 893–928.
104. (a) A. Bauzá, D. Quiñonero, P. M. Deya and A. Frontera, *CrystEngComm*, 2013, **15**, 3137–3144; (b) M. Iwaoka and N. Isozumi, *Molecules*, 2012, **17**, 7266–7283; (c) G. Sánchez-Sanz, C. Trujillo, I. Alkorta and J. Elguero, *ChemPhysChem*, 2012, **13**, 496–503; (d) L. M. Azofra and S. Scheiner, *J. Chem. Phys.*, 2014, **140**, 034302; (e) J. Fanfrlík, A. Přáda, Z. Padělková, A. Pecina, J. Macháček, M. Lepšík, J. Holub, A. Růžička, D. Hnyk and P. Hobza, *Angew. Chem., Int. Ed.*, 2014, **53**, 10139–10142; (f) J. Fanfrlík, A. Přáda, Z. Padělková, A. Pecina, J. Macháček, M. Lepšík, J. Holub, A. Růžička, D. Hnyk and P. Hobza, *Angew. Chem.*, 2014, **126**, 10303–10306.
105. (a) F. T. Burling and B. M. Goldstein, *J. Am. Chem. Soc.*, 1992, **114**, 2313–2320; (b) Y. Nagao, T. Hirata, S. Goto, S. Sano, A. Kakehi, K. Iizuka and M. Shiro, *J. Am. Chem. Soc.*, 1998, **120**, 3104–3110; (c) J. C. Taylor and G. D. Markham, *J. Biol. Chem.*, 1999, **274**, 32909–32914; (d) W. Brandt, A. Golbraikh, M. Tager and U. Lendeckel, *Eur. J. Biochem.*, 1999, **261**, 89–97; (e) S. Wu and A. Greer, *J. Org. Chem.*, 2000, **65**, 4883–4887.
106. (a) M. Iwaoka, S. Takemoto, M. Okada and S. Tomoda, *Chem. Lett.*, 2001, 132–133; (b) R. E. Rosenfield Jr, R. Parthasarathy and J. D. Dunitz, *J. Am. Chem. Soc.*, 1977, **99**, 4860–4862.

107. R. E. Rosenfield, R. Parthasarathy and J. D. Dunitz, *J. Am. Chem. Soc.*, 1977, **99**, 4860–4862.
108. A. Kucsman and I. Kapovits, *Organic Sulphur Chemistry*, Elsevier, Amsterdam, 1985, pp. 191–245.
109. V. Nziko and S. Scheiner, *J. Phys. Chem. A*, 2014, **118**, 10849–10856.
110. J. T. Goettel, P. Chaudhary, P. Hazendonk, H. P. A. Mercier and M. Gerken, *Chem. Commun.*, 2012, **48**, 9120–9122.
111. (a) J. E. Del Bene, I. Alkorta, G. Sánchez-Sanz and J. Elguero, *J. Phys. Chem. A*, 2012, **116**, 3056–3060; (b) M. Solimannejad, V. Ramezani, C. Trujillo, I. Alkorta, G. Sánchez-Sanz and J. Elguero, *J. Phys. Chem. A*, 2012, **116**, 5199–5206; (c) I. Alkorta, G. Sánchez-Sanz, J. Elguero and J. E. Del Bene, *J. Chem. Theory Comput.*, 2012, **8**, 2320–2327; (d) I. Alkorta, G. Sánchez-Sanz, J. Elguero and J. E. Del Bene, *J. Phys. Chem. A*, 2013, **117**, 193–194; (e) I. Alkorta, J. Elguero and J. E. Del Bene, *J. Phys. Chem. A*, 2013, **117**, 10497–10503; (f) J. E. Del Bene, I. Alkorta and J. Elguero, *J. Phys. Chem. A*, 2013, **117**, 6893–6903; (g) G. Sánchez-Sanz, C. Trujillo, M. Solimannejad, I. Alkorta and J. Elguero, *Phys. Chem. Chem. Phys.*, 2013, **15**, 14310–14318; (h) J. E. Del Bene, I. Alkorta, G. Sánchez-Sanz and J. Elguero, *J. Phys. Chem. A*, 2013, **117**, 3133–3141; (i) I. Alkorta, J. Elguero and M. Solimannejad, *J. Phys. Chem. A*, 2014, **118**, 947–953.
112. (a) G. Sánchez-Sanz, I. Alkorta, C. Trujillo and J. Elguero, *ChemPhysChem*, 2013, **14**, 1656–1665; (b) G. Sánchez-Sanz, C. Trujillo and J. Elguero, *ChemPhysChem*, 2016, **18**, 9148–9160.
113. (a) S. Scheiner, *J. Phys. Chem. A*, 2011, **115**, 11202–11209; (b) S. Scheiner, *Acc. Chem. Res.*, 2013, **46**, 280–288; (c) D. Setiawan, E. Kraka and D. Cremer, *Chem. Phys. Lett.*, 2014, **614**, 136–142; (d) D. Setiawan, E. Kraka and D. Cremer, *J. Phys. Chem. A*, 2015, **119**, 1642–1656; (e) Q. Zhao, *J. Mol. Model*, 2016, **22**, 1–7; (f) M. D. Esrafili and F. Mohammadian-Sabet, *Chem. Phys. Lett.*, 2015, **638**, 122–127.
114. S. Ghosh, S. Biswas, A. Bauzá, M. Barceló-Oliver, A. Frontera and A. Ghosh, *Inorg. Chem.*, 2013, **52**, 7508–7523.
115. P. Batail, D. Grandjean, F. Dudragne and C. Michaud, *Acta Crystallogr., Sect. B: Struct. Crystallogr. Cryst. Chem.*, 1975, **31**, 1367–1372.
116. (a) M. R. Sundberg, R. Uggla, C. Vinas, F. Teixidor, S. Paavola and R. Kivekäs, *Inorg. Chem. Commun.*, 2007, **10**, 713–716; (b) A. C. Marr, M. Nieuwenhuyzen, C. L. Pollock and G. C. Saunders, *Organometallics*, 2007, **26**, 2659–2671; (c) C. Ganesamoorthy, M. S. Balakrishna, J. T. Mague and H. M. Tuononen, *Inorg. Chem.*, 2008, **47**, 7035–7047; (d) S. Bauer, S. Tschirschwitz, P. Lçnnecke, R. Frank, B. Kirchner, M. L. Clarke and E. Hey-Hawkins, *Eur. J. Inorg. Chem.*, 2009, **2009**, 2776–2788; (e) M. Bìhl, P. Kilian and J. D. Woollins, *ChemPhysChem*, 2011, **12**, 2405–2408.
117. E. V. Avtomonov, K. Megges, S. Wocadlo and J. Lorberth, *J. Organomet. Chem.*, 1996, **524**, 253–261.
118. S. J. Grabowski, *Chem.–Eur. J.*, 2013, **19**, 14600–14611.

119. (a) M. Marín-Luna, I. Alkorta and J. Elguero, *Struct. Chem.*, 2015, **26**, 639–645; (b) L. F. Ji, A. Y. Li and Z. Z. Li, *Chem. Phys. Lett.*, 2015, **619**, 115–121.
120. G. Müller, J. Brand, S. E. Jetter, F. Chemie, U. Konstanz and D. Konstanz, *Z. Naturforsch., B: Chem. Sci.*, 2001, **56**, 1163–1171.
121. S. Scheiner, *Chem. Phys.*, 2011, **387**, 79–84.
122. M. M. Watt, M. S. Collins and D. W. Johnson, *Acc. Chem. Res.*, 2013, **46**, 955–966.
123. (a) W. J. Vickaryous, R. Herges and D. W. Johnson, *Angew. Chem., Int. Ed.*, 2004, **43**, 5831–5833; (b) W. J. Vickaryous, E. R. Healey, O. B. Berryman and D. W. Johnson, *Inorg. Chem.*, 2005, **44**, 9247–9252; (c) V. M. Cangelosi, A. C. Sather, L. N. Zakharov, O. B. Berryman and D. W. Johnson, *Inorg. Chem.*, 2007, **46**, 9278–9284; (d) V. M. Cangelosi, L. N. Zakharov, S. A. Fontenot, M. A. Pitt and D. W. Johnson, *Dalton Trans.*, 2008, **37**, 3447–3453; (e) V. M. Cangelosi, L. N. Zakharov, J. L. Crossland, B. C. Franklin and D. W. Johnson, *Cryst. Growth Des.*, 2010, **10**, 1471–1473; (f) N. R. Lindquist, T. G. Carter, V. M. Cangelosi, L. N. Zakharov and D. W. Johnson, *Chem. Commun.*, 2010, **46**, 3505–3507; (g) M. A. Pitt, L. N. Zakharov, K. Vanka, W. H. Thompson, B. B. Laird and D. W. Johnson, *Chem. Commun.*, 2008, **44**, 3936–3938; (h) V. M. Cangelosi, M. A. Pitt, W. J. Vickaryous, C. A. Allen, L. N. Zakharov and D. W. Johnson, *Cryst. Growth Des.*, 2010, **10**, 3531–3536; (i) W. J. Vickaryous, L. N. Zakharov and D. W. Johnson, *Main Group Chem.*, 2006, **5**, 51–59; (j) V. M. Cangelosi, L. N. Zakharov and D. W. Johnson, *Angew. Chem., Int. Ed.*, 2010, **49**, 1248–1251; (k) V. M. Cangelosi, T. G. Carter, J. L. Crossland, L. N. Zakharov and D. W. Johnson, *Inorg. Chem.*, 2010, **49**, 9985–9992; (l) S. A. Fontenot, V. M. Cangelosi, M. A. W. Pitt, A. C. Sather, L. N. Zakharov, O. B. Berryman and D. W. Johnson, *Dalton Trans.*, 2011, **40**, 12125–12131.
124. (a) W. Frank, J. Schneider and S. Müller-Becker, *J. Chem. Soc., Chem. Commun.*, 1993, 799–800; (b) D. Mootz and V. Händler, *Z. Anorg. Allg. Chem.*, 1986, **533**, 23–29.
125. A. Bauza, T. J. Mooibroek and A. Frontera, *Angew. Chem., Int. Ed.*, 2013, **52**, 12317–12321.
126. (a) S. J. Grabowski, *Phys. Chem. Chem. Phys.*, 2014, **16**, 1824–1834; (b) A. Bauza, T. J. Mooibroek and A. Frontera, *Chem.–Eur. J.*, 2014, **20**, 10245–10248; (c) A. Bauzá, R. Ramis and A. Frontera, *Comput. Theor. Chem.*, 2014, **1038**, 67–70.
127. I. Alkorta, I. Rozas and J. Elguero, *J. Phys. Chem. A*, 2001, **105**, 743–749.
128. I. Alkorta, *J. Organomet. Chem.*, 2001, **625**, 148–153.
129. (a) R. S. Ruoff, T. Emilsson, A. I. Jaman, T. C. Germann and H. S. Gutowsky, *J. Chem. Phys.*, 1992, **96**, 3441–3446; (b) A. R. Rossi and J. M. Jasinski, *Chem. Phys. Lett.*, 1990, **169**, 399–404; (c) R. D. Urban, G. Rouillé and M. Takami, *J. Mol. Struct.*, 1997, **413–414**, 511–519.

130. (a) C. J. Levy and R. J. Puddephatt, *J. Am. Chem. Soc.*, 1997, **119**, 10127–10136; (b) J. Langer, S. Matejcik and E. Illenberger, *Phys. Chem. Chem. Phys.*, 2000, **2**, 1001–1005; (c) J. Mikosch, S. Trippel, C. Eichhorn, R. Otto, U. Lourderaj, J. X. Zhang, W. L. Hase, M. Weidemìller and R. Wester, *Science*, 2008, **319**, 183–186; (d) M. Sohail, R. Panisch, A. Bowden, A. R. Bassindale, P. G. Taylor, A. A. Korlyukov, D. E. Arkhipov, L. Male, S. Callear, S. J. Coles, M. B. Hursthouse, R. W. Harrington and W. Clegg, *Dalton Trans.*, 2013, **42**, 10971–10981.
131. D. Mani and E. Arunan, *Phys. Chem. Chem. Phys.*, 2013, **15**, 14377–14383.
132. S. P. Thomas, M. S. Pavan and T. N. Guru Row, *Chem. Commun.*, 2014, **50**, 49–51.
133. Q. Li, X. Guo, X. Yang, W. Li, J. Cheng and H. B. Li, *Phys. Chem. Chem. Phys.*, 2014, **16**, 11617–11625.
134. P. G. Taylor, A. R. Bassindale, Y. El Aziz, M. Pourny, R. Stevenson, M. B. Hursthouse and S. J. Coles, *Dalton Trans.*, 2012, **41**, 2048–2059.
135. (a) X. Dai, D. L. Schulz, C. W. Braun, A. Ugrinov and P. Boudjouk, *Organometallics*, 2010, **29**, 2203–2205; (b) X. Dai, S. B. Choi, C. W. Braun, P. Vaidya, S. Kilina, A. Ugrinov, D. L. Schulz and P. Boudjouk, *Inorg. Chem.*, 2011, **50**, 4047–4053.
136. J. Tillmann, F. Meyer-Wegner, A. Nadj, J. Becker-Baldus, T. Sinke, M. Bolte, M. C. Holthausen, M. Wagner and H. W. Lerner, *Inorg. Chem.*, 2012, **51**, 8599–8606.
137. (a) K. Ghosh, K. Harms, A. Bauzá, A. Frontera and S. Chattopadhyay, *Dalton Trans.*, 2018, **47**, 331–347; (b) S. Roy, M. G. B. Drew, A. Bauzá, A. Frontera and S. Chattopadhyay, *ChemistrySelect*, 2017, **2**, 10586–10594; (c) M. Hussain, A. Bauzá, A. Frontera, K. M. Loc and M. M. Naseer, *CrystEngComm*, 2018, **20**, 150–154.
138. (a) D. Mani and E. Arunan, *J. Phys. Chem. A*, 2014, **118**, 10081–10089; (b) P. R. Varadwaj, A. Varadwaj and B. Y. Jun, *Phys. Chem. Chem. Phys.*, 2014, **16**, 17238–17252; (c) M. M. Azofra and S. J. Scheiner, *Chem. Phys.*, 2015, **142**, 034307.
139. X. Dai, S. B. Choi, C. W. Braun, P. Vaidya, S. Kilina, A. Ugrinov, D. L. Schulz and P. Boudjouk, *Inorg. Chem.*, 2011, **50**, 4047–4053.
140. E. C. Escudero-Adán, A. Bauzá, A. Frontera and P. Ballester, *ChemPhysChem*, 2015, **16**, 2530–2533.
141. (a) A. Bauzá and A. Frontera, *ChemPhysChem*, 2016, **17**, 3181–3186; (b) A. Bauzá and A. Frontera, *Theor. Chem. Acc.*, 2017, **136**, 37; (c) A. Bauzá, X. García-Linás and A. Frontera, *Chem. Phys. Lett.*, 2016, **666**, 73–78.
142. (a) S. J. Grabowski, *ChemPhysChem*, 2014, **15**, 2985–2993; (b) S. J. Grabowski, *ChemPhysChem*, 2015, **16**, 1470–1479.
143. F. Weinhold and C. Landis, *Valency and Bonding, a Natural Bond Orbital Donor–acceptor Perspective*, Cambridge University Press, Cambridge, UK, 2005.
144. M. Weinmann, M. Kroschel, T. Jaschke, J. Nuss, M. Jansen, G. Kolios, A. Morillo, C. Tellaeche and U. Nieken, *J. Mater. Chem.*, 2008, **18**, 1810–1818.

145. K. Anton, H. Noth and H. Pommerening, *Chem. Ber.*, 1984, **117**, 2479–2494.
146. A. Bauzá, T. J. Mooibroek and A. Frontera, *Chem. Commun.*, 2015, **51**, 1491–1493.
147. A. Bauzá, A. Frontera and T. J. Mooibroek, *Nat. Commun.*, 2017, **8**, 14522.
148. A. Bauzá, A. Frontera and T. J. Mooibroek, *Cryst. Growth Des.*, 2016, **16**, 5520–5524.
149. A. Bauzá, A. V. Sharko, G. A. Senchyk, E. B. Rusanov, A. Frontera and K. V. Domasevitch, *CrystEngComm*, 2017, **19**, 1933–1937.
150. X. Guo, L. Cao, Q. Li, W. Li and J. Cheng, *J. Mol. Model.*, 2014, **20**, 1.

Subject Index